SAGE Publishing: **Our Story**

Founded in 1965 by 24-year-old entrepreneur Sara Miller McCune, SAGE continues its legacy of making research accessible and fostering **CREATIVITY** and **INNOVATION**. We believe in creating fresh, cutting-edge content to help you prepare your students to thrive in the modern world and be **TOMORROW'S LEADING SOCIAL SCIENTISTS**.

- By partnering with **TOP SOCIOLOGY AUTHORS** with just the right balance of research, teaching, and industry experience, we bring you the most current and applied content.
- As a **STUDENT-FRIENDLY PUBLISHER**, we keep our prices affordable and provide multiple formats of our textbooks so your students can choose the option that works best for them.
- Being permanently **INDEPENDENT** means we are fiercely committed to publishing the highest-quality resources for you and your students.

Seeing Social Problems

For Nancy, Arielle, and Benjamin—your love and support continually fuel my conviction about the importance of the ideas expressed in this book.

Seeing Social Problems

The Hidden Stories Behind Contemporary Issues

Ira Silver

Framingham State University

Los Angeles | London | New Delhi
Singapore | Washington DC | Melbourne

FOR INFORMATION:

SAGE Publications, Inc.
2455 Teller Road
Thousand Oaks, California 91320
E-mail: order@sagepub.com

SAGE Publications Ltd.
1 Oliver's Yard
55 City Road
London EC1Y 1SP
United Kingdom

SAGE Publications India Pvt. Ltd.
B 1/I 1 Mohan Cooperative Industrial Area
Mathura Road, New Delhi 110 044
India

SAGE Publications Asia-Pacific Pte. Ltd.
18 Cross Street #10-10/11/12
China Square Central
Singapore 048423

Acquisitions Editor: Jeff Lasser
Editorial Assistant: Tiara Beatty
Content Development Editor: Tara Slagle
Production Editor: Laureen Gleason
Copy Editor: Shannon Kelly
Typesetter: C&M Digitals (P) Ltd.
Proofreader: Theresa Kay
Indexer: Wendy Allex
Cover Designer: Anupama Krishnan
Marketing Manager: Will Walter

Printed in Canada

Library of Congress Cataloging-in-Publication Data

Names: Silver, Ira, author.

Title: Seeing social problems : the hidden stories behind contemporary issues / Ira Silver, Framingham State University.

Description: Los Angeles : SAGE, [2021] | Includes bibliographical references and index.

Identifiers: LCCN 2019033248 | ISBN 9781506386812 (paperback ; alk. paper) | ISBN 9781544398624 (epub) | ISBN 9781544398631 (epub) | ISBN 9781544398648 (pdf)

Subjects: LCSH: Social problems. | Sociology.

Classification: LCC HN18.3 .S584 2021 | DDC 361.1—dc23
LC record available at https://lccn.loc.gov/2019033248

This book is printed on acid-free paper.

20 21 22 23 24 10 9 8 7 6 5 4 3 2 1

• Brief Contents •

• Detailed Contents •

• Preface •

To the Instructor

Think back to the best classes you took in college and why you remember them. For me, it was because the professors demonstrated how the material we were learning built upon familiar topics. My experience certainly isn't unique. As an instructor, you may have discovered that one of the best ways to motivate and inspire students is by showing them exciting new ways to think about something they already know.

A course about social problems is conducive to this type of learning. Unlike our colleagues who teach about subjects unfamiliar to undergraduates, students enter our classes having heard about—and often having experienced—many of the issues the course explores. We don't have to spend much class time illustrating that drugs are addictive, bullying is hurtful, or obesity poses health risks. Social media, TV shows, movies, and personal experiences familiarize students with this sort of information. Instead, we can focus on discussing how sociological explanations of social problems further what students already know and find interesting about them. In my twenty-five years of teaching undergraduates, I've found that being mindful about how to build on students' existing knowledge is a terrific way to engage them in thinking about the world sociologically.

That's the approach of *Seeing Social Problems: The Hidden Stories behind Contemporary Issues*. It addresses topics and concepts that are fitting for an entry-level sociology course, employing a writing style and tone that respond to a sentiment readers often express about textbooks: They're boring! This book will captivate your students by tapping into curiosities they have about the world around them and encouraging them to connect new ideas with what they already know. The narrative stimulates students' critical thinking instead of presenting encyclopedic content, which rarely ignites passion. Therefore, this book offers a refreshing alternative to others you may have used in teaching social problems. I have written it *to* students rather than *for* them.

There are twelve case studies, each of which begins with a vignette that enables students to recognize the conventional wisdom concerning particular social problems and motivates them to learn more. The chapters then discuss snapshots of pertinent studies highlighting hidden sociological perspectives that illuminate these issues in new ways. While the case studies are about different topics, this is not a "problem-of-the-week" book. Instead, it develops themes across the case studies by addressing cross-cutting sociological concepts like class, race, and gender.

With an eye toward tapping student interest and curiosity, I have given catchy titles to the chapters and use lucid, jargon-free prose. Many visuals illuminate the narrative, referencing trends in popular culture and presenting accessible tidbits of social science data. I also draw upon relevant experiences I've had as a teacher, researcher, parent, husband, pet owner, and sports fan. In using these personal anecdotes to highlight how my life connects to the topics students are learning, I encourage them to think about how theirs do too.

The chapters are written as stand-alone discussions and can be assigned in any order. Early on in each chapter, there is a list of questions that taps students' first impressions about the topic. At the end, there's another list of questions that encourages students to think about how what they know now compares with these initial impressions.

• Acknowledgments •

Books mask the time and energy that goes into writing them. While most of the labor was my own, I benefited in tremendous ways from the help of friends, colleagues, and students. Without their support, it's unimaginable how I ever would have completed *Seeing Social Problems.*

When it was just a kernel of an idea, David Shulman spent considerable time reading an early draft of the proposal and discussing with me ways to improve it. I thank him for impressing upon me that I had worthy ideas and for affirming that I should keep developing them.

Through numerous phone conversations, Pam Bauer inspired me to discover the richness in these ideas. Elizabeth and Thad Robey often engaged with my thoughts and motivated me to continue refining them. Graham Peck has, over many years, validated my efforts to write in a lucid and engaging style that aims to make a difference in readers' lives.

In 2017 and 2018, I got an incredible opportunity to attend Framingham State University's annual writing retreat. Both times, Sandy Hartwiger and Belinda Walzer offered me a wealth of invaluable insights. They provided significant feedback on several draft chapters and helped me brainstorm ways to begin writing others. Their advice echoed in my mind as I worked toward completing the manuscript.

I appreciate that Debbie Feldman read and offered her thoughts on several parts of the book. Marian Cohen provided input about the mental illness chapter. Beth Whaley compiled a list of studies for me to consult before starting to write the gender violence chapter, and Rachel DiBella made thoughtful comments on an early draft. When I presented material from this book at Framingham State's 2018 Lyceum Lecture, the following people offered support, encouragement, and advice: Corri Taylor, Diane Pecora, Mitch Rosenberg, Stefane Cahill-Farella, Jason Giannetti, Dave Ripp, Lisa Eck, Alison Butterfield, Lina Rincon, and Mary-Ann Stadtler-Chester.

Since *Seeing Social Problems* is written for students, their help has been the most invaluable of all. Frank Legere and Tiffany Agostini assisted me early in the project by identifying vignettes I could use to begin each case study. Danielle DiNardo compiled a list of suggested readings to accompany the chapters. Kiara Davis spent many hours during the summer identifying images to include that would make the book visually engaging. Benjamin Silver offered thoughtful comments about several draft chapters and, through innumerable daily conversations, he significantly sharpened my ideas. His relentless support and encouragement have motivated me to keep developing my ideas to their fullest potential.

My deepest appreciation goes to Aviva Rosenberg. She provided detailed feedback about every chapter, with a particular eye to how I could include more examples that would be relatable to students. She also suggested ideas to incorporate into the book and offered tips about engaging ways to discuss them. I cannot thank Aviva enough for her dedication and support. This book is dramatically better because of the energy and intellect she gave to it.

Over the three years that I wrote *Seeing Social Problems*, I have been continually reminded of a line by my Framingham State colleague and friend, Virginia Rutter: "Books will break your heart." She is right! However, I'm happy to say that it has been worth all of the effort and sacrifice. In addition to her wisdom, I'm thankful to Virginia for introducing me to Jeff Lasser at SAGE. He and Tara Slagle have been terrific editors, offering thoughtful and constructive comments about how to frame my ideas and present them lucidly. I also appreciate the help Tiara Beatty provided in guiding me through the permission process, Laureen Gleason's thoroughness in explaining and carrying out the production editing, and Shannon Kelly's thoughtful and detailed copy editing.

Finally, I am grateful to the peer reviewers listed below, as well as five anonymous others. At different stages of the writing process, each of them offered valuable comments about this book's merits.

Faezeh Bahreini, University of South Florida

Sherry Cooke, Grayson College

Joanna Kempner, Rutgers University

Jesse Klein, Florida State University

Rick Matthews, Carthage College

Debarashmi Mitra, Central New Mexico Community College

Jane Nielsen, College of Charleston

Charles Quist-Adade, Kwantlen Polytechnic University

Deborah Riddick, Elizabeth City State University

Amy Ruedisueli, Tidewater Community College

Suzanne Schneider, Manhattan College

Steven Smith, South Piedmont Community College

Erik T. Withers, University of South Florida

• About the Author •

Ira Silver is Professor of Sociology at Framingham State University, where he has taught since 2002. He also teaches during the summer at Wellesley College. He received his B.A. *summa cum laude* from Amherst College and his master's and Ph.D. in sociology from Northwestern University.

In addition to Social Problems, Ira also teaches Nonprofit Giving; Death & Dying; Animals & Society; Society, Technology & the Future; Sport in Society; and Issues & Influences in Education.

He has authored or edited four previous books. *Giving Hope: How You Can Restore the American Dream* (CreateSpace, 2013) is a guide for charitable giving that addresses the opportunity divide in the United States. *Social Problems: Readings* (Norton, 2008) is a collection of essays about contemporary issues. *Academic Street Smarts: Informal Professionalization of Graduate Students in Sociology* (American Sociological Association, 2008) is a collection of essays that provide tips for career success in the field. *Unequal Partnerships: Beyond the Rhetoric of Philanthropic Collaboration* (Routledge, 2006) is a study of power relations between grantors and grantees.

1

Looking beyond What You Already Know

Becoming Curious about Social Problems

Learning Objectives

1. Explain what social problems are and discuss the importance of conventional wisdom in how society views them.
2. Define sociology and discuss its importance.
3. Describe the sociological perspective and its impacts on social problems.
4. Explain the importance of sociological case studies in understanding social problems.

Social Problems Are Everywhere

1.1 Explain what social problems are and discuss the importance of conventional wisdom in how society views them.

Think for a moment how often you hear about issues like police brutality, teenage pregnancy, discrimination against LGBTQ people, or mass shootings. Maybe you and your friend recently had a conversation about struggling with anxiety or depression, a student who's been bullied, your mother's coworker who's addicted to prescription painkillers, or the deportation of undocumented immigrants. Perhaps you know someone who doesn't have health insurance or who has experienced sexual assault. Maybe there was a racist incident recently on your campus. Each of these examples is a **social problem**—a harm in our society that people believe should and can be fixed.

Much of what you know about social problems is at your fingertips.
iStockphoto.com/oneinchpunch

I wrote this book to show you fascinating new ways to understand these important topics that may directly impact your life. Rest assured—this is *not* a boring textbook filled with lots of facts for you to memorize and master. You've probably had enough experience with such books over the years that the very thought of having to read another one would produce yawns. Instead, my goal is to tap into what you know about drug addiction, cyberbullying, police brutality, and other timely issues because they're mentioned on social media or dramatized in YouTube videos, Hollywood movies, and TV shows. Therefore, this book will engage the curiosities you already have about the world around you, motivating you to connect new ideas with what you currently understand. Each chapter highlights thought-provoking ways to see a different social problem, offering you a lens for making sense of the world that you likely haven't considered before or even knew existed.

iStockphoto.com/Eoneren

You've learned many valuable things about social problems from teachers. However, it's useful to do your own investigations too, by looking beyond what teachers have told you.
iStockphoto.com/skynesher

Chapters begin with a story that encourages you to become aware of your **conventional wisdom**, or what you think is true about a topic based on information you've acquired over the course of your life. For example, it's commonly believed that teenage pregnancy results from unsafe sex, bullies exploit kids who don't fit in, and online predators harm those who are not careful about whom they friend on social media. Such beliefs give you a basic understanding of events and issues that shape the world and your place in it. Yet if you've never had the chance to explore beyond your conventional wisdom, you might think what you know about poverty, obesity, mass shootings, or any other social problem is all you need to know. Since this knowledge comes from authority figures—police officers, therapists, scientists, journalists, politicians, and other experts—it's easy to take for granted that they're giving you the whole story. In reading this book, you'll discover that often they're not.

Has anyone ever suggested that you get a second opinion on a medical treatment? While doctors recommend what they view as the best course of action, getting multiple perspectives illustrates that there is usually more than one option. The same can be said for how we view social problems.
iStockphoto.com/Cecilie_Arcurs

This book offers you a lens for expanding upon your conventional wisdom. Maybe you knew a boy in high school who attempted suicide. Chapters 12 and 13 explore reasons why. Perhaps he suffered from depression. He may also have felt that it wouldn't be "manly" to share his pain with others. Taking a comprehensive approach highlights that there's no simple explanation for suicide. The same is true for other social problems. In building upon what you already know about these familiar topics, you will recognize how you can see them in eye-opening new ways.

Throughout this book, I will share with you how my life connects to the topics at hand in order to encourage you to think about how yours does too. Let's start with the irony of the title *Seeing Social Problems*, given that I have a congenital visual impairment. Focusing clearly on the world around me is a daily challenge. You might be wondering, therefore, how I could possibly be qualified to show you ways to acquire a greater perception of the social world. I had similar thoughts before I took my first sociology course in college. Little could I have imagined at the time that this subject would enhance my vision in a way no medical intervention ever could. In fact, being a person who doesn't see well is, in part, what led me to embrace a way of thinking that has enabled me to see better. My aim in this book is to share that vision with you.

This book will show you that seeing isn't something we do with our eyes, but with our minds.
iStockphoto.com/metamorworks

FIRST IMPRESSIONS?

1. Jot down a social problem that interests you. What do you know about the problem? Where have you gotten that knowledge?
2. Why do people typically put trust in what experts say about social problems?

The Virtue of Sociology: Opening Your Eyes to the Hidden World around You

1.2 Define sociology and discuss its importance.

Many years ago, a Unitarian Universalist minister named Robert Fulghum wrote a book with a catchy title that people continue to find inspiring. *All I Need to Know I Learned in Kindergarten* struck a chord with millions of readers because of its simple, yet powerful message. Fulghum is right in that we acquire a lot of our most important knowledge very early in life. But when it comes to the workings of our society, there's a lot we cannot know unless we learn how to look for it.

Most college students born after the mid-1990s are part of Generation Z. From a young age, you likely developed an expectation that people should be readily available via mobile devices. What are other examples of how your beliefs have been shaped by being a member of this generation?
Zoonar GmbH/Alamy Stock Photo

Sociology offers tools for seeing the world in refreshing new ways and opening your eyes to the intricate workings of society. You're already familiar with the basic rules that guide daily life, and you know which behaviors violate those rules. Yet things are often not as they appear. Sociology can enable you to push past your own blinders. Learning about social problems will uncover hidden truths about how our society operates.

Sociologists conduct research by directly observing people doing their daily activities, interviewing them, or asking survey questions. These research techniques produce **sociological data**—evidence about how people behave or think—that reveal the ways groups shape individuals' thoughts and behaviors. If you stop and think for a moment, you can name many groups to which you belong. I don't mean just the ones you've officially joined, but all of your sources of identity. These include your gender, race, social class, family, sexual orientation, religion, where you live, and when you were born. And this is hardly a complete list.

NASCAR events illustrate the core sociological idea that each of us is shaped by others with whom we spend significant time. Fans don't just share a love of auto racing; they also often work in similar types of jobs and have common political views.
iStockphoto.com/Onfokus

Because golf enables people to be with others who can similarly afford to spend money and time playing this sport, it contributes to their tastes about how to enjoy leisure time.
iStockphoto.com/LightFieldStudios

This book shows you how to discover the hidden workings of our society.
iStockphoto.com/bowie15

Let's consider other groups that may be important to you. For example, it can be sociologically meaningful whether you are a "dog person," a "cat person," or dislike pets altogether. The same goes for people who prefer Instagram versus Snapchat versus don't use social media at all (such people exist!). Even more significant are the differences between those who have served in the military and those who have lived an entirely civilian life, between religious fundamentalists and atheists, and between pro-choice and pro-life abortion activists.

A person must do substantial research, often lasting several years, in order to earn a Ph.D. and become a sociologist. However, you need no prior training to begin thinking like one. All that's required is an interest in exploring how data about human behavior can expand your current understanding about the workings of society. I recall having this insight when I took my first sociology course in college. During the ensuing thirty plus years that I've studied and taught sociology, I've learned again and again how to think in refreshing ways I didn't know were possible. One of my greatest joys has been seeing students come to this same realization.

Viewing Our Society through Different Lenses

1.3 Describe the sociological perspective and its impacts on social problems.

The topics in this book may lead you to revisit social media posts you've read or written, movies you've seen, and TV shows you've watched. Your mind may wander across diverse subjects such as the overweight person who orders supersize meals at McDonald's, the runner who takes performance-enhancing drugs and breaks a world record, or the gunman who shoots randomly at the mall. These examples highlight the **individual perspective** toward social problems, which focuses on the person who commits wrongdoing. From this perspective, their behavior is **deviance**. It violates **social norms**—rules people have agreed upon for appropriate conduct—and therefore highlights flaws in the person who acts this way. The individual perspective is an appealing way to understand social problems because it places blame on people whom you or I can easily cast as distinct from the rule followers we believe we are. Because this perspective is widespread in American society, it's understandable if you assumed it to be the only possible way to understand social problems.

Seeing Social Problems shows how you can also understand these issues through a **sociological perspective** by recognizing the **social forces** that shape individual behavior (see Figure 1.1). Just as gravity is a physical force causing objects to fall downward instead of flying into the sky, groups are social forces that influence a person's likelihood of following or deviating from social norms. Paying attention to social

FIGURE 1.1 ● Uncovering the Hidden Story

The sociological perspective exposes social forces that influence how each of us behaves, thinks, and feels.

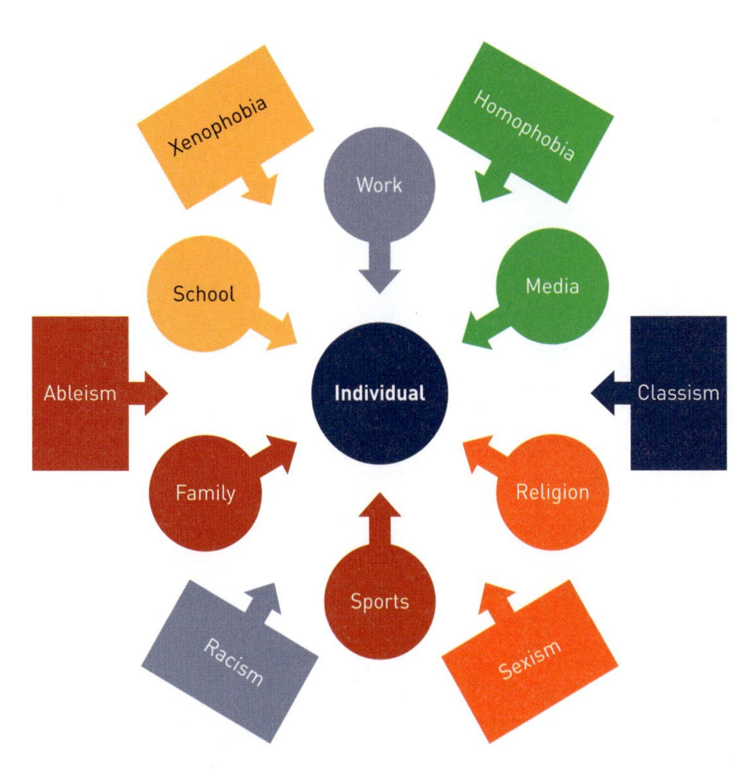

forces will enable you to get to the root of why some people bully their classmates, abuse drugs, or commit sexual assault. The sociological perspective exposes the contexts that give rise to deviance. This way of thinking doesn't make excuses for wrongdoing but explains why some individuals make rational decisions to break the rules.

One context that produces deviance is **social inequality**—the disparities among people in their amount of money, power, or status. For example, Chapter 11 reveals that the teens most likely to ignore stranger-danger warnings and forge relationships with online sexual predators are kids who've grown up with significant emotional vulnerabilities stemming from family violence. Another context is **culture**, or the beliefs, values, and behaviors that a particular group of people shares in common. Chapter 7 shows that cheating often stems from being in a workplace or school environment that puts a premium on doing whatever it takes to beat the competition. Cheaters, therefore, may not see their actions as shortcuts to success but as ways to level the playing field. Highlighting the significance of social inequality and culture isn't a justification for irresponsible behavior but a recognition that actions that appear deviant from the perspective of our own personal experiences are explainable when we acknowledge the social forces underlying them.

Let's illustrate what this shift in perspective looks like by focusing on a social problem that's become more pronounced during your lifetime. In the United States, one out of every three adults—and one in six children—is obese. Many people have a narrow understanding of those who are overweight. My guess is that your beliefs probably align with the message echoed in *The Biggest Loser*, a popular TV show from 2004 to 2016: Obese people choose to overeat and prefer being sedentary to exercising. In other words, the conventional wisdom is that they are personally responsible for the health problems they experience.

The sociological perspective enables you to build upon the conventional wisdom that obesity and other social problems stem from personal irresponsibility.
iStockphoto.com/Fertnig

It's a sensible viewpoint in that people make choices about how much they eat and exercise. However, there's more to this picture than meets the eye. If you have a friend who's overweight, seeing obesity sociologically can enable you to look upon that person with greater awareness of what is entailed in the struggle to lose pounds. It's hardly as simple as being a more disciplined eater or getting up early every morning to go to the gym. We can see why by considering data indicating that obesity is concentrated among lower-income Americans. Poorer people face constraints in their ability to make healthy food choices and in their access to fitness opportunities that those who live in wealthier communities seldom experience.

The sociological perspective recognizes that while each of us makes choices that affect our body size, these choices reflect the constraints or opportunities of living in an unequal society. Whereas many people are inclined to see the obese as having failed to conform to social norms about thinness, obese individuals behave in ways that mirror the variable distribution of resources in American society. Both individual and sociological perspectives are useful ways to understand obesity. Overweight people may irresponsibly care for their bodies *and* also may experience income-related challenges in accessing nutritious foods and fitness opportunities. Chapter 5 offers a fuller discussion of this topic.

The particular perspective people use to understand obesity or any other social problem informs the strategies they believe can remedy it. Those who see weight strictly through an individual perspective favor diets and vigorous exercise. Embracing the sociological perspective doesn't preclude exerting greater self-discipline, but it does highlight why this often isn't enough. A sociologically informed remedy would be to promote greater access to nutritious foods and fitness opportunities. This might involve the government creating incentives for supermarket chains to open stores in low-income neighborhoods, where fresh fruits and vegetables are often unavailable to residents.

Recognizing That Social Problems Have Diverse Explanations

1.4 Explain the importance of sociological case studies in understanding social problems.

In my Social Problems course, on the first day of the semester I ask people to jot down the sorts of topics they expect to learn about in the course. All of the issues students mention are included in this book: poverty, racism, sexism, drug addiction, mass shootings, cyberbullying, obesity, and teenage pregnancy. There are also chapters about subjects that students rarely anticipate yet find fascinating: mental illness, animal cruelty, online predators, and cheating.

Each chapter is a **case study**—an example that exposes and illustrates a broader theme. This book's twelve case studies highlight the rich ways the sociological perspective builds on what you already know about social problems. It's important to explore multiple case studies because they collectively uncover fascinating features of our society that you may not have ever given much thought to before. As you read in subsequent chapters about diverse topics, you'll discover how these case studies speak to one another in teaching you about the hidden workings of our society.

The case studies each begin with a vignette highlighting the sorts of beliefs you may currently hold about a particular social problem. This vignette is the springboard for discussion of research aimed at expanding your understanding of the issue. Because they will open your eyes to new ways of understanding timely issues, the case studies as a whole offer models for how you can better understand *any* social problem.

Studying social problems may be the gateway to the major you choose or it may be your only exposure to sociology during college. Either way, this book is written for you. By offering a set of tools for applying the sociological perspective to your own life, it aims to instill in you a lifelong curiosity about the hidden workings of our society. You will discover how groups that seem to have little influence on you significantly impact the choices you make and the beliefs you hold close to your heart.

This perspective opens your eyes to noticing how your life connects to the broader society around you. Therein lies the beauty of sociology.

What Do You Know Now?

1. Jot down three different groups that you are part of, as well as specific ways each group has shaped who you are.
2. What is the difference between the individual and sociological perspectives toward social problems?
3. Why is it important to investigate multiple case studies of social problems?

Key Terms

Social problem 1
Conventional wisdom 2
Sociology 3
Sociological data 3
Individual perspective 5
Deviance 5
Social norms 5
Sociological perspective 5
Social forces 5
Social inequality 6
Culture 6
Case study 7

Visit **edge.sagepub.com/silver** to help you accomplish your coursework goals in an easy-to-use learning environment.

iStockphoto.com/Eoneren

2

Opportunity for Few

The Withering of the American Dream

Learning Objectives

1. Discuss how economic inequality is impacted by the American Dream.
2. Describe the opportunity divide that exists in the United States.
3. Explain why some students succeed in school and others do not.
4. Recognize why poor people may be unmotivated to better their lives.
5. Discuss how the same social forces that enable some people's successes contribute to others' hardships.
6. Explain why Americans are often resistant to providing aid to those living in poverty.
7. Describe the shared societal costs of economic inequality.

Struggling to Get Ahead in the Land of Opportunity

2.1 Discuss how economic inequality is impacted by the American Dream.

Linda Tirado used to work two low-wage jobs and live in a weekly rental motel. Although her life was tough, she acknowledged in a viral blog post that she had also made some bad choices. She smoked, ate junk food, and got pregnant. Therefore, you'd have reason to wonder whether these choices contributed to her being poor or living in **poverty**, which the federal government defines as having a household income below an amount annually adjusted for inflation. In 2018 this amount was about $25,000 a year for a family of four, and less for a single mother like Tirado (see Table 2.1). Whereas the majority of poor people in the U.S. are White, Blacks and Latinos are more likely than Whites to live in poverty (see Figure 2.1 on page 12).

TABLE 2.1 • Defining Who Is Poor

Whether Americans live in poverty depends on the size of their family and their annual household income.

Size of Family Unit	Annual Household Income
One person	$12,752
Two person	$16,414
Three people	$19,173
Four people	$25,283
Five people	$30,490
Six people	$35,069

Source: U.S. Census Bureau, "Poverty Thresholds," 2018, https://www.census.gov/data/tables/time-series/demo/income-poverty/historical-poverty-thresholds.html.

FIGURE 2.1 ● Race and Poverty

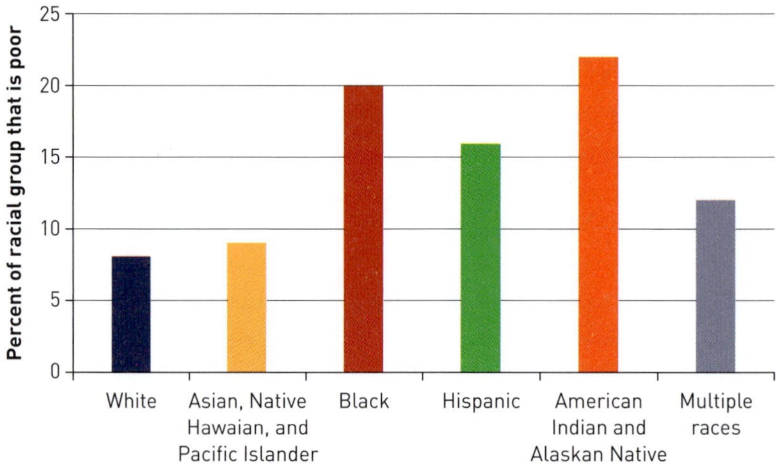

Source: Adapted from Henry J. Kaiser Family Foundation, "Poverty Rate by Race/Ethnicity," 2017, https://www.kff.org/other/state-indicator/poverty-rate-by-raceethnicity/?currentTimeframe=0&sortModel=%7B%22colId%22:%22Location%22,%22sort%22:%22asc%22%7D#note-1.

Based on Tirado's own words, several portions of her story appear to confirm the conventional wisdom that poor people contribute to their own struggles:

- "Junk food is a pleasure that we are allowed to have; why would we give that up?"
- "I smoke. It's expensive."
- "I make a lot of poor financial decisions."[1]

The thirty-something Michigan native's life, indeed, seems emblematic of the way many Americans think about poor people—as having themselves to blame for their inability to overcome hardship.

This individual perspective is deeply rooted in American culture. There's a long-standing belief that our nation is the land of opportunity, where hard work and self-discipline lead to **upward mobility**, which is a long-term increase in a person's income and status. This belief, commonly known as the **American Dream**, is the core of our founding narrative about colonists coming to these shores with aspirations to create better lives than the ones they left behind in Europe. They regarded determination and sacrifice as the keys to upward mobility. The view that anyone can "pull themselves up by their bootstraps" remains prevalent. Although this view has waned a bit over the past couple of decades, over 60 percent of Americans still believe it's possible for someone who begins life poor to become rich. This belief offers a justification for seeing someone like Linda Tirado who's stuck in poverty as deserving blame for their struggles.[2]

Because of this belief, there's considerable opposition to the government providing **welfare**—food stamps, health insurance, affordable housing, and other types of support to enable low-income people to have a basic standard of living. Look at the right side of Figure 2.2. These data indicate that whereas in 2014 fewer than 20 percent of Americans thought the government spent too little on welfare, nearly 50 percent thought it spent too much. Many believe welfare prevents people from getting ahead; for example, because they use their food stamps to buy soda, juice, and other sugary drinks with little nutritional value. Opposition to welfare is strongest among Whites who see it as giving Blacks unfair advantages and making them lazy. Yet these same Whites make up part of the overwhelming support that exists for assistance to the poor (see the left side of Figure 2.2). Labels matter; assisting the poor doesn't fuel public resentment even though it's the same as welfare.[3]

Publicity about Oprah Winfrey and other celebrities with rags-to-riches stories reinforces the idea that anyone can succeed in the U.S.
Kevin Winter/Getty Images

FIGURE 2.2 ● Words Matter

Americans' attitudes toward *assistance to the poor* and toward *welfare* substantially differ from one another.

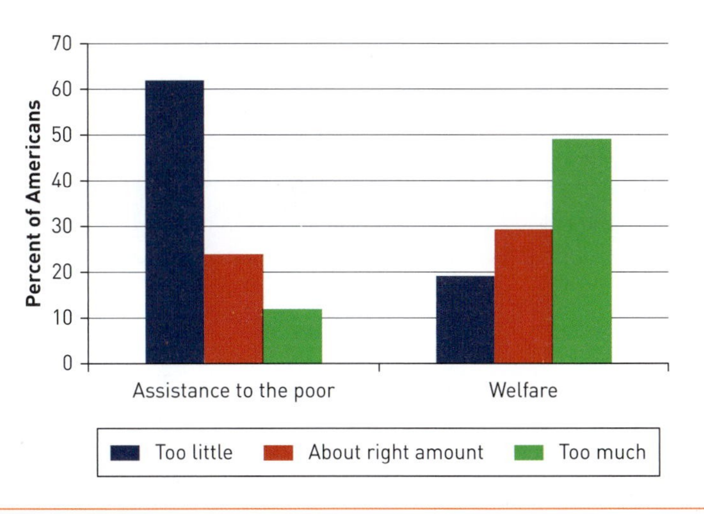

Source: The Associated Press-NORC Center for Public Affairs Research, "Inequality: Trends in Americans' Attitudes,"2015,http://www.apnorc.org/projects/Pages/HTML%20Reports/inequality-trends-in-americans-attitudes0317-6562.aspx.

During the 1980s, President Ronald Reagan often talked about a woman who allegedly used fake names, addresses, and telephone numbers to collect over $150,000 in Social Security, food stamps, and veterans' benefits for deceased husbands who didn't exist.[4]
IanDagnall Computing/Alamy Stock Photo

Each year, the U.S. Census Bureau releases a report containing data about America's poor.[5] Over the past several years, roughly one in seven Americans has lived in poverty. Millions more are modestly better off yet also face daily struggles. Nearly *half* the population has annual family earnings under twice the poverty line. Low-income people who live in expensive cities like New York, San Francisco, Boston, and Washington, D.C., especially experience hardship. Table 2.2 paints a picture of **economic inequality**, or gaps among people based on their financial worth. Whereas 60 percent (the bottom three quintiles) of the population collectively earns just a quarter of all income, 20 percent (the highest quintile) earns over half. Figure 2.3 reveals, moreover, that these earnings overwhelmingly go to a very tiny slice of the population. Inequalities in **wealth**, which includes total assets a person has in addition to their earnings, are even greater. The four hundred richest Americans collectively have more wealth than the 150 million people who make up the bottom 60 percent of the U.S. population.[6]

Some low-income people may bear partial responsibility for their struggles. However, we'll see in this chapter that the individual perspective cannot account for the difficulties such a vast segment of Americans experiences, nor for the successes of others who have good jobs and earn decent incomes. By highlighting the social forces that create disparities in attainment of the American Dream, the sociological perspective explains these disparities.

TABLE 2.2 ● Economic Inequality in the U.S.

	Percentage of Total U.S. Income	Median Income	Income Range
Highest quintile	51.1	$202,366	More than $117,003
Fourth quintile	23.2	$92,031	$72,002 to $117,002
Third quintile	14.3	$56,832	$43,512 to $72,001
Second quintile	8.2	$32,631	$22,801 to $43,511
Lowest quintile	3.1	$12,457	Less than $22,800

Source: Bernadette D. Proctor, Jessica L. Semega, and Melissa A. Kollar, "Income and Poverty in the United States: 2015," 2016, U.S. Census Bureau, https://www.census.gov/content/dam/Census/library/publications/2016/demo/p60-256.pdf.

FIGURE 2.3 ● Strikingly Unequal Earnings

In the U.S., the lion's share of income goes to just .01 percent of the population.

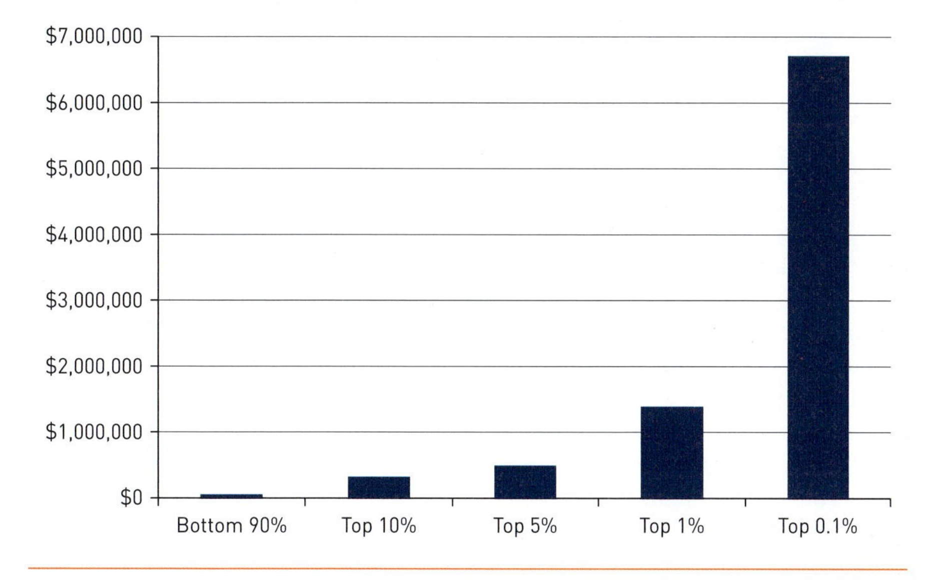

Source: Emmanuel Saez, "Income Inequality," Institute for Policy Studies, 2017, https://inequality.org/facts/income-inequality.

FIRST IMPRESSIONS?

1. Where does your knowledge about poverty come from—social media, TV, news reports, encountering panhandlers on the street, firsthand experience, or somewhere else? How do these sources influence the ways you look upon poor people?
2. How does Linda Tirado's story lend support to the individual perspective toward poverty?
3. Why are many Americans opposed to welfare?

The Opportunity Divide: How American Society Produces Economic Inequality

2.2 Describe the opportunity divide that exists in the United States.

Did you play musical chairs when you were a child? You may recall that kids circle around a row of chairs and when the music stops, everyone tries to sit down. Since there's one fewer chair than the number of players, someone always comes up short. In the end, only one person is left sitting. I remember when I played that some kids

In addition to being a popular children's game, musical chairs is emblematic of how our society produces inequalities in jobs, education, housing, and other opportunities.
iStockphoto.com/SerrNovik

would try as aggressively as they could to win. I never liked those kids; they took this simple and fun game way too seriously.

Regardless of how well people play musical chairs, all but one person eventually loses. That reality is built into the structure of the game. In a similar way, poverty is carved into how our society is organized. Upward mobility is certainly possible, and the success of those who achieve it may stem from individual characteristics like hard work and self-discipline. However, for millions of others stuck in poverty, it's typically not because they haven't embraced these traits. Instead, they're constrained by social forces beyond their control.[7]

To see why, we need to examine the **opportunity divide**: People born into higher-income families have lots of chances to better their lives, while those born into lower-income families have comparatively few. Paying attention to this divide focuses our attention on how economic inequality is eroding the American Dream. Consider the opening passage of a book social critic Robert Putnam wrote about the opportunity divide:

> My hometown was, in the 1950s, a passable embodiment of the American Dream, a place that offered decent opportunity for all the kids in town, whatever their background. A half century later, however, life in Port Clinton, Ohio, is a split-screen American nightmare, a community in which kids from the wrong side of the tracks that bisect the town can barely imagine the future that awaits the kids from the right side of the tracks.[8]

Putnam's words highlight our country's failure to live up to its image as a place where anyone who works hard can achieve upward mobility.

The withering of the American Dream is evident in towns and cities across the country. Consider that 92 percent of people born in 1940 earned more as adults than did their parents, with the greatest gains occurring among the poor and middle class. Yet only half of those born in 1984 outearned their parents. Figure 2.4 illustrates poor kids' dim chances of doing better financially as adults than their parents. Among people born into the bottom quintile (20 percent) of families based on income, 43 percent stay there as adults; 70 percent either remain or move up only slightly. A person's family income at birth has increasingly become an indicator of their family income as an adult. Biology is predictive of destiny.[9]

Since low-income people are often unable to achieve upward mobility no matter how hard they try, the individual perspective cannot adequately explain these inequalities. A better explanation explores why a society that trumpets "opportunity for all" provides many more chances to the haves than to the have-nots.

FIGURE 2.4 • Stuck at the Bottom

Despite efforts to achieve the American Dream, people born into low-income families often fare no better financially as adults.

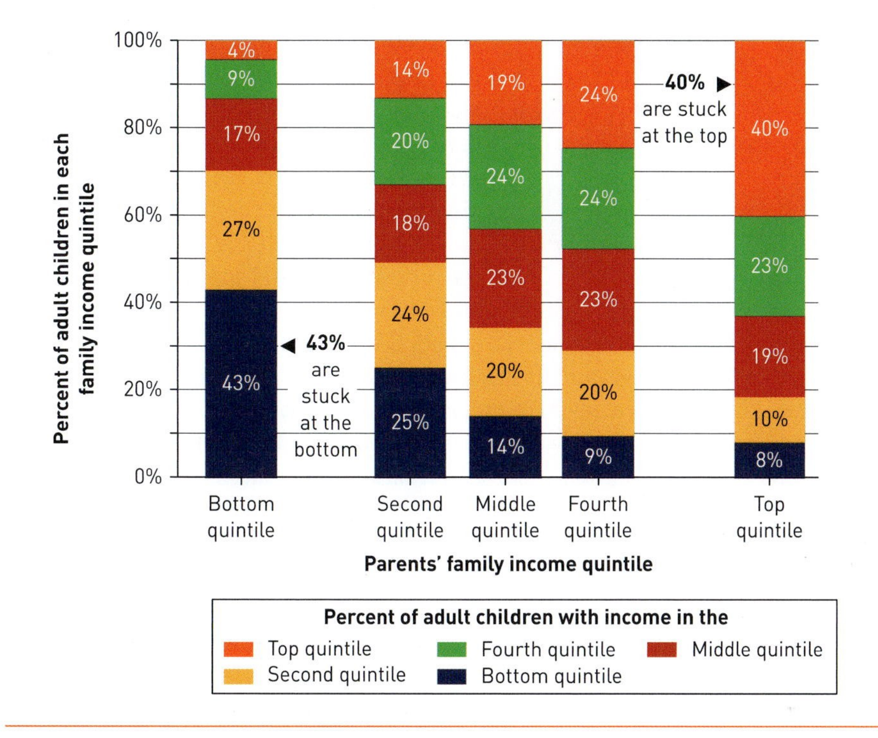

Source: Pew Charitable Trusts, "Pursuing the American Dream: Economic Mobility across Generations," July 2012, http://www.pewtrusts.org/~/media/legacy/uploadedfiles/wwwpewtrustsorg/reports/economic_mobility/pursuingamericandreampdf.pdf.

Let's develop this sociological perspective by chronicling stories of people whose lives have had very different measures of success and failure. Despite variation in their experiences, we'll see how similar social forces explain the directions these Americans' lives have taken.

Blocked Opportunities: How Social Forces Impede Low-Income Youth from Living the Dream

2.3 Explain why some students succeed in school and others do not.

Think about the pleasure you have felt whenever you've read a book so riveting that you couldn't put it down. That was my experience with Jay MacLeod's *Ain't*

In addition to the examples mentioned in this chapter, what other forms of cultural capital are important for academic success?
iStockphoto.com/PeopleImages

No Makin' It, which is what got me hooked on sociology when I was in college. The author studied a group of Black teenage boys living in a housing project near Boston. Despite their poor upbringing, these boys—whom MacLeod called the Brothers—believed someday they'd have good jobs and own houses. MacLeod followed up with them twice, seven years and twenty-two years after he conducted the study I read in college. What he discovered shattered popular beliefs about the American Dream. Though they had followed the script expected of low-income kids striving to achieve a better life—working hard and staying out of trouble—the Brothers fared only marginally better financially than their parents had.

One major reason is because the Brothers lacked **cultural capital**, or the particular types of knowledge one needs in order to achieve upward mobility. Kids with this knowledge are most likely to have acquired it at home. The Brothers' parents, however, didn't impress upon them that there are various steps one needs to take in order to apply to college. As you know, these include taking college prep courses, forging relationships with high-school teachers who can write letters of recommendation, and researching universities that match your skills and interests.

Even though the Brothers were industrious, cultural capital isn't something they could gain simply by working harder. After all, not having it meant they didn't know what they needed to know. Therefore, lacking cultural capital wasn't the Brothers' fault. Nor was it their parents' fault; since the parents didn't have this knowledge, they couldn't transmit it to their sons.

In order to explain why poor students are often disadvantaged by a lack of cultural capital, take a look at Figure 2.5. It presents data about kids' performance on the SAT, which of course is a critical factor in college admissions. You can see that those from lower-income families typically do not fare as well as their higher-income peers. It's not that richer kids are born with greater intellectual gifts or a more disciplined work ethic. Rather, various social forces give them a leg up in being prepared to do well on high-stakes tests. The sociological perspective enables you to discover the factors that play into these kids' advantages. In what ways are higher-income parents better positioned than lower-income parents to give their kids resources for success on the SAT?

The Brothers' lack of cultural capital was a key reason why none of them earned a four-year degree after graduating from high school. While having limited funds to pay for college is certainly a barrier, we're seeing that the obstacles low-income kids face in attending college go well beyond cost. Several Brothers did attend community college, but since they didn't have a clear sense of how this might lead them toward a four-year degree or directly into a good job, they dropped out. The one person who enrolled at

FIGURE 2.5 ● Test Scores Reflect Family Background

How well a student does on the SAT can be predicted by their parents' income.

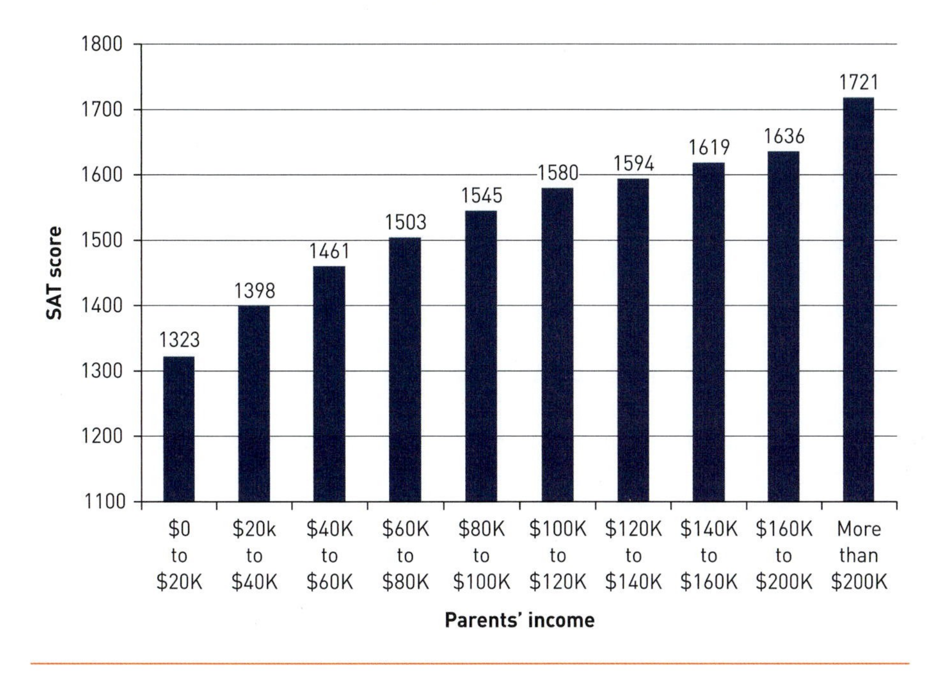

Source: Daniel H. Pink, "How to Predict a Student's SAT Score: Look at the Parents' Tax Return," 2012, https://www.danpink.com/2012/02/how-to-predict-a-students-sat-score-look-at-the-parents-tax-return.

Note: This graph is based on data from 2011, when a perfect SAT score was 2400.

a state university also left. He felt ostracized, not knowing how to fit in with his many peers from suburbs and small towns.[10]

The Brothers' story is emblematic of the experience of other low-income youth. Whereas 81 percent of kids who come from the highest-earning quintile of families (the top 20 percent) are enrolled in a two- or four-year college, just 51 percent of kids from the bottom quintile are. There's also a gap in graduation rates. Two-thirds of kids from the top quartile (the top 25 percent) earn a bachelor's degree, yet fewer than half as many kids from the bottom quartile do. These data, as a whole, highlight the significance of **social class** in shaping opportunity. Social class refers to a

Even if low-income students know about financial aid and other resources for applying to college, these students may still lack the cultural capital needed to persevere through academic and personal challenges so they can graduate.
iStockphoto.com/ericsphotography

person's financial standing and how it relates to other types of resources and opportunities they have or don't have, such as their level of educational attainment, type of occupation, the people they know, and their cultural capital.[11]

Becoming familiar with the Brothers' story offers you a mirror for examining your own situation in ways you probably have never done before. If you grew up with a clear understanding of the college process, you now understand why inequalities in cultural capital are so consequential in explaining why some kids pursue higher education and others don't. If your childhood more closely resembled the Brothers', you're now aware of a rarely advertised payoff of college: It offers opportunities for gaining knowledge that can help you succeed. I don't mean formal knowledge that comes from taking courses, but informal knowledge you can pick up along the way about how to advance yourself. Ask your peers—especially those who are close to graduating—for tips about how you can access opportunities during the rest of your time in school and beyond. If you do so, you may discover that acquiring cultural capital is one of the most valuable benefits of college.

The Brothers also faced racial barriers in their efforts to get ahead. Employers were often biased against hiring them because they were Black. Even though they were well-mannered and respectful, employers didn't necessarily infer this demeanor from their appearance. The ways they dressed and spoke signaled that they were lower class, and some employers clung to the prejudice that low-income Black males are thugs who can't be trusted. (See Chapter 3 for a fuller discussion of this racial bias and its effects.)[12]

The Brothers' story exposes how factors outside a person's control may impede poor people from fulfilling their aspirations for success. Having a strong work ethic hardly guarantees that someone will get ahead. Moreover, for low-income minority youth to believe they can succeed when the odds are stacked against them can be damaging to their self-esteem. They're prone to blame themselves for their failure to get ahead and may also believe that negative stereotypes about their racial or ethnic group are a valid explanation for their failure.[13]

Trying hard in school but not achieving upward mobility may lead poor minority kids to believe they're failures.
iStockphoto.com/asiseeit

The Brothers' experiences highlight how scarring it can be for poor minority kids to believe they have personal control over their lot in life. These boys' story exposes why the individual perspective is shortsighted. It's unreasonable simply to blame high-aspiring youth for their struggles to escape poverty. These boys did exactly what our society expects of kids who want to achieve the American Dream, and yet they still came up short. Their story underscores why it's crucial that we shine a spotlight on the social forces causing the United States to fail to live up to its promise as the land of opportunity.

Tarnished Hopes for the Future: How Poverty Impedes Low-Income People's Motivations to Get Ahead

2.4 Recognize why poor people may be unmotivated to better their lives.

Let's now return to the story of Linda Tirado, the woman we met at the beginning of this chapter. Whereas the Brothers believed that hard work would propel their lives forward, poverty stymied Tirado's ability to plan for the future. She described in her blog post her fatigue from working two low-wage jobs while going to school and raising a child:

> You have to understand that we know that we will never not feel tired. We will never feel hopeful. We will never get a vacation. Ever. We know that the very act of being poor guarantees that we will never not be poor. It doesn't give us much reason to improve ourselves.
>
> We don't plan long-term because if we do we'll just get our hearts broken. It's best not to hope.[14]

These words underscore how poverty can strip a person of the motivation to overcome it. A person's brain responds to chronic hardship by producing stress and fear, both of which impair the capacity to think beyond the immediate situation. Given the neurological obstacles they face in setting goals and completing tasks, it's often hard for poor people to imagine a better future.[15]

The sociological perspective reveals why those who can least afford the cost of smoking are the most likely to do so.[16]
iStockphoto.com/Mac99

Because of these obstacles, someone living paycheck to paycheck may make choices that seem foolish to those who haven't experienced similar hardships. For a poor person to spend $10 or more for a pack of cigarettes may strike you as irresponsible. But to Linda Tirado, smoking offered relief from her daily grind. It also helped her stay awake and alert. The same was true for consuming chips, candy, soda, and other junk foods. They were not only affordable; they provided one of the few forms of satisfaction available to her. Spending money on these habits enabled Tirado to cope with the challenges of making ends meet.

The manufacturers of cigarettes and junk food understand the constraints poor people face and seek to exploit them. Tobacco companies concentrate their storefront and billboard advertising in poor neighborhoods and also sometimes offer low-income consumers coupons for future purchases. Likewise, companies like Frito-Lay and Coca-Cola aggressively market junk foods that are high in salt, sugar, and fat to low-income minority communities. Given the daily struggles of living in poverty and the power of tobacco and food companies to promote products that offer cheap and addictive pleasures, these products are practically irresistible to low-income consumers.[17]

Living hand to mouth can inhibit a person from believing they can have a better life, prompting them to spend their limited income on immediate forms of gratification, like junk food.
iStockphoto.com/nkbimages

Given the particular constraints that poor women face, this sight is familiar among America's homeless population.
iStockphoto.com/AvailableLight

We're seeing how the sociological perspective explains why someone like Linda Tirado may make choices that compound her hardships. Her story offers a different sociological explanation for the opportunity divide than we saw with the Brothers. They had a strong work ethic yet were unable to achieve college success or land good jobs. Social forces blocked them from *achieving* upward mobility. Tirado's account highlights that being poor can impede a person from developing a strong work ethic in the first place. Social forces blocked her from *believing* upward mobility was even possible.

These forces particularly impact poor women. In addition to working in low-wage jobs, they typically shoulder greater parenting responsibilities than the fathers of their children. Like other women, they're also susceptible to gender bias. Because of these constraints, in 2016 women were 38 percent likelier than men to be poor. Among families headed by single mothers, 35.6 percent lived in poverty—over twice the poverty rate (17.3 percent) for families headed by single fathers and over five times the rate (6.6 percent) for two-parent families. Sociologists use the term **feminization of poverty** to refer to women's greater likelihood than men to be poor.[18]

Linda Tirado's story took an improbable turn after she wrote her blog post. Not only did the post go viral but people came out of the woodwork with offers of help. So Linda created a GoFundMe, which raised more money than she'd been earning in her two jobs. Moreover, a top publisher invited her to write a longer account of her experiences living in poverty. Her book became a commercial success. The odds that a woman of her background with dim hopes for the future would become upwardly mobile were about as likely as her winning the lottery.[19]

A Personal Tale from the Other Side of the Opportunity Divide: How Social Forces Bolster Success

2.5 Discuss how the same social forces that enable some people's successes contribute to others' hardships.

My life is night and day compared to the people discussed so far in this chapter. I earn an income that puts me among the top 15 percent of Americans. I have a stable career and a happy marriage. I have two well-adjusted kids and a cute dog. I don't say these things to gloat. My aim is instead to bring humility to my achievements—to show you that the sociological perspective can explain the good fortune of someone like me just as it can deepen your understanding of poor people's struggles.

From an individual perspective, my accomplishments are the result of effort and perseverance. I did indeed work very hard during my ten years spent in college and graduate school. But, if I'm being honest—and a good sociologist!—I have to acknowledge that this perspective only partially explains my achievements. Although I've worked hard, that's not the whole story.

Inequalities across school districts in the amount of money they spend per student produce inequalities in kids' chances to do well academically.
Sangoiri/Shutterstock.com

Consider that I grew up in a spacious, single-family house in Scarsdale, a suburb of New York City. The kids I knew spanned the spectrum of the middle class. Some families were well-off, but that level of wealth now pales in comparison to how the community has changed over the past few decades. No longer can any but the superrich afford to live in Scarsdale anymore. With a median family income of $241,453, it's the wealthiest town in the U.S.[20]

Long before Scarsdale came to epitomize the very top of the opportunity divide, there were distinct advantages to living there. I attended Edgemont High School, which for years has been among the top-ranked in the country. Even though I graduated in the 1980s, data from 2015–16 offer a snapshot of the perks I enjoyed. That year, the school district spent $28,551 per student, which was 22 percent more than the average for New York as a whole. The poorest kids in the state attended schools that allocated less than 50 percent of the amount per student as Edgemont.[21]

It's not hard to see disparities between the social class disadvantages these kids had in school and the advantages available to me. Jonathan Kozol, a longtime activist for equity in public education who studied several poorly funded schools in New York City, offered a riveting account of one of these schools:

> I had made repeated visits to a high school where a stream of water flowed down one of the main stairwells on a rainy afternoon and where green fungus molds were growing in the office where the students went for counseling. A large blue barrel was positioned to collect rainwater coming through the ceiling.[22]

This kind of environment obviously hampers learning. It makes kids think that if their school doesn't place value on their education, then why should they?

The scene Kozol described starkly differs from Edgemont High School, which is located on a sprawling campus lined with trees and gardens. My classes were small,

Dilapidated schools have to spend funds for repairs that could otherwise be used to purchase instructional materials or pay higher salaries to more qualified teachers.
Getty/ranplett

Extracurricular activities enable kids to develop self-confidence that contributes to academic success, yet many parents can't afford to pay for their kids to participate in these activities.
iStockphoto.com/adamkaz

and teachers gave me lots of individualized attention. This was critical since my visual impairment made it difficult for me to see notes written on the blackboard. Whereas I could have floundered academically, the extra help teachers provided enabled me to reach my potential.

The support I received at school multiplied the benefits of having a family invested in my educational success. Higher-class parents can most afford to make these investments. Even though lower-income parents may want what's best for their children, these parents often don't have the resources to provide what can enable their kids to thrive. Therefore, while some kids have a relatively easy time maintaining the social class advantages they received at birth, others struggle to become upwardly mobile.[23]

My parents both went to college, held professional jobs, and could afford to pay for my expenses at a private, four-year institution. I graduated without loans, which meant I could start saving money for a house long before this was possible for my peers who had incurred considerable debt. Moreover, my parents contributed to the down payment on the home that my wife Nancy and I own.

It's also significant that I'm White. The house I grew up in was my parents' second suburban address. They had bought their prior house in 1966 when there was still **redlining**—a practice that barred Blacks from purchasing real estate in certain communities. Although this practice legally ended six years before my parents bought the Scarsdale house in 1974, the unequal effects of redlining lingered. By being able to buy a starter home in the mid-1960s, my parents were able to make a profitable investment at a time when many Black families could not. When they sold that house at a much higher price, my parents had enough money to make a down payment on the Scarsdale house.

Historic housing discrimination is a leading reason for racial inequalities in wealth. A person with an annual income of $40,000 and a house valued at $400,000 is wealthier than a person who earns $100,000 a year and owns a house worth $300,000. Moreover, wealth typically grows more rapidly than income, as my parents experienced by buying a starter house that significantly appreciated in value. Figure 2.6 highlights the stark wealth gap between Whites and Blacks: $110,729 versus $4,955. This is over 22:1. Whereas 70 percent of White adults own homes, the homeownership rate for Blacks is just 41.2 percent.[24]

Comparing my story to the two others discussed in this chapter highlights that in order to explain inequalities in achievement of the American Dream, we must take into account **intersectionality**, which is the idea that a person's identities are interwoven. My social class and race combine to create advantages for me in the same ways that they produce disadvantages for low-income Black people. Moreover, being male is another key piece of my success. We saw earlier in relation to Linda Tirado's story that women are likelier to be poor than men. Figure 2.7 (on page 26) highlights, moreover, that being White and male advantage me compared to women of color. Race, gender, and social class combine to produce significant inequalities of opportunity.

Despite the many advantages I've had, they still haven't guaranteed my successes. Take another look at Figure 2.4, discussed on page 17. Eight percent of kids born into the top quintile of families by income (the top 20 percent) ended up in the bottom quintile as adults—an indicator that downward mobility is possible. A person may lack initiative or experience unexpected hardship. Since I could have struggled to launch a successful career, the individual perspective is certainly a useful way to understand my story, but it hardly captures the whole picture.

FIGURE 2.6 ● Wealth Inequality

There is a huge disparity in the U.S. between Whites and Asians as compared to Blacks and Hispanics.

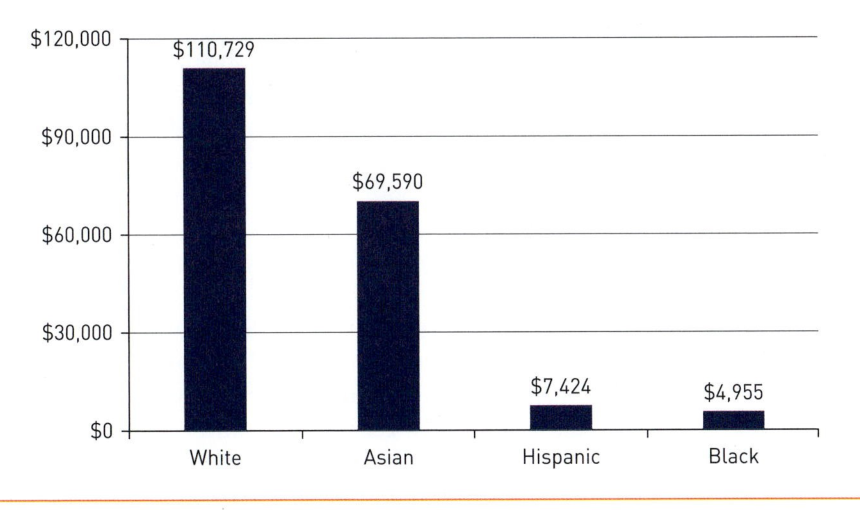

Source: Tami Luhby, "Worsening Wealth Inequality by Race," CNN Money, June 21, 2012, http://money.cnn.com/2012/06/21/news/economy/wealth-gap-race/index.htm. Data source: U.S. Census Bureau.

FIGURE 2.7 ● White Male Advantage

There is significant inequality in earnings based on race and gender.

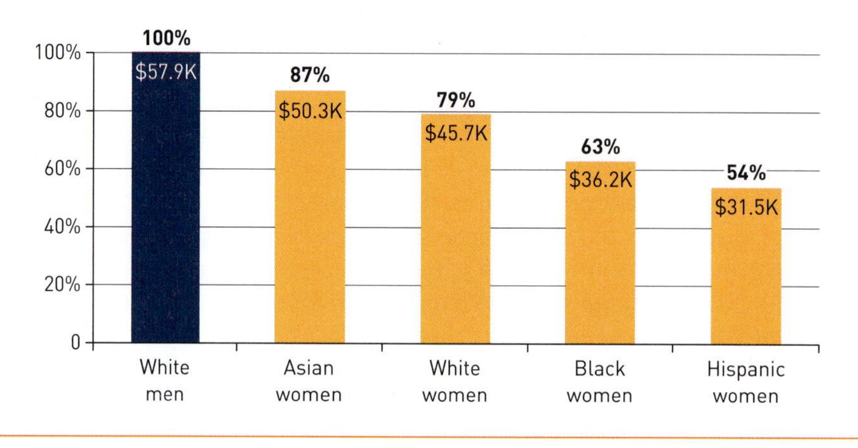

Source: Sonam Sheth, Shayanne Gal, and Skye Gould, "Six Charts Show How Much More Men Make Than Women," *Business Insider*, August 27, 2018, https://www.businessinsider.com/gender-wage-pay-gap-charts-2017-3/#the-gender-wage-gap-varies-widely-depending-on-the-state-1. Data source: U.S. Census Bureau, "2017 American Community Survey."

The sociological perspective alternatively emphasizes how the advantages I grew up with turned the tides in my favor. Having a supportive family, financial security, access to a high-quality education, and racial privilege gave me both the drive to work hard over many years and the confidence that trying to advance myself was worth the effort. Notice how these details contrast with the earlier stories you read about Linda Tirado and the Brothers. Whereas working two low-wage jobs to make ends meet impeded Tirado from thinking about how to achieve a better life, I had lots of reasons to believe my chances for success were high. I also had the cultural capital to know how to advance myself educationally, which the Brothers lacked. Indeed, many social forces influence how far a person is likely to advance educationally. And, as Figure 2.8 highlights, there is a strong relationship between one's level of educational attainment and their employment opportunities.

Moreover, the particular circumstances of my upbringing enabled me to see beyond the conventional wisdom that people create their own lot in life. My parents worked in a very different type of neighborhood than the bucolic setting of our suburban home. They were both lawyers, partners in work as well as marriage. When I wasn't in school, I'd occasionally help them by filing legal papers and doing other clerical work. What I remember most about going to their office was the surrounding neighborhood. It was two blocks from Yankee Stadium in one of the poorest sections of New York City. There were boarded-up buildings and panhandlers standing on street corners.

FIGURE 2.8 ● Credentials Matter

As a person attains more formal education, they're likelier to earn a better income and avoid unemployment.

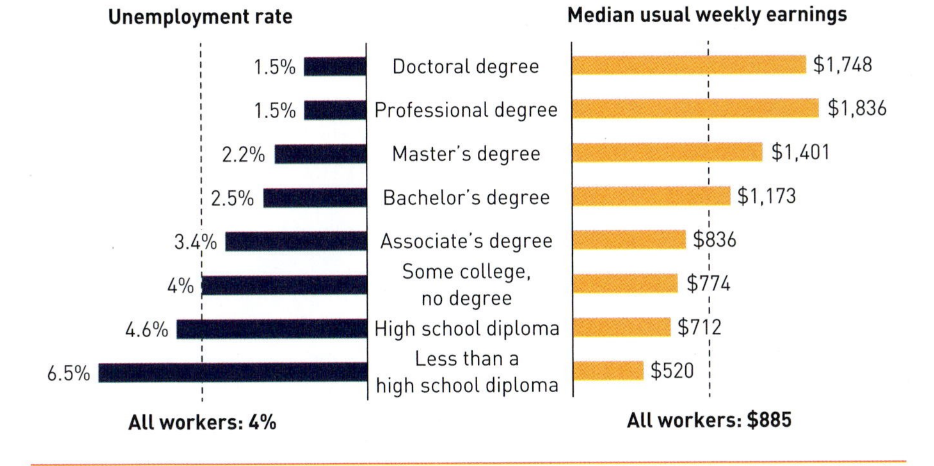

Source: Bureau of Labor Statistics, 2017.

Sociology enables me to recognize that scenes of neighborhood disparity I noticed as a child but didn't understand are telling indicators of broader inequalities in American society.
iStockphoto.com/artisticco

I was just a kid, so I had no idea why my eyes focused on these contrasts. It wasn't until I took my first sociology course in college that I started to connect the dots. I began to understand how a combination of individual and social forces explains the opportunities a person has in life. I can now see that the successes I've enjoyed and the struggles of people like the Brothers and Linda Tirado share commonalities. Whereas the advantages I grew up with have enabled me to reap ongoing benefits, the deck has been stacked against these low-income people enjoying similar successes.

Rethinking Welfare: How the Sociological Perspective Can Widen Your Understanding of Who Deserves Help

2.6 Explain why Americans are often resistant to providing aid to those living in poverty.

Ten years after my parents closed their law practice in 1999, their neighbors opened a new ballpark a block from where the old Yankee Stadium had been. For several years before construction plans were finalized, owner George Steinbrenner had threatened to move the Yankees across the Hudson River to New Jersey. Since this would have been a huge loss for the city, it agreed to subsidize $528 million of the overall building costs in partnership with the State of New York. A **subsidy** is government money—which comes from taxpayers—given to a recipient that reduces the recipient's out-of-pocket price for goods or services. Including subsidies from the federal government, the total public expense of the new ballpark rose to a whopping $1.2 billion. This meant taxpayers funded more than half of the $2.3 billion overall cost. And this wasn't the only publicly financed ballpark to open in New York City in 2009; the crosstown Mets (my favorite team!) unveiled Citi Field the same year.[25]

Over the past few decades, nearly every Major League Baseball franchise has gotten a new stadium, with taxpayers footing most of the bill. The strong public support for these construction projects sits in contrast to the sentiment toward government spending for programs that help poor people. We saw at the beginning of this chapter that in American society there is significant opposition to welfare. On the other hand, politicians have been able to convince urban residents that new ballparks are a worthy investment because they generate tax revenue to make up for the subsidies, and then some. Welfare recipients, on the other hand, are allegedly moochers who offer society nothing in return for the assistance our tax money provides them. We've already exposed one reason to call into question this way of thinking: It masks the social forces that keep people in poverty and therefore in need of help. The other shoe is about to drop.

The high regard baseball has as the "national pastime" contributes to public support for subsidizing new stadiums.
iStockphoto.com/AlexLMX

One of the most eye-opening facts I learned in doing the research for this book is how common it is for Americans to experience poverty. A study by sociologist Mark

Rank of people's annual earnings over the course of their careers found that 58.5 percent of those studied had been poor for at least a year, 45 percent for two or more years, 38.6 percent for three or more, and 33.9 percent for four or more. Therefore, a majority of Americans experience poverty at some point in their lives. "For most of us," Rank writes, "the question is not whether we will experience poverty, but when."[26] Therefore, one of the assumptions fueling opposition to welfare—that poor people's values are at odds with the mainstream—often isn't true. For many low-income individuals who at one time experienced financial security, their lives spiraled downward because of a job loss, illness, or divorce—*not* because they suddenly and unexplainably became irresponsible.

Contrary to popular belief, any of us may unexpectedly fall into a situation where we temporarily need help from the Supplemental Nutrition Assistance Program (SNAP) or other welfare programs in order to get back on our feet.
iStockphoto.com/jetcityimage

Moreover, consider how often adults turn to the government for help. Rank found that between ages twenty and sixty-five, nearly two-thirds of Americans benefit from welfare. Over 58 percent use it for two or more years and well over half (52.7 percent) for three or more years. These statistics dispel negative stereotypes surrounding welfare. Those who accept Medicaid or food stamps do not reject traditional values like hard work and self-discipline.[27]

On what basis does it make sense to think about welfare in connection with Major League Baseball?
iStockphoto.com/adamkaz

In fact, low-income people are much *less* likely to receive help from the government than Major League Baseball teams and other corporations. These companies collectively get hundreds of billions of dollars in federal subsidies and tax breaks each year. I say this with no axe to grind against baseball, but as an avid fan. However, my love of the game is beside the point. Given how often people question poor Americans' deservedness of help, it's reasonable to ask whether the Yankees, Mets, and other teams with subsidized ballparks are providing as much public value as taxpayers have doled out to build them.[28]

Taking Collective Responsibility: Why the Withering of the American Dream Is Everyone's Problem

2.7 Describe the shared societal costs of economic inequality.

Each school day, over thirty million kids receive a free or reduced-price meal from the federally subsidized National School Lunch Program. This program enables low-income children to learn about healthy eating habits and ensures that they're adequately nourished so they can have a chance to do well in school. Subsidizing lunches seems like a worthy use of government funds—which come from our tax dollars—since schools cannot possibly be engines of the American Dream if some kids are unable to concentrate due to hunger.[29]

Don't we have a responsibility to promote equal opportunity given the disadvantages that stem from economic inequality? These disadvantages include the following:

- Low-income kids experience higher rates of asthma and lead poisoning than kids living in higher-income neighborhoods.
- Poor children are less likely to receive routine medical care, and when they do, it tends to be from a neighborhood or hospital-based clinic rather than a private doctor.
- By the time kids enter kindergarten, test scores of literacy and math on average tend to be 60 percent lower among the poorest children than among the most affluent.
- This gap grows during elementary and middle school, to the point that students from low-income families are six times more likely to drop out of high school than kids from high-income families.

You might wonder how any of these inequalities affect those who don't directly experience them. Let's consider what happens when a vast segment of the population grows up unhealthy and undereducated. They go to the emergency room for costly procedures that could have been avoided if they'd had better primary care, and they don't productively contribute to the economy. Moreover, they're likely to spend time in prison, which has a *higher* per capita annual expense ($31,286) than it costs to attend many universities. In all, childhood poverty costs our society a staggering half a trillion dollars a year.[30]

School is where you've heard the strongest message that how hard you work influences how much you succeed. Embracing the sociological perspective, therefore, requires critically reevaluating this core belief you've been taught since you were very young.
iStockphoto.com/AntonioGuillem

Nonetheless, it may be tempting to ignore the huge social and economic costs of poverty. Such denial is a by-product of the individual perspective, which views the hardships poor people experience as their own fault. Personalizing the reasons for other people's struggles can validate one's own successes. From a very young age, we hear the message that hard work is the key to getting ahead in life. This may incline a person to see their achievements as wholly earned. As a result, it becomes easy to feel contempt toward those who have little and convenient to believe that if they only worked harder, they'd be better off.

The sociological perspective elevates your understanding of these inequalities by exposing the social forces that influence people's choices. Integrating sociological thinking into how you understand the withering of the American Dream doesn't make excuses for people's struggles; it offers a clearer picture of why a person's life takes a particular course. Start paying attention to how the individual and sociological perspectives jointly enable you to understand the chances you had growing up and the shape your life may take in the years to come. The goal isn't to feel guilty about having had advantages or self-pity for having struggled but rather to gain insight about the realities of America's opportunity divide.

What Do You Know Now?

1. In what sense are explanations for inequality of opportunity in American society analogous to the reasons why people win or lose at musical chairs? Explain how, just like in the classic children's game, being successful in life is the result of *both* individual and sociological factors.
2. How did the Brothers' lack of cultural capital hurt them in their efforts to achieve a better life?
3. How does Linda Tirado's account of living in poverty challenge the individual perspective that the reason people like her struggle to get ahead is because of bad choices they've made?
4. Draw on details about the author's life to illustrate how his achievements have been the result of both personal efforts and social advantage.
5. Why is it significant that more than half of Americans experience poverty at some point in their lives?
6. How does your life story fit within this chapter? In other words, how do the opportunities you've had compare to those of the people you read about?

Key Terms

Poverty 11
Upward mobility 12
American Dream 12
Welfare 13
Economic inequality 14
Wealth 14
Opportunity divide 16
Cultural capital 18
Social class 19
Feminization of poverty 22
Redlining 24
Intersectionality 25
Subsidy 28

Visit **edge.sagepub.com/silver** to help you accomplish your coursework goals in an easy-to-use learning environment.

Notes

1. Linda Tirado, "This Is Why Poor People's Bad Decisions Make Perfect Sense," *Huffington Post*, November 22, 2013, https://www.huffingtonpost.com/linda-tirado/why-poor-peoples-bad-decisions-make-perfect-sense_b_4326233.html.
2. Andrew Ross Sorkin and Megan Thee-Brenan, "Many Feel the American Dream Is out of Reach, Poll Shows," *New York Times*, December 10, 2014, https://dealbook.nytimes.com/2014/12/10/many-feel-the-american-dream-is-out-of-reach-poll-shows/.
3. Nutrition Assistance Program Report, U.S. Department of Agriculture, "Foods Typically Purchased by Supplemental Nutrition Assistance Program (SNAP) Households," November 2016, https://fns-prod.azureedge.net/sites/default/files/ops/SNAPFoodsTypicallyPurchased.pdf; Martin Gilens, *Why Americans Hate Welfare: Race, Media, and the Politics of Antipoverty Policy* (Chicago: University of Chicago Press, 1999), 1–10.
4. Reagan started referring to "welfare queens" during his unsuccessful presidential campaign in 1976. He returned to this image after winning the presidency in 1980. See Ian Haney-Lopez, "The Racism at the Heart of the Reagan Presidency," *Salon*, January 11, 2014. http://www.salon.com/2014/01/11/the_racism_at_the_heart_of_the_reagan_presidency/.
5. In 2016, the Census Bureau reported that there were approximately forty-three million poor Americans; see U.S. Census Bureau, "Income, Poverty and Health Insurance Coverage in the United States: 2015," September 13, 2016, https://census.gov/newsroom/press-releases/2016/cb16-158.html.
6. Tim Worstall, "Well of Course 1 in 2 Americans Are Poor or Low Income, Naturally, It's Obvious," *Forbes*, December 16, 2011, https://www.forbes.com/sites/timworstall/2011/12/16/well-of-course-1-in-2-americans-are-poor-or-low-income-naturally-its-obvious/#4f83842e3895. The gap between the four hundred wealthiest Americans versus 150 million others is reported in Gabriel Zucman, "Global Wealth Inequality," Working Paper 25462, National Bureau of Economic Research, January 2019, https://papers.nber.org/tmp/38195-w25462.pdf.
7. Mark Rank, "Rethinking American Poverty," *Context* 10, no. 2 (2011): 16–21.
8. Robert D. Putnam, *Our Kids: The American Dream in Crisis* (New York: Simon & Schuster, 2015), 1.
9. Raj Chetty et al., "The Fading American Dream: Trends in Absolute Income Mobility since 1940," *Science* 356, no. 6336 (April 2017): 398–406; David Leonhardt, "Our Broken Economy, in One Simple Chart," *New York Times*, August 7, 2017, https://www.nytimes.com/interactive/2017/08/07/opinion/leonhardt-income-inequality.html?emc=edit_ty_20170808&nl=opinion-today&nlid=73017964&te=1&_r=0.
10. Jay MacLeod, *Ain't No Makin' It: Aspirations and Attainment in a Low-Income Neighborhood* (Boulder, CO: Westview Press, 2009), 216–8.
11. Drew Desilver, "College Enrollment Among Low-income Students Still Trails Richer Groups," Pew Research Center, January 15, 2014, http://www.pewresearch.org/fact-tank/2014/01/15/college-enrollment-among-low-income-students-still-trails-richer-groups/;

Matt Bruenig, "The College Graduation Gap," Demos, May 18, 2004, https://www.demos.org/blog/5/18/14/college-graduation-gap.

12. Other sociological research reveals evidence of this racial bias that Jay MacLeod observed. See Kathryn M. Neckerman and Joleen Kirschenman, "Hiring Strategies, Racial Bias, and Inner-City Workers," *Social Problems* 38, no. 4 (1991): 433–47 and Devah Pager, Bruce Western, and Bart Bonikowski, "Discrimination in a Low-Wage Labor Market: A Field Experiment," *American Sociological Review* 74 (2009): 777–99.
13. Melinda D. Anderson, "Why the Myth of Meritocracy Hurts Kids of Color," *The Atlantic*, July 27, 2017, https://www.theatlantic.com/education/archive/2017/07/internalizing-the-myth-of-meritocracy/535035/.
14. Tirado, "This Is Why Poor People's Bad Decisions Make Perfect Sense."
15. Ibid.; Tara Garcia Matthews, "How Poverty Changes the Brain," *The Atlantic*, April 19, 2017, https://www.theatlantic.com/education/archive/2017/04/can-brain-science-pull-families-out-of-poverty/523479/.
16. Elizabeth M. Barbeau, Nancy Krieger, and Mah-Jabeen Soobader, "Working Class Matters: Socioeconomic Disadvantage, Race/Ethnicity, Gender, and Smoking in NHIS 2000," *American Journal of Public Health* 94, no. 2 (February 2004): 269–78.
17. Diana P. Hackbarth, Barbara Silvestri, and William Cosper, "Tobacco and Alcohol Billboards in 50 Chicago Neighborhoods: Market Segmentation to Sell Dangerous Products to the Poor," *Journal of Public Health Policy* 16, no. 2 (1995): 213–30; Cati G. Brown-Johnson et al., "Tobacco Industry Marketing to Low Socioeconomic Status Women in the USA," *Tobacco Control* 23 (2014): 139–46; Rudd Center for Food Policy & Obesity, "Food Advertising Targeted to Hispanic and Black Youth: Contributing to Health Disparities," August 2015, http://www.uconnruddcenter.org/files/Pdfs/272-7%20%20Rudd_Targeted%20Marketing%20Report_Release_081115%5B1%5D.pdf.
18. Tara Culp-Ressler, "How the United States' Growing Income Inequality Is Hurting Women's Mental Health," *ThinkProgress*, October 31, 2013, https://thinkprogress.org/how-the-united-states-growing-income-inequality-is-hurting-women-s-mental-health-ded33666da8/; Kayla Patrick, "National Snapshot: Poverty among Women & Families, 2016," National Women's Law Center Fact Sheet, September 2017, https://nwlc.org/wp-content/uploads/2017/09/Poverty-Snapshot-Factsheet-2017.pdf.
19. Linda Tirado's memoir is *Hand to Mouth: Living in Bootstrap America* (New York: Berkley Books, 2014).
20. Scarsdale, New York, ranks as the richest town in the U.S. based on 2016 data; see Devin Alessio, Steele Marcoux, and Christina Oehler, "These Are the 25 Richest Towns in America," *Veranda*, June 7, 2019, http://www.veranda.com/luxury-lifestyle/g1563/10-richest-towns-in-the-us/.
21. Sarah Moses, "Spending per Student NYS School Districts, 2015: Look Up, Compare Any District, Rank," Syracuse University, May 25, 2015, http://www.syracuse.com/schools/index.ssf/2015/05/spending_per_student_nys_school_districts_2015_lookup_compare_any_district_rank.html.
22. Jonathan Kozol, "Still Separate, Still Unequal," *Harper's Magazine* (September 2005): 41–54, 44.
23. Annette Lareau, *Home Advantage: Social Class and Parental Intervention in Elementary Education* (New York: Rowman and Littlefield, 2000).
24. Janelle Jones, "The Racial Wealth Gap: How African-Americans Have Been Shortchanged out of the Materials to Build Wealth," Working Economic Blog, February 13, 2017, http://www.epi.org/blog/the-racial-wealth-gap-how-african-americans-have-been-shortchanged-out-of-the-materials-to-build-wealth/; Mark Whitehouse, "Homeownership and the White-Black Wealth Gap," *Bloomberg News*, February 27, 2017, https://www.bloomberg.com/view/articles/2017-02-27/home-equity-and-the-white-black-wealth-gap.

25. Eliot Brown, "IBO: New Yankee Stadium Costing City, State $528 M," *Observer*, January 14, 2009, http://observer.com/2009/01/ibo-new-yankee-stadium-costing-city-state-528-m/.
26. Mark Rank, *One Nation, Underprivileged: Why American Poverty Affects Us All* (New York: Oxford University Press, 2005), 94; Mark Rank, "Poverty in America Is Mainstream," *New York Times*, November 2, 2013, https://opinionator.blogs.nytimes.com/2013/11/02/poverty-in-america-is-mainstream/?_r=0.
27. Rank, *One Nation*, 105.
28. Bill Quigley, "Ten Examples of Welfare for the Rich and Corporations," *Huffington Post*, January 14, 2014, http://www.huffingtonpost.com/bill-quigley/ten-examples-of-welfare-for-the-rich-and-corporations_b_4589188.html.
29. Data from 2012 indicate that more than 31.6 million kids received a free or reduced-price lunch each school day. See U.S. Department of Agriculture, "National School Lunch Program Fact Sheet," September 2013, https://www.fns.usda.gov/sites/default/files/cn/NSLPFactSheet.pdf.
30. Neal Halfon and Paul W. Newacheck, "Childhood Asthma and Poverty: Differential Impacts and Utilization of Health Services," *Pediatrics* 91, no. 1 (1993): 56–61; Brett Drake and Shanta Pandey, "Understanding the Relationship between Neighborhood Poverty and Specific Types of Child Maltreatment," *Child Abuse & Neglect* 20, no. 11 (1996): 1003–18; David Wood, "Effect of Child and Family Poverty on Child Health in the United States," *Pediatrics* 122, supp. 3 (2003): 707–11; Valerie E. Lee and David T. Burkam, *Inequality at the Starting Gate: Social Background Differences in Achievement as Children Begin School* (Washington, DC: Economic Policy Institute, 2002); Mark Santora, "City Annual Cost per Innate Is $168,000, Study Finds," *New York Times*, August 23, 2013, https://www.nytimes.com/2013/08/24/nyregion/citys-annual-cost-per-inmate-is-nearly-168000-study-says.html; Harry J. Holzer et al., "The Economic Costs of Childhood Poverty," *Journal of Children and Poverty* 14, no. 1 (2008): 41–61.

iStockphoto.com/duncan1890

3

"I Can't Breathe"

Policing, Race, and Violence

Learning Objectives

1. Recognize the effects of police brutality on the livelihood of Black Americans.
2. Discuss the role of implicit bias in how individuals view police brutality.
3. Explain how police brutality is a reflection of the fears of crime that pervade American society.
4. Identify the social forces that lead some Black males to engage in deviant behavior.
5. Discuss the importance of the sociological perspective in understanding the relationship between race and police brutality.

Violent Encounters between Cops and People of Color

3.1 Recognize the effects of police brutality on the livelihood of Black Americans.

Eric Garner gasped "I can't breathe" eleven times and then became unconscious. About half an hour earlier that afternoon, July 17, 2014, two White New York City police officers had approached the forty-three-year-old Black man outside a beauty supply store. They'd noticed him selling individual untaxed cigarettes, known as "loosies." Twice that spring, cops had arrested and released Garner for the same offense. Earlier in July, they'd issued him a warning. This time, when the two officers approached Garner, he begged them to leave him alone. Amidst his pleas, one of the cops restrained him around the neck and held on for about fifteen seconds. Garner suffocated to death. A friend standing nearby recorded the tragic scene.[1]

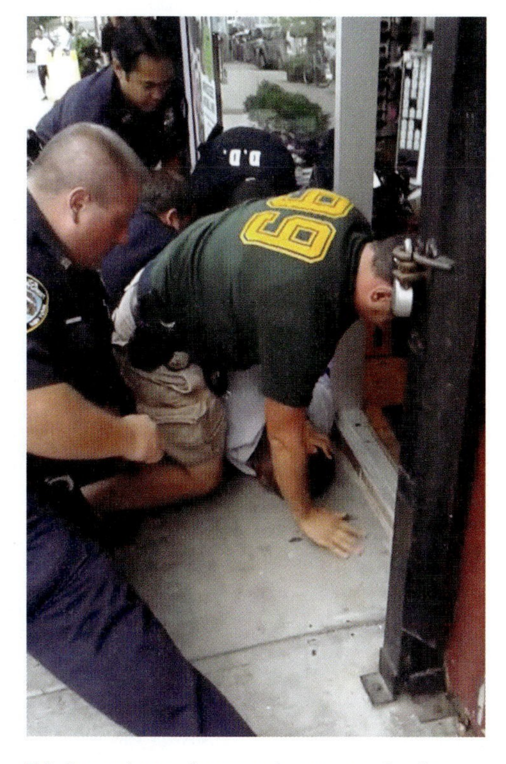

Eric Garner's story is among the many tragic tales publicized in recent years of suspects dying in police custody.
NY Daily News via Getty Images

This video and many others on the internet might give you the impression that cops killing unarmed suspects is a social problem that's spiraling out of control. Actually, what's changed is our capacity to know about this problem. The graphic footage available online and often featured in news reports makes it possible for anyone to get an up-close glimpse of these tragedies. However, they're not new. They've existed for as long as cities and towns have been employing law enforcement agencies to control crime—a practice that began in the 1830s amidst significant urban population growth.[2]

Scott Olson/Getty Images

Videos of White cops killing unarmed minority suspects paint the police in a negative light. That's unfair to the majority of officers who do their jobs responsibly. Some choose, moreover, to spend time outside of work volunteering for police athletic/activities leagues—coaching sports teams and mentoring kids in valuable life skills. Such officers, while off duty, give of themselves in order to better the lives of kids growing up in disadvantaged minority neighborhoods. It's vital that we keep these "good" cops in mind as we begin to focus our attention on the ones like Daniel Pantaleo—who choked Eric Garner to death—that fuel negativity about the police.

When I ask my students about interactions they've had with cops, glaring racial inequalities become apparent. It's rare for a Black or Brown person to have lived eighteen or more years without having been stopped by a police officer, but it's common for a White student. A disproportionate percentage of the twenty million annual stops of motor vehicles in the U.S. involves a minority, even though minorities aren't any likelier than Whites to break traffic laws. Cops pull over Hispanics 7 percent more frequently and Blacks 31 percent more frequently than Whites. Police officers also search people of color more often despite their being no more likely to possess illegal drugs or weapons.[3]

Stories like Eric Garner's are the most egregious examples of police officers' mistreatment of people based on race. Whereas Blacks make up 13 percent of the U.S. population, they comprise 31 percent of the people killed by cops.[4] Media images often focus on male victims; however, many Black women whose names are likely unfamiliar to you, such as Renisha McBride, Shereese Francis, Malissa Williams, and Meaghan Hockaday, have been killed by the police too.[5] All of these tragedies highlight that some officers view Black people as dangerous regardless of whether or not they're carrying a weapon. Unarmed Blacks like Eric Garner are 500 percent likelier than unarmed Whites to die at the hands of the police. Although such homicides are relatively rare compared to all Black victims of homicide, these killings demand investigation. After all, the perpetrators are public employees whom our society invests with the authority to protect all people justly and fairly.[6]

The slogan "Driving while Black" highlights racial inequalities in police stops and searches.
Ammentorp Photography/Alamy Stock Photo

Homicides by cops of unarmed Black people defy simple explanation. What is the line between situations where cops must resort to violence in order to avert danger versus situations of **police brutality**—officers unnecessarily, and therefore excessively, using force? While there is law pertaining to this question, the most interesting answers come from the court of public opinion. A poll conducted in 2014—the year Eric Garner was killed—found that while just 7 percent of Whites lacked confidence in the police to exercise their power responsively and avoid using excessive force, an astounding 33 percent of Black respondents felt that way. Other opinion data indicate that Blacks are about half as likely as Whites to believe that the police have an interest in protecting

FIGURE 3.1 ● Differing Perceptions of Cops

Blacks are more than twice as likely as Whites to view police officers as behaving too harshly toward suspects.

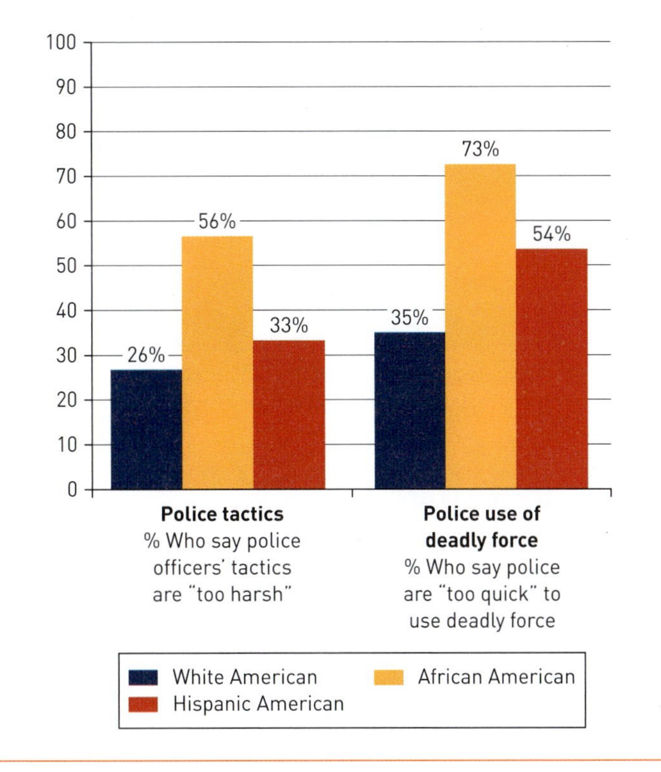

Source: Emiliy Ekins, "Policing in America: Understanding Public Attitudes toward the Police: Results from a National Survey," Cato Institute, December 7, 2016, https://www.cato.org/survey-reports/policing-america.

them. Figure 3.1 reveals that Blacks are much likelier than either Hispanics or Whites to believe that police practices are often harsh or violent. These data, as a whole, highlight that how a person understands police violence toward unarmed suspects is colored both by their background and how they view race relations in the U.S.[7]

Police violence toward unarmed Blacks is a social problem that demands investigation because it undermines our society's credo that all people deserve fair treatment under the law.
BrunoWeltmann/Alamy Stock Photo

You may be a person of color who's experienced mistreatment by the police or who lives in fear that someday you will. Perhaps you have relatives or friends who are cops, or you aspire to become one yourself after you graduate from college. Regardless of your biography and how it influences your perspective toward police killings of unarmed Black people, this chapter will expand how you think about these killings. Although they're reprehensible, they *are* still explainable.

We'll expose that officers who brutalize Black suspects aren't merely "bad" cops; they are responding to racial fears that pervade our society. Likewise, we'll see why it's shortsighted to view Black people who commit crime—and therefore come under police surveillance—simply as having acted irresponsibly. Their illegal behavior is, rather, a response to the limited opportunities available to them. The sociological perspective doesn't excuse cops who abuse their power or Blacks who break the law; these people are accountable for their actions. This perspective instead exposes hidden stories that can deepen your understanding of tragic encounters between cops and criminal suspects.

FIRST IMPRESSIONS?

1. What types of interactions have you had with the police throughout your life? How do these experiences contribute to your views about police homicides of unarmed people of color? How much trust do you place in the police to use force only when it's absolutely necessary?
2. Go to https://mappingpoliceviolence.org/ and read two accounts of cops killing unarmed people of color. What similarities do these accounts have with the homicide of Eric Garner?

Assigning Personal Blame, Either to Bad Cops or to Dangerous Black Men

3.2 Discuss the role of implicit bias in how individuals view police brutality.

Twenty years from now when you hear about an unarmed Black person dying at the hands of the police, the death of Eric Garner or a similar tragedy may come to mind. That was my experience when I watched the video of him gasping for his life. It resurrected images of a grisly scene I saw repeatedly during my last semester of college: the beating of Rodney King. On March 3, 1991, a Los Angeles man named George Holliday heard sirens outside his apartment and proceeded to record four White police officers repeatedly beating the twenty-five-year-old Black man after they'd stopped him for a traffic violation. Though this was over a decade before it was common for people to post videos on YouTube, millions saw Holliday's footage on TV.[8]

The beating influenced how my generation thought about violent encounters between White cops and Black suspects, just as more recent videos have shaped yours. In watching the raw footage, each of us makes sense of this violence by drawing on preconceptions we have about policing, race, and crime. Most often, we embrace an individual perspective and attribute responsibility either to the cops or to the suspects.

Many people see videos of White officers mistreating unarmed Black suspects as testimony of **explicit bias**, or the overt prejudices a person holds toward a particular group of people—in this case, a racial bias toward African Americans. This used to be how I saw these videos. In my view, the four White L.A. police officers who repeatedly beat Rodney King abused their power by crossing the line between legitimate and excessive force. From this individual perspective, White cops who injure or kill unarmed Black suspects are deviant because they violate the rules for appropriate police conduct. Tainted by explicit racial bias, they neglect their duty to act justly toward all Americans. It's eye-opening to consider that there are situations where the police are deviant since their work is supposed to control and reduce deviance among civilians.

After an all-White jury acquitted four White Los Angeles cops of excessively beating Rodney King, riots broke out in the city's South-Central neighborhood.
US Air Force Photo/Alamy Stock Photo

Data about law enforcement agencies' varying homicide rates of unarmed Black suspects offer support for this individual perspective. From 2013 to 2017, the homicide rate per one million Blacks living in Buffalo was 2.3, but this rate was 113 percent higher in New York City; one of these instances was Eric Garner's death. This comparison suggests that racism may have been common within the New York Police Department during this period. The bias, however, was more implicit than explicit. **Implicit bias** refers to the deep-seated, often unconscious prejudices a person holds toward a particular group of people. Blacks are 46 percent more likely than Whites to see police violence toward unarmed people of color as a reflection of implicit racial bias. This sentiment was instrumental in the creation of **Black Lives Matter**, an organization that's been a leading voice in exposing racial injustices in policing.[9]

After Eric Garner was choked to death, "I can't breathe" became a rallying cry among activists exposing the horrors of police brutality.
M. Stan Reaves/Alamy Stock Photo

Even as Black Lives Matter has attracted massive support, many people embrace a different individual perspective to explain police officers' mistreatment of unarmed minority suspects. In these people's eyes, blame rests not with the officers but the suspects. An indicator of this perspective's widespread influence is that only in relatively few of the approximately one thousand annual instances of police killings of minority suspects are the officers held criminally accountable. Just 35 percent of those who are charged receive a conviction, and many are never even charged. The grand jury that investigated Daniel Pantaleo, the officer shown on video putting Eric Garner in an illegal chokehold and maintaining it for fifteen seconds while Garner gasped for his life, didn't indict him. Since significant public presumption rests with cops, blame often falls instead on the victims of police violence.[10]

Some supporters of Blue Lives Matter see Black Lives Matter as giving people like Abdullah Brinsley the license to murder police officers. The photos at each end are the two cops he killed.
PACIFIC PRESS/Alamy Stock Photo

Some people who subscribe to this view see Black Lives Matter as belittling the courageous work cops do to maintain public safety. Another organization, **Blue Lives Matter**—which "seeks to honor and recognize the actions of law enforcement"—formed in 2014 after Abdullah Brinsley, a twenty-eight-year-old Black man, killed two New York City cops in broad daylight.[11] According to a post Brinsley made on Instagram, the grisly act was revenge for Eric Garner's death:

> I'm Putting Wings on Pigs Today. They Take 1 Of Ours. . .Let's Take 2 of Theirs #ShootThePolice #RIPErivGardner #RIPMikeBrown.[12]

In the minds of Blue Lives Matter supporters, Brinsley's ambush of officers in the line of duty justifies cops' aggressive policing of Black men—because they're dangerous. One of the most outspoken proponents of this view is former New York City mayor Rudy Giuliani. He's claimed that cops get an unfair rap because 99 percent of Black homicide victims are killed by Black civilians, not cops. This claim suggests Blacks have only themselves to blame for the violence that often plagues their own communities and insinuates that cops are justified in mistreating Black suspects because of the dangers they pose.[13]

Police officers most certainly deserve honor and respect for the heroic work they do by putting themselves in unpredictable and often risky situations. However, the underhandedness Giuliani and other Blue Lives Matter supporters display in diverting blame for violence against Blacks away from the police and toward other Black people is troubling. Aren't *both* types of violence significant social problems? Rather than get mad when I first learned about this underhandedness, I became motivated to research the hidden story behind it. An explanation for police violence toward unarmed Black suspects that lays the blame on these suspects clearly begs for an analysis of the social forces that lead many low-income Black males to commit crime and thereby elicit cops' suspicions. Before we go there, let's first see how the sociological perspective can deepen our view of cops who act violently toward unarmed Black suspects.

Responding to Public Fears: Police Brutality as a By-product of Getting Tough on Crime

3.3 Explain how police brutality is a reflection of the fears of crime that pervade American society.

Take a few minutes and jot down what comes to mind when you think of crime. Describe the scene. Where is it taking place? Who's breaking the law? What is this person's background? When I ask my students these questions, they typically mention violent crimes on the street or in other public places. They're likely to imagine the criminal as a young male from a low-income background. Many of them see the perpetrator as Black or Latino.[14]

You may hold similar views, particularly if you watch crime shows or mainstream movies. These media often reinforce **images**, or mental pictures, people have about low-income Black and Brown males. These images paint such males as committing most violent crimes in the United States.[15] Criminal images contain kernels of truth. The murder rate in Chicago, for example, is highest in predominantly Black neighborhoods.[16] Moreover, Blacks are by far the racial group that is most likely to be incarcerated—about four times more than Hispanics and six times more than Whites. Still, fewer than 2 percent of Blacks are in prison at any given time, and many of those who are incarcerated are there for nonviolent offenses. Even when we take into account other Blacks who have served time behind bars, it's clear that criminal images distort the reality of being Black in America.[17] "If you walked into a group of 1,000 randomly selected Blacks," wrote journalist Jamelle Bouie, "the vast majority—upward of 998—would never have had anything to do with violent crime."[18]

When Hollywood directors cast Black men as thugs, it reinforces criminal images about these men. What movies have you seen that portray Black men in this way?
iStockphoto.com/manley099

In their efforts to control crime, cops of any race may treat people unfairly because they fit popular images of dangerous criminals.
Pulsar Imagens/Alamy Stock Photo

By considering the significance of criminal images, we can begin to see how police brutality reflects implicit racial bias. While most police officers who kill minority suspects are White, that's a reflection of most cops being White. Not only do minority officers also sometimes use excessive force against minority suspects, they're proportionally *more* likely to do so than White cops. Consider that Blacks, Hispanics, and Asians collectively comprise 31.7 percent of the Cincinnati Police Department, which from 2013 to 2017 had a homicide rate of 11.8 Blacks per one million Blacks living in the city. The Baltimore Police Department is more racially diverse (49 percent minority) and yet had a higher Black homicide rate: 13.6 per one million Blacks living in the city. These data reveal that, like their White counterparts, minority cops also internalize the implicit biases contained in criminal images. And since minority cops are often assigned to patrol communities of color, it's no wonder they're more likely than White officers to use excessive violence.[19]

By the same token, it's not just White residents fearful of becoming crime victims who have prejudices toward Black males. People living in the very places most impacted by police brutality may carry criminal images too. As a result, they may advocate for more cops to patrol their neighborhoods. Historian Michael Javen Fortner, who writes about high-crime Black communities that have advocated for stronger policing, offers a recollection of his childhood growing up in New York City:

> I remember Black folks constantly worrying about keeping their children, homes, and property safe. These working- and middle-class families did not express much sympathy for and empathy with the perpetrators of crime in the neighborhood. I recall hearing "that's what he gets" every time one of "our youngsters" was arrested.[20]

Because hardworking and law-abiding parents living in Black neighborhoods are concerned about the safety of their children, it makes sense why these parents would support the police cracking down on violent crime.

Therefore, situations where cops excessively use force reflect the spread of criminal images across our society. The police officers who mistreat Black suspects do so because they are following law-enforcement practices implemented in accordance with these images. Consider **stop and frisk**, a policy instituted by Mayor Rudy Giuliani in New York City during the early 2000s. It gave cops the authority to question and search anyone they suspected of carrying weapons or drugs. Given the implicit bias within criminal images, it's unsurprising that stop and frisk disproportionately targeted people of color. It was consistent with the **broken windows** theory that had influenced policing since the 1980s. This theory contends that highly visible nonviolent offenses can give people the license to commit more serious crimes. Officers target loitering, panhandling, and other low-level offenses in order to stamp out evidence of disorder that may encourage violent crimes.[21]

The militarization of policing is a key factor in cops getting tough on crime.
Edwin Remsberg/Alamy Stock Photo

Knowing that the two White New York City cops who confronted Eric Garner for selling loose cigarettes were practicing broken windows policing deepens our understanding of why they acted violently toward him. Whereas from an individual perspective this brutality appears to be explicitly biased policing, there's a more nuanced sociological explanation for their deviant behavior. Because these officers were responding to a public and political mandate to remove the seeds of violent crime, they saw Garner as a threat to public safety. It didn't matter that he was standing at a busy intersection in the middle of the afternoon unarmed. The officers had internalized an implicit bias that since Garner was a large Black man, he posed a threat. So they subdued him. The sociological perspective, therefore, highlights how criminal images fuel the adoption of law-enforcement practices that can give rise to racially unjust behavior by cops.

Before reading this chapter, you may have believed that violent behavior by White police officers toward unarmed Black suspects is a sign of explicit racial bias by rogue cops. If so, it may have also been hard for you to imagine that there could be a different way to see police violence. When I was in college and watched the video of Rodney King being beaten by four White cops, this individual perspective seemed like the only way to make sense of what I was seeing. Even though I was about to graduate with a bachelor's degree in sociology, I had never thought about the social forces that lead cops to use excessive violence against Black suspects. My aim here has been for you to discover earlier in your life than I did in mine the value of seeing police brutality from the sociological perspective, which highlights why cops may hold the same implicit racial biases as do many other people.

Given that policing reflects public and political mandates to be tough on crime, in some situations it may be difficult for cops to distinguish between exercising force as a necessary response to danger versus an abuse of their power.
tom carter/Alamy Stock Photo

The Hidden Roots of Crime: How Race Shapes Opportunity

3.4 Identify the social forces that lead some Black males to engage in deviant behavior.

The sociological perspective is equally valuable if, prior to reading this chapter, you believed that blame lies with the victims of police violence since people of color often commit violent crimes. To develop this perspective, let's revisit Eric Garner's story. At the beginning of this chapter, we explored the events surrounding his death. Now we'll consider the conditions in which he lived. Doing so exposes the social forces that led him to sell loose cigarettes on the street, which put him in a situation where he risked being approached by the cops and becoming another minority victim of police violence.

After graduating high school, Garner earned an associate's degree in automotive technology. Figure 3.2 (on page 46) illustrates that the unemployment rate for Blacks with this degree is 60 percent higher than for Whites with the same credential. Blacks with no postsecondary education—both high school graduates and dropouts—are more than twice as likely to be unemployed as Whites with the same schooling. Given the disadvantages Blacks experience in converting their educational attainment into steady and decent-paying work, they are vulnerable to turning to criminal activities in order to make ends meet.

Though some employers have a preference for White jobseekers over Black jobseekers with similar educational credentials, we mustn't presume that this preference exclusively reflects **racial discrimination**, which is the systemic disadvantaging of people based on the color of their skin. Employers might also base

FIGURE 3.2 ● Excluded from the Labor Market

At all levels of educational attainment, Blacks are less likely to be employed than Whites.

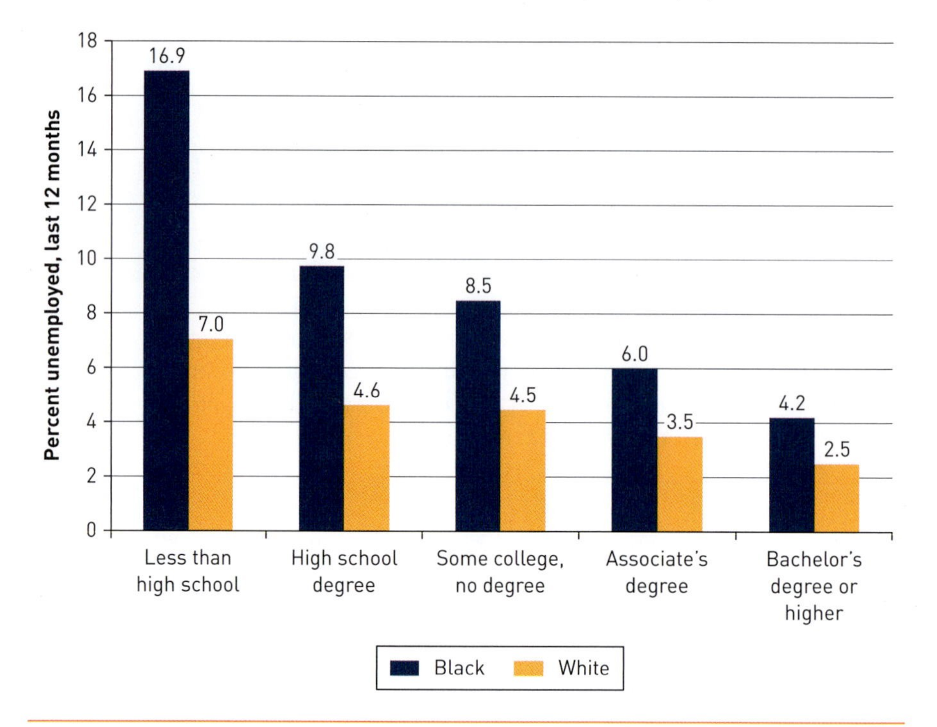

Source: Nick Buffie, "The Problem of Black Unemployment: Racial Inequalities Persist Even amongst the Unemployed," Center for Economic and Policy Research, November 4, 2015, http://cepr.net/blogs/cepr-blog/the-problem-of-black-unemployment-racial-inequalities-persist-even-amongst-the-unemployed.

their hiring decisions on the appearance of an applicant's resume or how they dress for an interview. Indeed, a combination of factors explains why a Black person may be disadvantaged in the search for employment.

In order to pinpoint the effect race has on job opportunity, sociologists Devah Pager, Bruce Western, and Bart Bonikowski created two fascinating experiments. In the first, they recruited people to apply in person for low-skill jobs in New York City and to take notes about their experiences. Applicants answered ads for movers, delivery drivers, stockers, telemarketers, and restaurant workers. Three different people applied for each job—one White, one Latino, and one Black. Except for race, the applicants were similar. All were males, ages twenty-two to twenty-six, and 5′10″ to 6′0″ tall. They were comparable in how well they made eye contact, dressed, and spoke. Their résumés similarly indicated having a high school degree and experience in low-skill jobs. None of them had a criminal record.

The researchers found the effects of race on hiring to be huge. Employers showed an interest in learning more about 31 percent of the White applicants' suitability for the jobs. Only 25.1 percent of Latinos and 15.2 percent of Blacks got a positive response, even though members of both groups had comparable credentials to the White applicants. In some cases, this disparity stemmed from an explicit bias against hiring Blacks—such as when the three jobseekers applied for a position in a warehouse. The woman collecting applications told them they could leave because there wouldn't be interviews that day. While they were waiting for the bus, she motioned for them to return but told the Black applicant he could go back to the bus stop because she only wanted to speak to the others. Then she asked the White and Latino applicants to come back later that day to start work.[22]

There is substantial evidence of racial discrimination in hiring.
iStockphoto.com/hyejin kang

At other times, employers displayed implicit bias. They drew upon deep-seated, unconscious racial prejudices by shifting their standards depending on whether the applicant was White, Black, or Latino. For example, when they applied for a position as a mover, all three jobseekers indicated on their résumés that they had experience stocking boxes for a storage company. The hiring manager told the Black and Latino applicants they didn't have the necessary skills. He said the same thing to the White applicant, but qualified that "because you've worked for a [storage company], that has a little to do with moving."[23] The White applicant got the job.

There were yet other instances of implicit bias when Black applicants received offers but for jobs with fewer customer service responsibilities than the positions they'd applied for—maintenance worker instead of retail sales, or delivery instead of counter person. Such downward channeling happened nine times for Blacks, five times for Latinos, and just once for Whites. In six cases, employers offered Whites positions with *greater* responsibilities than the ones listed in the job ad. This upward channeling also happened twice for Latinos, yet not a single time for Black applicants. These findings are further evidence that employers often implicitly devalue Black applicants relative to jobseekers of other racial backgrounds with comparable qualifications.

In addition to facing explicit racial discrimination, Blacks even more often experience implicit bias.
ARTPUPPY/Getty Images

For Black job applicants, having a criminal record makes them practically unemployable. A study conducted in Milwaukee found that employers showed interest in just 5 percent of Black male jobseekers who listed on their applications that they'd been convicted of a nonviolent drug felony. In contrast, employers responded positively to 17 percent of White felons who

Approximately 8 percent of Black men between the ages of twenty and twenty-four are in prison, versus just 1.3 percent of their White peers.[26]
Greg Smith/CORBIS/Corbis via Getty Images

applied for these same jobs.[24] It's no wonder that Black male inmates are more prone (80 percent) than the prison population as a whole (75 percent) to **recidivism**, which occurs when an ex-convict commits another crime. Being unable to find work inclines Black males toward illegal activities in order to earn income.[25]

The most startling finding from the study of racial discrimination in New York City's low-skill labor market came from the second experiment the researchers did. The Black and Latino jobseekers again indicated on their résumés having no criminal record, but this time the White applicants listed a nonviolent drug felony conviction. Not surprisingly, the latter received a positive response less frequently than the 31 percent rate for White applicants with a clean record in the first experiment. As you can see in Figure 3.3, however, the 17.2 percent positive response rate for White felons

FIGURE 3.3 ● Minorities Need Not Apply

In the New York City experiment, employers expressed a 12 percent greater preference for White felons than for law-abiding Latinos and a 32 percent greater preference for White felons than for law-abiding Blacks.

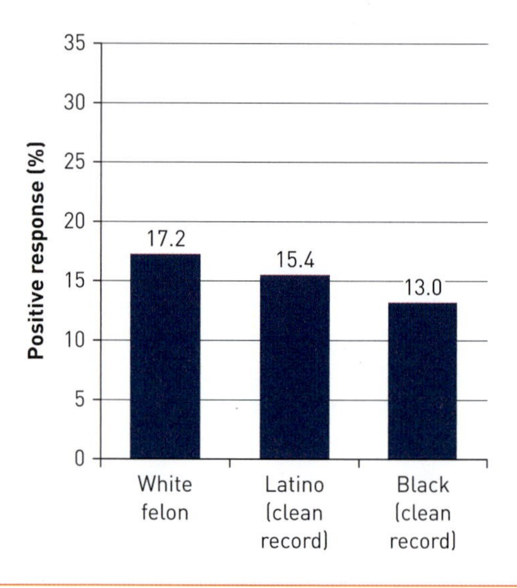

Source: Devah Pager, Bruce Western, and Bart Bonikowski, "Discrimination in a Low-Wage Labor Market: A Field Experiment," *American Sociological Review* 74 (2009): 777–99, https://scholar.harvard.edu/files/bonikowski/files/pager-western-bonikowski-discrimination-in-a-low-wage-labor-market.pdf. Reprinted by permission of American Sociological Association.

was *higher* than for Latinos (15.4 percent) or Blacks (13 percent) with no criminal record. Some employers believed a White felon was more reliable and trustworthy than a Black or Brown person who had never committed a crime.[27]

How could this be? Consider that whereas committing a crime is a matter of fact, "being a criminal" is a judgment of character. Employers may think that a Black person is a criminal even if they haven't broken the law and also may think that a felony conviction doesn't necessarily make a White person untrustworthy. Such employers, after all, carry the same implicit racial biases as do the cops who use excessive violence toward unarmed people of color.

Because Ban the Box laws remove questions about a person's criminal history from job applications, some employers may presume that a Black jobseeker must have a rap sheet simply because of their race.
iStockphoto.com/Hailshadow

"The outrage over relatively rare police killings," wrote journalist Kelefa Sanneh, "should remind us just how much everyday violence—and just how much everyday inequality—we have learned to ignore."[28] Indeed, the biases that produce racial inequality pervade our society, coloring the minds of many otherwise seemingly fair and open-minded people. Implicit racial biases are deep-seated, stemming from the criminal images we see in TV shows, movies, and social media posts.

Because of these implicit biases, laws prohibiting employers from asking jobseekers to check a box indicating whether they have a criminal record may not have their intended effect—and may even have the opposite effect. These so-called Ban the Box laws aim to reduce the disadvantages of having a rap sheet, particularly for Black job applicants. Yet a study comparing before and after the passage of such laws in New York and New Jersey found otherwise. Whereas White applicants as a whole fared slightly better after passage than they had before, Black applicants fared significantly worse.[29]

We're seeing how racial discrimination in the low-wage labor market disadvantages Black jobseekers, particularly those with a felony conviction. Whether discrimination is based on an applicant's criminal record or on an employer's belief that being Black is a sign of criminality, this bias can create a **self-fulfilling prophecy**, which occurs when a person acts in ways that confirm how others label them. A self-fulfilling prophecy can be positive; for example, if you work harder in a class and earn a better grade after your teacher tells you that you're smart. In this case, however, Black men who can't find work because they're presumed to be criminals may turn to unlawful employment—such as Eric Garner did by selling loose, untaxed cigarettes—as a way to earn income. And if they get a rap sheet for doing so, they become even less employable.[30]

Earning income unlawfully often involves violence. The nearly 50 percent of Black males who grow up without a father are especially susceptible to going down this road. A major reason for the absence of Black fathers is that 30 percent of Black men are unemployed or in prison. Roughly one in nine Black children grows up with an absent father behind bars—about twice as many as White children (see Figure 3.4).[31] This disparity is a key reason why Black kids lag significantly behind their White peers in academic achievement. The lack of a present, law-abiding father figure and poor school performance lead males to grow up feeling they can't live up to what our society expects of them. Some turn to violent crime not only because it's a source of income but also because it bolsters their fragile sense of masculinity.[32] Here again there is a self-fulfilling prophecy: For Black males who internalize the societal expectation that being manly means earning respect, they may fulfill that expectation by emulating the criminal behavior of the males they revere most, such as their fathers, brothers, or friends.[33]

To see how workplace discrimination and absent fathers push Black males into criminal activities, consider a story sociologist Alice Goffman recounted in her study of a low-income, predominantly Black neighborhood in Philadelphia. Chuck, age eighteen, was driving his eleven-year-old brother Tim to school. It was Chuck's girlfriend's car, and he had no idea it was hot. The police stopped them, charging Chuck with receiving stolen property and Tim with being an accessory to a crime. It was Tim's first offense; Chuck already had a rap sheet.

FIGURE 3.4 ● Missing Dads

Nearly half of Black children grow up without a father compared to 15 percent of White children.

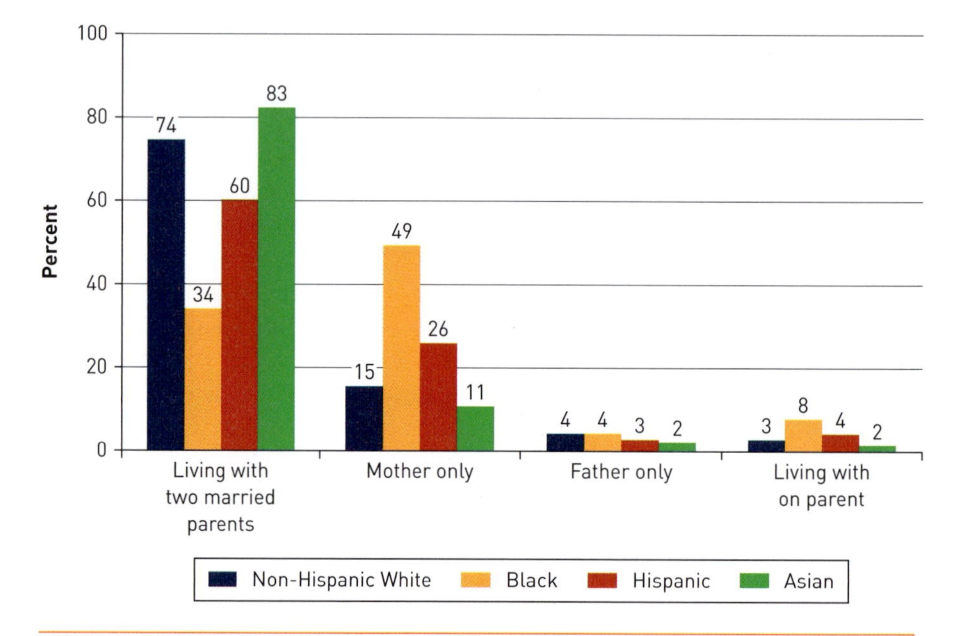

Source: Child Trends, "Family Structure: Indicators of Child and Youth Well-Being," December 2015, https://www.childtrends.org/wp-content/uploads/2015/12/59_Family_Structure.pdf.

Goffman subsequently observed Chuck and his friends repeatedly coming up empty in their search for work. After several months trying, Chuck turned to selling drugs. Here's an excerpt of a conversation Goffman had with him:

Goffman: "Are you OK?"

Chuck: "I hate this. I seen what [crack] did to my mother. I hate doing that to other people's mothers. Like I'm causing their pain."

Goffman: "Yeah."

Chuck: "And I know I'm probably going right back to jail."

Goffman: "Yeah."

Chuck: "But what am I supposed to do? I need to eat. Tim needs to eat."[34]

Because he was unable to find legitimate work and needed to support his younger brother, Chuck turned to drug dealing as a way to earn desperately needed income.

One way to understand Chuck's choice to sell crack is that he could have—and should have—stuck it out longer in his search for a legitimate job. From this individual perspective, it's easy to feel unsympathetic toward Black males like Chuck who get stopped by the police for unlawful behavior and therefore are put at risk of experiencing mistreatment. If these men hadn't broken the law, after all, they wouldn't have come under police suspicion in the first place. Moreover, there are many Black males growing up in neighborhoods like Chuck's who manage to make ends meet without turning to crime. The ones who stay out of trouble appear to be testimony that others like Chuck bear responsibility for how drugs and violence have plagued their communities.

Like other successful Black males, Barack Obama is often heralded for staying out of trouble as a kid instead of giving in to the temptation of criminal activities.
RHONA WISE/AFP/Getty Images

However, Alice Goffman's research highlights that in order to come to a thorough understanding of why a low-income Black male might deal drugs, we have to situate their decision to break the law within the broader social context in which they live. In embracing this sociological perspective, Goffman learned something eye-opening about the constant police surveillance in the neighborhood she studied: In addition to controlling unlawful behavior, cops' presence led residents to identify with crime. She came to this discovery by observing the role cops played in people's daily lives:

> I watched the police stop and search young men in the street, chase them, make arrests, raid houses in the middle of the night and threaten girlfriends and mothers who refused to cooperate. I saw the police take young men into custody, not only on the streets, but at their jobs, in their mothers' homes, at funerals and even in a hospital delivery room.[35]

The sociological perspective exposes hidden forces that contribute to Black males' decision to commit crime.
iStockphoto.com/keeweeboy

Given these realities of crime control, it's telling that Goffman often saw children playing games of chase where they role-played criminals and cops. "Cops" would act out what kids had often seen during their short lives: They'd push "criminals" to the ground and handcuff them. These kids were growing up with an understanding that constant surveillance was normal—a simple fact of life. Given that a cop might approach them or someone they knew at any time and in any place, they learned to see themselves and their neighbors as criminals—people marked as guilty just for being who they were and living where they lived. They expected that the police would confront and mistreat them, just as cops did to Eric Garner, Rodney King, and countless other "dangerous" Black men.

The Sociological Stories behind the Footage: Taking a Deeper Look at Police Violence Videos

3.5 Discuss the importance of the sociological perspective in understanding the relationship between race and police brutality.

A friend of mine who's a film professor is skilled at teaching people how to watch movies in ways they may never have before. Because of his guidance, the story on the screen becomes richer for me. This holds just as true for raw footage depicting police homicides of unarmed Black suspects as it does for Hollywood films. Indeed, our aim in this chapter has been to uncover the hidden stories in videos of police violence. These stories highlight invisible social forces that can enrich your understanding of this violence, regardless of your prior beliefs about who was at fault and why.

Let's start with the police officer. We've seen that cops of all racial backgrounds respond to public demands—from communities of color and Whites alike—to be tough on crime. Criminal images fuel support for policing practices that can blur the line between the necessary use of violence and brutality. Reasonable concerns you may have about feeling safe in public may give cops the license to behave in abusive ways. These concerns place expectations on cops so that snap judgments made in risky situations may lead them to use excessive force against Black suspects.[36]

Therefore, the responsibility for this social problem doesn't rest solely on the shoulders of individual officers or even entire police departments. Each of us also contributes to police brutality. Focusing on collective responsibility isn't a cover for cops who behave egregiously; they are still accountable for their actions. However, this sociological perspective suggests that unjust killings of unarmed Black civilians are likely to continue as long as criminal images inform law enforcement practices.

Just as the individual perspective is often the starting point for making sense of police brutality, the same holds true for the way many people think about Black males who commit crime. When I first heard about the killing of Eric Garner, amidst my outrage I also wondered whether he might still be alive if he hadn't illegally been selling loose cigarettes and hadn't spoken to the cops in ways that led them to believe he was dangerous. Many people embrace this individual perspective, which casts aspersions on Black men like Garner for their criminal choices.

The sociological perspective offers an alternative lens for understanding why these men commit crime. It situates their behavior within the broader social contexts that shape their lives. We've seen that males like Alice Goffman's friend Chuck face significant obstacles in the search for low-wage jobs as well as constant suspicion from the police. Many of them also grow up fatherless. These factors collectively explain why such males are prone to violence and account for why Black gunmen perpetrate most homicides of Black victims in the United States.

Whereas college is your ticket to getting ahead, for many people with fewer opportunities than you the most viable pathway for earning income is criminal activities that will likely land them in prison.
iStockphoto.com/jaminwell

Our aim in embracing the sociological perspective is to see these homicides not as a Black problem but as an *American* problem. Consider that Chuck lacked the very opportunities that propel other young people toward higher education. You may have experienced some of the same disadvantages he did. However, being in college means you likely have also had key supports in your life that he lacked. The reality is our society affords only some people the opportunity to pursue higher education—a significant point to bear in mind given that postsecondary training is essential nowadays for attaining good jobs.[37]

Those without access to this opportunity distinctly differ from two groups of people: low-income college students for whom higher education is a pathway to upward mobility and middle-class students who are likely to move even further up the social ladder after college. If one of these groups describes you, taking responsibility for the inequalities that lie at the heart of violent police encounters with unarmed Black suspects means acknowledging that if you had grown up with similarly bleak opportunities, you too might have turned to illegal activities to earn income and therefore been at risk of becoming a victim of such violence.[38]

What Do You Know Now?

1. Discuss the ways that criminal images contribute to police brutality. How does this explanation highlight that cops who use excessive force toward minority suspects are not the only people responsible for their actions?

2. How can you show concern about police brutality while still acknowledging the crucial, often dangerous work cops do to maintain public safety? In other words, how can you be against police brutality without being against the police?
3. How does the discussion of racial discrimination in the labor market challenge the conventional wisdom that Black males who commit crime have chosen to break the law?
4. The next time you hear about a cop killing an unarmed Black or Brown suspect, how might you react differently than before you read this chapter?
5. Given that beliefs about police violence toward unarmed Black suspects are often racially polarizing, how can reading this chapter foster a common ground of understanding?

Key Terms

Police brutality 38
Explicit bias 40
Implicit bias 41
Black Lives Matter 41
Blue Lives Matter 42
Images 43
Stop and frisk 44
Broken windows 44
Racial discrimination 45
Recidivism 48
Self-fulfilling prophecy 49

Visit **edge.sagepub.com/silver** to help you accomplish your coursework goals in an easy-to-use learning environment.

Notes

1. Al Baker, J. David Goodman, and Benjamin Mueller, "Beyond the Chokehold: The Path to Eric Garner's Death," *New York Times*, June 13, 2015, https://mobile.nytimes.com/2015/06/14/nyregion/eric-garner-police-chokehold-staten-island.html.
2. Eliott C. McLaughlin, "We're Not Seeing More Police Shootings, Just News Coverage," CNN, April 21, 2015, https://www.cnn.com/2015/04/20/us/police-brutality-video-social-media-attitudes/index.html; Katie Nodjimbadem, "The Long, Painful History of Police Brutality in the U.S," *Smithsonian*, July 27, 2017, https://www.smithsonianmag.com/smithsonian-institution/long-painful-history-police-brutality-in-the-us-180964098/.
3. Lynn Langton and Matthew Durose, "Police Behavior during Traffic and Street Stops,"

U.S. Department of Justice, October 27, 2016, https://www.bjs.gov/content/pub/pdf/pbtss11.pdf; Rod K. Brunson, "'Police Don't Like Black People': African American Young Men's Accumulated Police Experiences," *Criminology & Public Policy* 6 (2007): 71–102.

4. German Lopez, "There Are Huge Racial Disparities in How U.S. Police Use Force," *Vox*, November 14, 2018, https://www.vox.com/identities/2016/8/13/17938186/police-shootings-killings-racism-racial-disparities.
5. Crystal Fleming, *How to Be Less Stupid about Race: On Racism, White Supremacy, and the Racial Divide* (Boston: Beacon Press, 2018), 60.
6. The statistics about racial disparities in police homicides of unarmed suspects are based on 2015 data and come from Mapping Police Violence, https://mappingpoliceviolence.org/Zunarmed.
7. Roper Center for Public Opinion Research "Black, White, and Blue: Americans' Attitudes on Race and Police," September 22, 2015, https://ropercenter.cornell.edu/black-white-blue-americans-attitudes-race-police/; Gallup, "Public Opinion Context: Americans, Race and Police," July 8, 2016, http://www.gallup.com/opinion/polling-matters/193586/public-opinion-context-americans-race-police.aspx.
8. Richard I. Kirkland, "What Can We Do Now?" *Fortune* (June 1992): 41–8.
9. Mapping Police Violence, "Rate of Police Killings per Population (Data from Jan. 2013 to Dec. 2017)," https://mappingpoliceviolence.org/cities/; John Gramlich, "Black and White Officers See Many Key Aspects of Policing Differently," Pew Research Center, January 12, 2017, http://www.pewresearch.org/fact-tank/2017/01/12/black-and-white-officers-see-many-key-aspects-of-policing-differently/.
10. Philip M. Stinson, "Charging a Police Officer in Fatal Shooting Case Is Rare, and a Conviction is Even Rarer," Criminology Commons, May 31, 2017, https://scholarworks.bgsu.edu/cgi/viewcontent.cgi?article=1079&context=crim_just_pub.
11. The Blue Lives Matter website can be found at http://archive.bluelivesmatter.blue/.
12. Quoted in Igor Volsky, "Two NYC Police Officers Assassinated 'Execution Style,'" *ThinkProgress*, December 21, 2014, https://thinkprogress.org/two-nyc-police-officers-assassinated-execution-style-f8a81379404f/.
13. Giuliani exaggerated the point. He was referring to data indicating that 93 percent of Black homicide victims die at the hands of Black suspects. See U.S. Department of Justice "Homicide Trends in the United States, 1980–2008," 2011, http://bjs.gov/content/pub/pdf/htus8008.pdf.
14. I got the idea for this exercise from "A Crime by Any Other Name" in Jeffrey Reiman and Paul Leighton, *The Rich Get Richer and the Poor Get Prison: Ideology, Class, and Criminal Justice* (New York: Routledge, 2017), 54–112.
15. The Sentencing Project, "Race and Punishment: Racial Perceptions of Crime and Support for Punitive Policies," 2014, http://sentencingproject.org/wp-content/uploads/2015/11/Race-and-Punishment.pdf.
16. Gary Lucido, "Chicago's Safest and Most Dangerous Neighborhoods by Murder Rate," *Chicago Tribune*, July 11, 2017, http://www.chicagonow.com/getting-real/2017/07/chicagos-safest-and-most-dangerous-neighborhoods-by-murder-rate/.
17. Equal Justice Initiative, "Study Finds Racial Disparities in Incarceration Persist," June 15, 2016, https://eji.org/news/sentencing-project-report-racial-disparities-in-incarceration.
18. Jamelle Bouie, "Black and Blue: Why More Diverse Police Departments Won't Put an End to Police Misconduct," *Slate*, October 13, 2014, http://www.slate.com/articles/news_and_politics/politics/2014/10/diversity_won_t_solve_police_misconduct_black_cops_don_t_reduce_violence.html.
19. Charles E. Menifield, Geiguen Shin, and Logan Strother, "Do White Law Enforcement Officers Target Minority Suspects?" *Public Administration Review* 75, no. 1 (2018): 56–68; "Police Department Race and Ethnicity

Demographic Data," *Governing*, http://www.governing.com/gov-data/safety-justice/police-department-officer-demographics-minority-representation.html; Mapping Police Violence, "Rate of Police Killings per Population"; German Lopez, "How Systemic Racism Entangles All Police Officers—Even Black Cops," *Vox*, August 15, 2016, https://www.vox.com/2015/5/7/8562077/police-racism-implicit-bias.

20. Michael Javen Fortner, *Black Silent Majority: The Rockefeller Drug Laws and the Politics of Punishment* (Cambridge, MA: Harvard University Press, 2015), xi.
21. Philip Bump, "The Facts about Stop and Frisk in New York City," *Washington Post*, September 26, 2016, https://www.washingtonpost.com/news/the-fix/wp/2016/09/21/it-looks-like-rudy-giuliani-convinced-donald-trump-that-stop-and-frisk-actually-works/?noredirect=on&utm_term=.bbd14ce20847.
22. Devah Pager, Bruce Western, and Bart Bonikowski, "Discrimination in a Low-Wage Labor Market: A Field Experiment," *American Sociological Review* 74 (2009): 777–99.
23. Quoted in Pager et al., "Discrimination in a Low-Wage Labor Market," 789.
24. Devah Pager, "The Mark of a Criminal Record," *American Journal of Sociology* 108, no. 5 (2003): 937–75.
25. William A. Galston and Elizabeth McElvein, "Reducing Recidivism Is a Public Safety Imperative," The Brookings Institution, March 25, 2016, https://www.brookings.edu/blog/fixgov/2016/05/25/reducing-recidivism-is-a-public-safety-imperative/.
26. Child Trends, "Young Adults in Jail or in Prison: Indicators of Child and Youth Well-Being," April 2012, https://www.childtrends.org/wp-content/uploads/2012/04/89_Young_Adults_In_Prison-1.pdf.
27. Pager et al., "Discrimination in a Low-Wage Labor Market," 786.
28. Quoted in Kelefa Sanneh, "Body Count," *New Yorker*, September 14, 2015, http://www.newyorker.com/magazine/2015/09/14/body-count-a-critic-at-large-kelefa-sanneh.
29. Amanda Y. Agan and Sonja B. Starr, "Ban the Box, Criminal Records, and Statistical Discrimination: A Field Experiment," University of Michigan Law & Economic Research Paper No. 16-012, August 24, 2016, https://law.yale.edu/system/files/area/workshop/leo/leo16_starr.pdf.
30. Congressional Budget Office, "Trends in the Joblessness and Incarceration of Young Men," May 2016, 14, https://www.cbo.gov/sites/default/files/114th-congress-2015-2016/reports/51495-YoungMenReport.pdf.
31. David Murphey and P. Mae Cooper, "Parents behind Bars: What Happens to Their Children," Child Trends, October 2015, https://www.childtrends.org/wp-content/uploads/2015/10/2015-42ParentsBehindBars.pdf.
32. Data about the effects of mass incarceration on inequalities in academic achievement come from a 2016 Economic Policy Institute report by Leila Morsy and Richard Rothstein titled "Mass Incarceration and Children's Outcomes: Criminal Justice Policy Is Education Policy," http://www.epi.org/publication/mass-incarceration-and-childrens-outcomes/. See also Robert J. Sampson, "Urban Black Violence: The Effect of Male Joblessness and Family Disruption," *American Journal of Sociology* 93, no. 2 (1987): 348–82.
33. Joseph Richardson and Christopher St. Vil, "Putting in Work: Black Male Youth Joblessness, Violence, Crime, and the Code of the Street," *Spectrum* 3, no. 2 (2015): 71–98.
34. Alice Goffman, "The Fugitive Life," *New York Times*, May 31, 2014.
35. Ibid.
36. Eli B. Silverman and Jo-Ann Della-Giustina, "Urban Policing and the Fear of Crime," *Urban Studies* 38, no. 5–6 (2001): 941–57; Frank Bruni, "Behind Police Brutality: Public Assent," *New York Times*, February 21, 1999,

http://www.nytimes.com/1999/02/21/weekinreview/ideas-trends-crimes-of-the-war-on-crime-behind-police-brutality-public-assent.html; Michael Sierra-Arrevalo, "American Policing and the Danger Imperative," November 1, 2016, http://dx.doi.org/10.2139/ssrn.2864104; Sandra Bass, "Policing Space, Policing Race: Social Control Imperatives and Police Discretionary Decisions," *Social Justice* 28, no. 1 (2001): 156–76.

37. Georgetown University Center on Education and the Workforce, "Job Growth and Education Requirements through 2020," https://cew.georgetown.edu/wp-content/uploads/2014/11/Recovery2020.ES_.Web_.pdf.
38. Ibid.

4

The Color of Drug Abuse

Handcuffs for Some Addicts, Help for Others

Learning Objectives

1. Describe the difference between criminalization and medicalization with regard to drug addiction.
2. Explain how drugs are a social construction.
3. Compare the unequal impact of drug laws on Whites and people of color.
4. Discuss how different drug scares throughout American history have been based on the scapegoating of minority groups.
5. Identify the social forces that influence a person's decision to do drugs.

Crack and Opioids: A Tale of Two Drug Epidemics

4.1 Describe the difference between criminalization and medicalization with regard to drug addiction.

In 2015, a judge told Ashley Radliff that if she didn't accept help, she'd return to prison. So the twenty-eight-year-old checked into The Next Step, an addiction treatment center for women in Albany, New York. As part of her rehab program, Radliff had to compute the dollar value of the drugs she'd used over the past eleven years. The realization that it was upward of $1 million injected a shot of sobriety through her bloodstream. She told a journalist, "You don't really get it until you see it in front of your face."[1]

Over twenty-one million people in the United States ages twelve and older have a **drug addiction**, which is characterized by physical and/or psychological dependence on a dangerous substance and the compulsive urge to use it.[2] Given how widespread addiction is, you may know people who've struggled with it. Ashley Radliff is among the roughly 2.5 million Americans addicted to **opioids**, which relieve pain by dulling the senses.[3] They are available by prescription in the form of oxycodone, hydrocodone, codeine, or morphine. Illegally, they either come from the opium poppy plant (e.g., heroin) or, like fentanyl, are produced synthetically in labs. Since the millennium, the only type of opioid use that hasn't led to a surge in overdoses is methadone, which is prescribed to help addicts detox from dependence on other, stronger opioids (see Figure 4.1 on page 60).

Over the past two decades, opioid abuse has become an **epidemic**—a harm that spreads rapidly across a larger and larger segment of people—particularly in states where a significant percentage of people live in rural areas (see Figure 4.2 on page 61). Every day, more than one thousand Americans receive treatment in hospital emergency rooms for opioid addiction and over ninety die of overdoses.[4] There is a greater number of annual deaths per year from opioids than from auto accidents. The mortality rate

FIGURE 4.1 ● Overdose Epidemic

The death rate in the U.S. from different types of opioids has spiked in recent years.

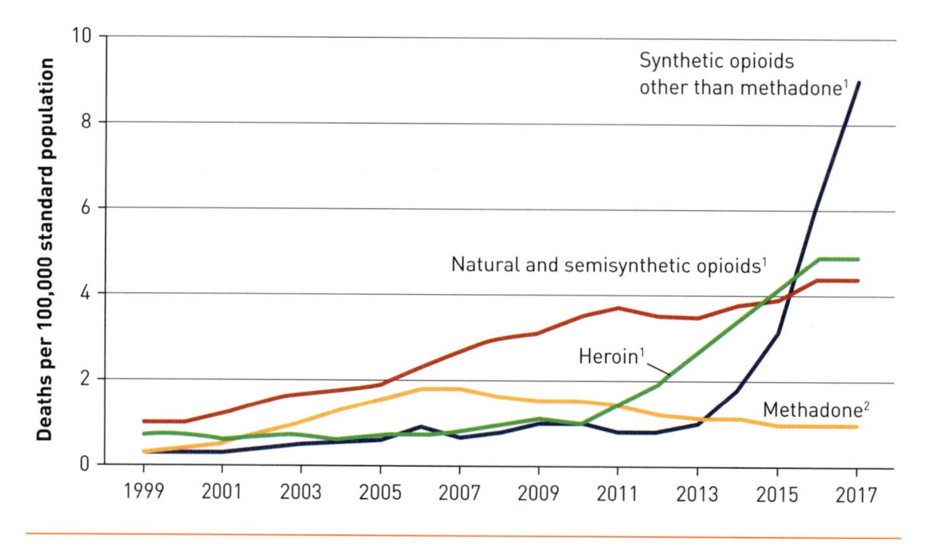

Source: Holly Hedegaard, Arialdi M. Miniño, and Margaret Warner, "Drug Overdose Deaths in the United States, 1999–2017," Centers for Disease Control and Prevention, November 2018, https://www.cdc.gov/nchs/products/databriefs/db329.htm.

from opioids exceeds the annual death totals from AIDS when that epidemic was at its peak in the mid-1990s.[5] And these numbers only begin to capture the devastation opioids have inflicted on addicts, their loved ones, and their communities.

Ashley Radliff's troubles started after a skateboarding accident, when her doctor prescribed oxycodone to manage the excruciating pain. Like so many other users of painkillers, she became hooked. Sometimes she would pop pills; other times she'd crush and snort them. As states began to restrict opioid prescriptions in the mid-2000s,[6] Radliff and many other addicts turned to heroin. It's a much cheaper drug that produces a similar high and is easier to get. As a result, it now causes more overdose deaths than prescription painkillers.[7] Radliff dropped out of college, lost her job, totaled five cars, and started stealing, even from her own parents. All she lived for was to get high with her girlfriend, a fellow addict. After they decided to have a baby, her girlfriend lied to her doctor about being clean so that he'd consent to her taking sperm injections. Their baby was born addicted to heroin.

A generation earlier, Beverly Black was part of a different drug epidemic. She became hooked on **crack**, a smokable crystal created by cooking powder cocaine with baking soda and water. Black started using in her early twenties amidst a long bout of depression that set in after her brother committed suicide. She continued while pregnant with her second child (she had given birth to her first before she got hooked), and, not surprisingly, her baby was born an addict. During her third pregnancy eight years later, Black's crack abuse induced early labor while she was

FIGURE 4.2 ● United States of Addiction

Drug overdose deaths, particularly from opioids, occur most frequently in states where a large percentage of people live in rural areas.

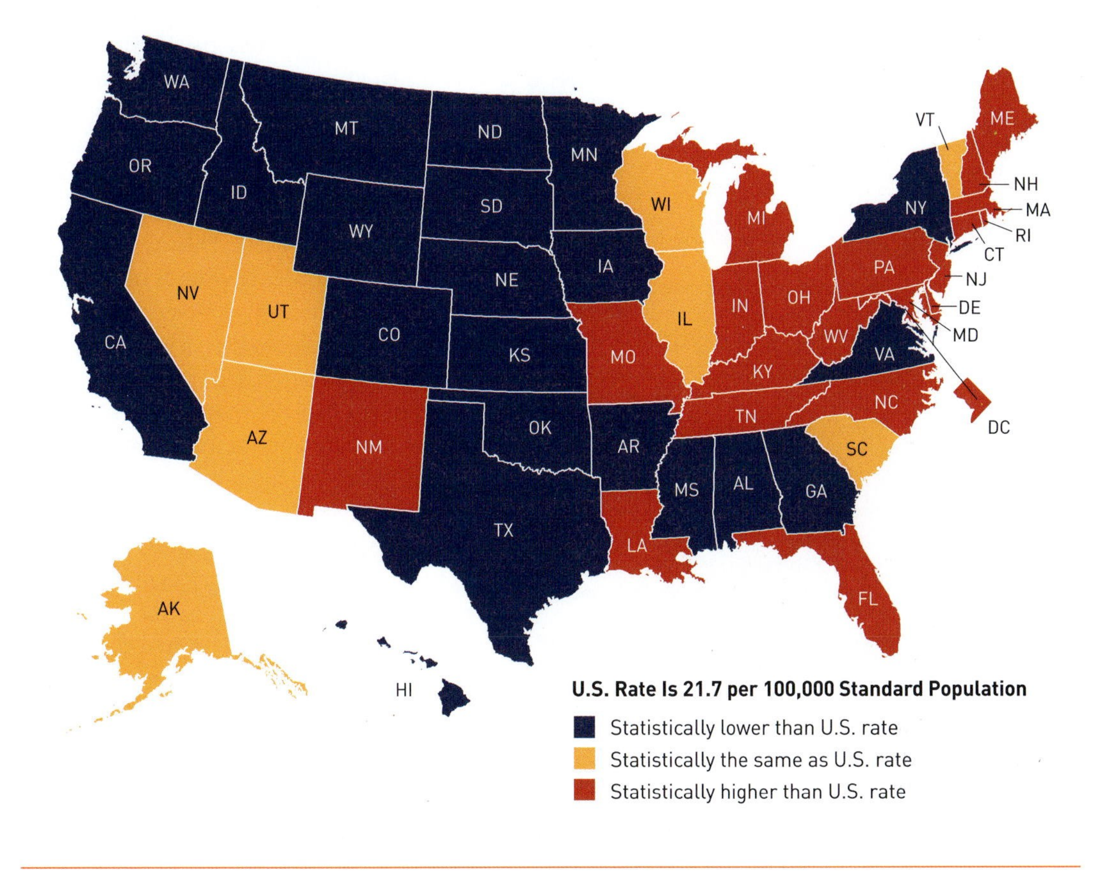

Source: Holly Hedegaard, Arialdi M. Miniño, and Margaret Warner, "Drug Overdose Deaths in the United States, 1999–2017," Centers for Disease Control and Prevention, November 2018, https://www.cdc.gov/nchs/products/databriefs/db329.htm.

still at home. After the birth, a firefighter escorted her and her newborn to the hospital for routine neonatal care. The fire station called the local media to ask them to chronicle her heroic impromptu delivery without medical assistance. Black agreed to allow a reporter to interview her, and one of his questions was why her baby had been born a crack addict. The next morning, a police officer came to the hospital to arrest her. He barred her from taking her newborn home and restricted her from seeing her two older kids.[9]

Beverly Black's story highlights the **criminalization** of drug abuse during the crack epidemic. This idea refers to the view that because a person has broken the law, they should be incarcerated. Although Ashley Radliff behaved similarly, instead of going to prison she received a court-mandated referral to a treatment center. This response

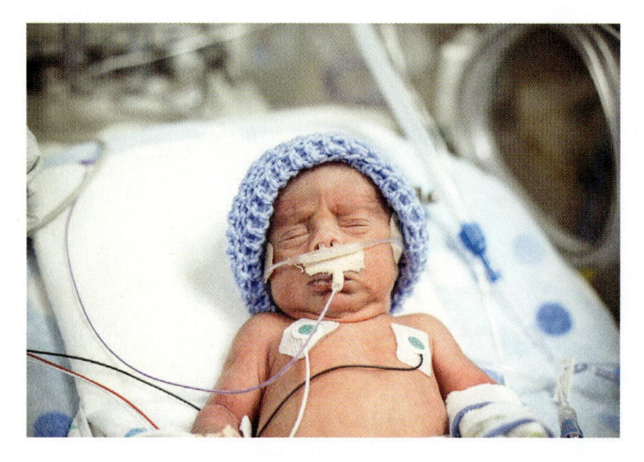

The rate of American babies born with neonatal abstinence syndrome, a postnatal condition of drug withdrawal often seen in infants exposed to opioids in utero, increased 300 percent between 1999 and 2013.[8]
iStockphoto.com/Yobro10

to her addiction was **medicalization**—when a problem experienced by a group of people is regarded as a treatable disease. The aim isn't to remove addicts from society but to help them kick their habit.

This chapter discusses the significance of these different ways of addressing drug abuse. First we'll trace why criminalization was the dominant response to the crack epidemic in the 1980s and 1990s, when there were a disproportionate percentage of African American addicts like Beverly Black. Then we'll consider why over the past twenty years as opioid abuse has surged, particularly among Whites like Ashley Radliff, there has been a growing preference for medicalization. Using the sociological perspective to explain these different understandings of substance abuse highlights how U.S. drug policy produces and reinforces racial inequalities.

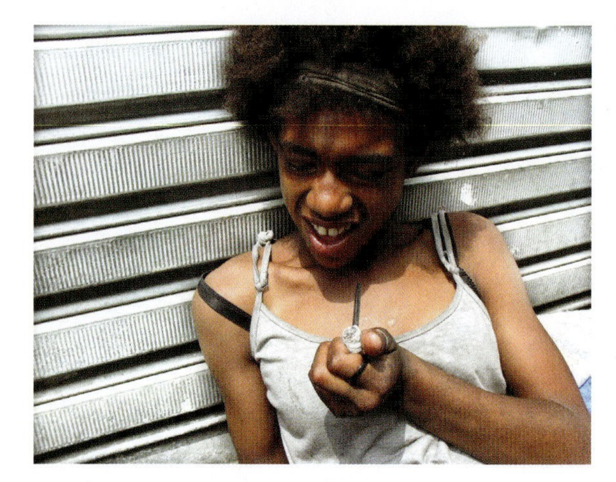

Crack produces a short, intense high and is extremely addictive.
Cavan/Alamy Stock Photo

This perspective also uncovers the social forces that give rise to drug epidemics. Beverly Black and many other African Americans have abused crack in response to broader constraints in their lives, such as bleak opportunities to get ahead and constant police surveillance. A similar sociological story explains why Ashley Radliff and so many other people have abused opioids. In most cases, their addiction started with a doctor's prescription. They began taking legal drugs that pharmaceutical companies had oversold by highlighting the drugs' capacity to relieve pain and by downplaying their dangers. The stories behind both of these epidemics highlight that drug abuse stems from factors far beyond a person's inability to exercise willpower and just say no.

FIRST IMPRESSIONS?

1. In what sense do Beverly Black's and Ashley Ratliff's experiences with addiction remind you of other stories you've heard of people's lives becoming upended by drugs?
2. After reading both of their stories, how sympathetic or unsympathetic are you to the plights they endured? Why?
3. What is significant about the difference between criminalized and medicalized responses to drug abuse?

Race and the Social Construction of Drugs

4.2 Explain how drugs are a social construction.

When I was in college, marijuana was illegal to use for any purpose. As of 2019, ten states had legalized the drug for recreational use and thirty-three for medical purposes. Moreover, think about the significance of its name change. Over the many years that marijuana was illegal, users referred to it in a number of different ways, such as *pot, grass, weed, bud,* and *reefer.* These terms seemingly added to the allure of this "forbidden fruit." In states that have legalized recreational usage, marketers have rebranded the drug as *cannabis* to differentiate it from its illicit past. It can be bought in brightly lit stores where it's packaged in beautiful containers and elegantly displayed in glass cases—quite a change from the days when people bought dime bags in the shady corners of city parks.[10]

The clean and pleasant appearance inside marijuana dispensaries is an indicator of how dramatically attitudes toward the drug have shifted.
BIOSPHOTO/Alamy Stock Photo

Cocaine has had the opposite story. Whereas until 1914 pharmacies sold it over the counter as a medicinal remedy for kids' toothaches, nowadays the drug is illegal and a person is likely to receive a prison sentence for possessing it. The drug hasn't changed; what's different is how our society defines its dangers and how to control them. Shifting attitudes toward marijuana and cocaine over time illustrate the **social construction of drugs**, the idea that people decide whether the possession, use, and sale of a particular drug are crimes and, if so, the penalty for breaking the law. The social construction of a drug doesn't simply reflect its chemical properties. The movement to legalize marijuana gained momentum during the same years that it became significantly more potent. Levels of THC—the psychoactive compound in marijuana—are on average about 30 percent nowadays, whereas in the 1980s the concentration was generally 10 percent or less. If the legality or illegality of a drug mirrored its chemical properties, we'd expect opposition to marijuana to have increased, not lessened, in recent years.[11]

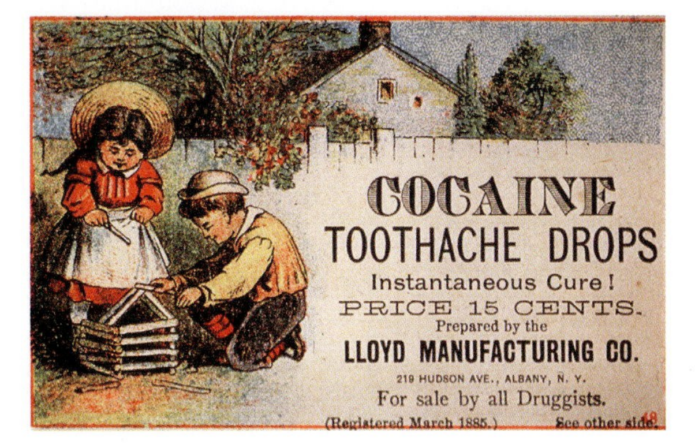

Given that nowadays people generally regard cocaine as a threat to kids, this ad from 1885 is revealing. How does it positively associate cocaine with childhood?
Smith Collection/Gado/Getty Images

Now that you are familiar with the social construction of drugs, let's consider how public perceptions of a drug's dangers reflect beliefs and prejudices about the racial group most associated with using it. Thinking about drugs in this way enables us to explain why

Beverly Black and other crack addicts in the 1980s and 1990s typically went to prison, whereas Ashley Radliff and other recent opioid addicts have more often received treatment for substance abuse.

Our story begins in 1971 when, during a press conference, President Richard Nixon labeled illicit drugs "America's public enemy number one."[12] He declared what became known as the **War on Drugs**, the government's substantial expansion of funding for the policing and punishment of drug crimes. Throughout the 1970s, expenditures on drug control grew fourfold, from a quarter of a percent of the federal budget to 1 percent. During the 1980s and early 1990s, spending for the War on Drugs went from just under $2 billion annually to just over $12 billion.[13] In a 1989 speech in which he called for doubling the allocation for drug enforcement, President George H. W. Bush held up a bag labeled "evidence" that appeared to be filled with crack. He said it was proof that drugs were "turning our cities into battle zones and murdering our children."[14] Piggybacking on a long-standing crusade against drugs, in 1994 President Bill Clinton signed the largest anticrime bill in American history. It substantially increased funding for hiring new police officers and building more prisons.[15]

President George H. W. Bush told the nation the crack he was holding had been seized in a park across from the White House. The deal to acquire it had actually been staged.
Bettmann/Getty Images

To understand the fervor behind the War on Drugs, we need to see how events that immediately preceded it made its timing significant. For most of American history, **racial discrimination**, which is the systemic disadvantaging of people based on the color of their skin, had been legal. In Southern states, where slavery had existed until 1865 and the end of the Civil War, there remained in place a system of **Jim Crow laws**—statutes enacted after the U.S. Civil War that permitted segregation in many aspects of everyday life, including busing, schooling, and public accommodations (hotels, restaurants, restrooms, and the like). Across the U.S., employers could refuse to hire Black job applicants and universities could bar students solely because of their race. Rosa Parks and Martin Luther King Jr. are among the many activists whose hard-fought efforts, beginning in the 1950s, enabled Blacks to

In the Jim Crow South until the 1950s, Blacks and Whites used separate water fountains.
Getty/kickstand

gain equal protection under the law. Their most significant victory was passage of the Civil Rights Act of 1964, which outlawed discrimination on the basis of race, color, religion, sex, or national origin.

The historic civil rights legislation President Johnson signed in 1964 sowed the seeds of a major shift in Southern voters' political loyalties.
Photo12/Universal Images Group via Getty Images

When President Lyndon Johnson signed this legislation, he supposedly turned to an aide and said, "I think we just delivered the South to the Republican Party for a long time to come."[16] Johnson proved to be right. The South, which for years had been a Democratic stronghold, shifted Republican in 1968 with the election of Richard Nixon and has remained that way to this day. Southern Democrats felt Johnson had betrayed them since the Democratic Party had previously been against extending civil rights to Blacks. Consequently, the War on Drugs was enticing to these voters. It tapped their racial prejudices through frequent mention of addicts living in the "inner city," "ghetto," or "slums." Each of these words is a **euphemism**, or a sanitized way of speaking about something dirty or unpleasant. When politicians used euphemisms in referring to the War on Drugs, they exposed their chief political aim: to get tough on Black people.

Since the Civil Rights Act of 1964 had inscribed racial equality into law, President Nixon thought that using explicit racist language to tap into Southern bigotry risked alienating the growing segment of voters who supported greater equality for Blacks. So his administration euphemistically focused public attention instead on urban neighborhoods, where Blacks were the most likely group to be abusing heroin. Coming across as "tough on drugs" was, therefore, a veiled way of appealing to this bigotry.[17]

John Ehrlichman, one of Nixon's top aides and a key figure in the Watergate scandal that led to Nixon's resignation from office in 1974, admitted in a 2016 interview that stoking racial fears lay at the heart of the War on Drugs:

> You want to know what this was really all about? The Nixon campaign in 1968, and the Nixon White House after that, had two enemies: the antiwar left and black people. You understand what I'm saying? We knew we couldn't make it illegal to be either against the war or blacks, but by getting the public to associate the hippies with marijuana and blacks with heroin, and then criminalizing both heavily, we could disrupt those communities. We could arrest their leaders, raid their homes, break up their meetings, and vilify them night after night on the evening news. Did we know we were lying about the drugs? Of course we did.[18]

Whereas the manifest function of religion is to feel a connection to a supernatural being, its latent function is to reinforce a sense of community. What other examples can you think of that illustrate the difference between manifest and latent functions?
iStockphoto.com/FatCamera

At the time that Nixon described drugs as public enemy number one, they were in truth a significant menace. Indeed, the **manifest function**—or stated intention—of the War on Drugs was to curb abuse and addiction. However, Ehrlichman's comment underscores that during the 1970s, 1980s, and 1990s, U.S. antidrug policy had a more significant **latent function**, or hidden intention: to reinforce racial inequalities.[19]

The Racial Inequalities Underlying Drug Enforcement

4.3 Compare the unequal impact of drug laws on Whites and people of color.

One day on her way to work, Michelle Alexander noticed a poster stapled to a telephone pole. It said "The drug war is the new Jim Crow." This comparison to the system of racial segregation that existed in the South following the Civil War seemed far-fetched to her:

> I thought that was hyperbole. I shook my head and said: "Yeah, the criminal justice system is racist in a lot of ways, but it doesn't help to make such absurd comparisons to Jim Crow. People will just think you're crazy."[20]

Like many Blacks of her generation, Alexander had benefited from the civil rights activism that occurred during the 1950s and 1960s. She had come of age believing our country was on a path toward continued racial progress. Alexander's parents had instilled in her that the Jim Crow era was a part of American history that Blacks had overcome.

As Alexander provided legal counsel to victims of police brutality over the ensuing years, the images in that poster came to mind whenever she saw evidence of racism. Her accumulating observations led her to begin a study of the effects of U.S. drug policy on different racial groups. Continuing to do criminal defense work while conducting her research led Alexander to an epiphany: "We hadn't ended racial caste in America. We had just re-designed it."[21] Her book, *The New Jim Crow,* presents rich evidence of how the War on Drugs has contributed to the resegregation of a sizable segment of Black America. Whereas Blacks comprise 12 percent of the U.S. population, they make up 33 percent of the prison population. More Blacks are under the control of the corrections system today than were enslaved in 1950.[22]

The War on Drugs led to **mass incarceration**—a surge in the percentage of people behind bars. Whereas about a tenth of a percent of Americans were locked up prior to the War on Drugs, the incarceration rate had risen to three-tenths of a percent by 1990, two-fifths of a percent by 1995, and half a percent by 2008 (see Figure 4.3). In terms of

FIGURE 4.3 ● Locked Up behind Bars

The War on Drugs led to a dramatic rise in the U.S. prison population.

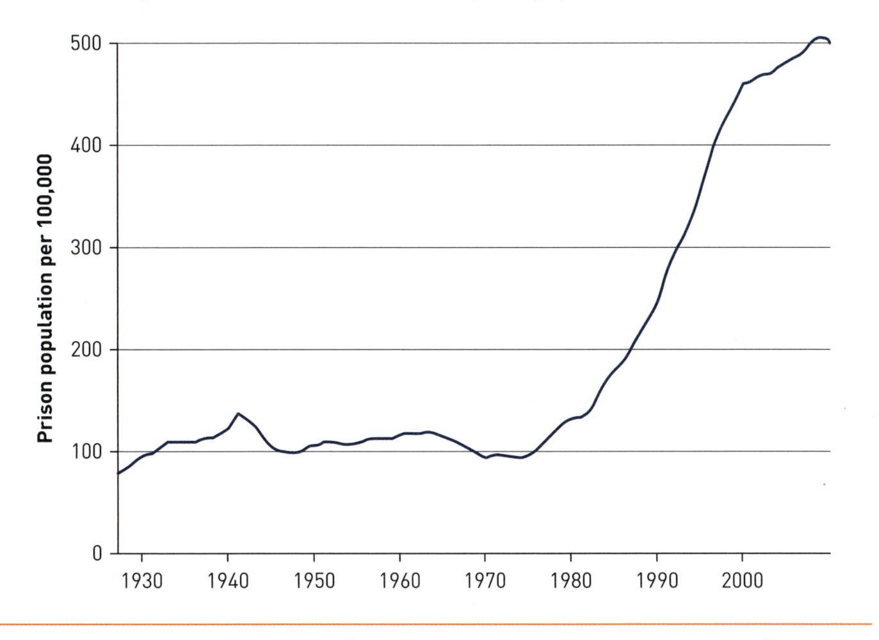

Source: Based on Prison Policy Initiative, "U.S. Incarceration Rate, 1925–2008," 2010, https://www.prisonpolicy.org/graphs/incarceration1925-2008.html.

how the War on Drugs reinforced racial inequalities, the most significant detail about this surge is that it occurred most dramatically among Black males without a high school diploma (see Figure 4.4 on page 68). Although the United States has less than 5 percent of the world's population, it has roughly 25 percent of all inmates, and it has the highest incarceration rate. This rate is 53 percent more than Russia's, the country that ranks second.[23]

Here's clear evidence of how the War on Drugs led to mass incarceration: Drug felons accounted for two-thirds of the surge in the federal prison population from the 1970s through the 1990s and half the growth in the number of people incarcerated in state penitentiaries. Many of these felons were first-time, nonviolent offenders—people like Beverly Black, the crack addict discussed at the beginning of this chapter. New mandatory minimum sentencing laws stipulated that even people convicted of possessing small amounts of drugs could be locked up for long stints.[24] Over 3,200 Americans are currently locked up for life because of nonviolent offenses, and 80 percent of those offenses were drug-related. Sixty-five percent of these lifers are Black, 18 percent are White, and 16 percent are Latino.[25]

By itself, the particular spike in the Black male prison population isn't proof of discriminatory drug enforcement. It could be the racial gap in incarceration for drug-related offenses reflected that Blacks committed proportionally more of these

FIGURE 4.4 ● Race and Incarceration Rates

During the War on Drugs, the portion of Black American men behind bars rose much more substantially than the White male prison population, especially among the least-educated men.

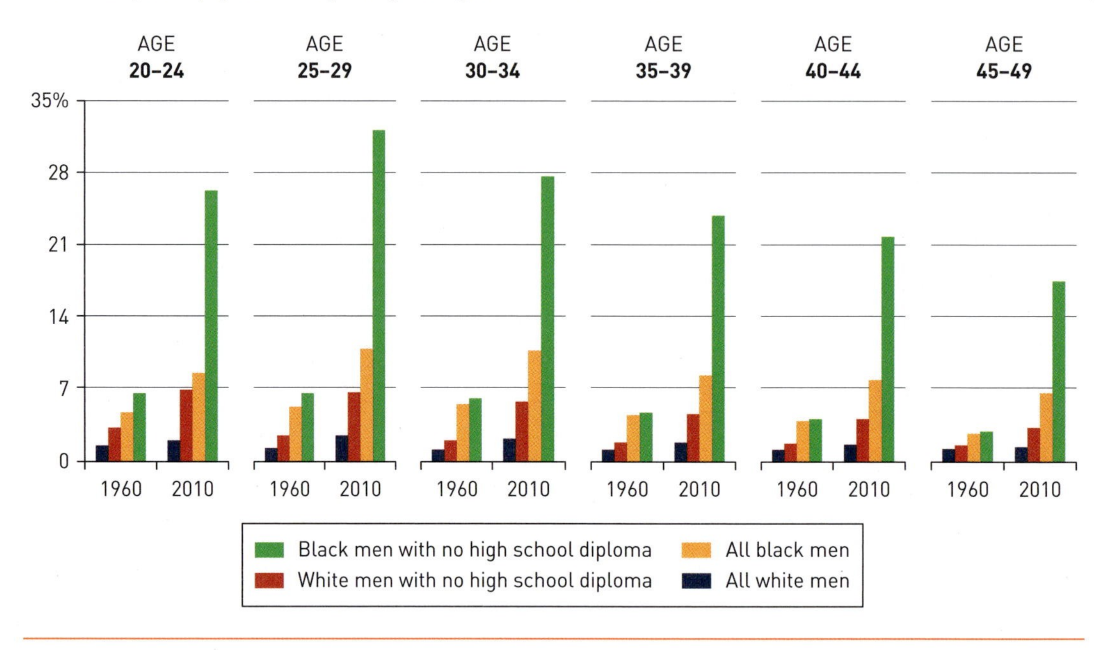

Source: Based on George Gao, "Chart of the Week: The Black-White Gap in Incarceration Rates," Pew Research Center, July 18, 2014, http://www.pewresearch.org/fact-tank/2014/07/18/chart-of-the-week-the-black-white-gap-in-incarceration-rates.

offenses. However, they didn't. A 1991 study by the National Institute on Drug Abuse found that people who'd admitted to using crack were 52 percent White, 38 percent Black, and 10 percent Hispanic—and yet much of the surge in the prison population was from Black crack users.[26] Additionally, even though crack and powder cocaine are essentially the same drug, a 1986 law stipulated that crack carried *one hundred times* the penalty as cocaine. In other words, someone possessing five grams of crack received the same sentence as someone possessing five hundred grams of powder cocaine. It's not a coincidence that lawmakers created this punishment disparity at a time when Blacks were much more likely to use crack and Whites were more likely to use powder cocaine. The disparity is clear evidence of racial discrimination in drug enforcement and punishment.[27]

Data about marijuana offer another case in point. During the 2000s, Blacks were arrested for possession at over three times the rate as Whites, even though the two groups were comparably likely to use the drug (see Figure 4.5).[28] This statistic may hit home for you since marijuana is the drug college students use most frequently. If you are White, it doesn't matter if you attend a university in a state where marijuana

FIGURE 4.5 • Busted for Being Black

In states where marijuana remains illegal, possession puts Blacks at a far greater risk of arrest than Whites.

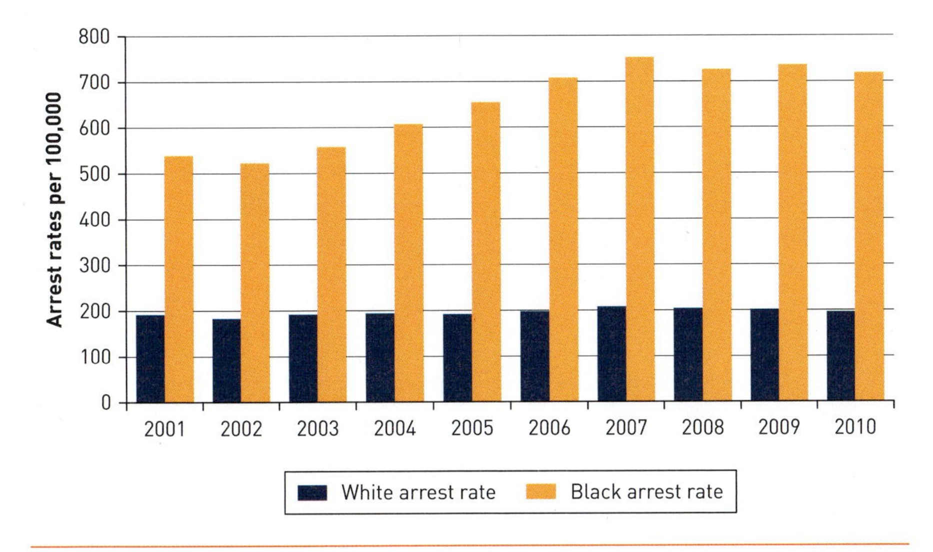

Source: American Civil Liberties Union, "The War on Marijuana in Black and White: Billions of Dollars Wasted on Racially Biased Arrests," June 2013, https://www.aclu.org/sites/default/files/field_document/1114413-mj-report-rfs-rel1.pdf. Data from FBI/UCR and U.S. Census.

remains illegal—you're unlikely to get arrested for possessing it. If you're a Black user, however, it's likelier you will be arrested and get a criminal record that carries adverse long-term effects.[29]

The War on Drugs has racially segregated a disproportionate segment of Blacks behind prison walls. Their exclusion from mainstream society continues long after inmates have served their sentences. People convicted of felonies may not live in public housing and aren't permitted to serve on juries. They are also denied the right to vote, which is one of several ways Blacks are disenfranchised and unable to exercise political influence comparable to their proportion of the population.[30]

Behind these hands lies the story of how the War on Drugs resegregated a large segment of Black Americans.
iStockphoto.com/oneword

The most devastating effect of having a criminal record occurs in the search for employment. If you've ever filled out a job application, you know that employers often ask about felony convictions. While any person with a record faces tough prospects in landing a job, an experiment conducted in Arizona revealed that Blacks have an especially hard time. Researchers found that for White male community college graduates who applied in person for low-skill jobs (customer service, manual labor, food service) and listed on their résumés employment experience gained while an inmate at a state prison, 13.3 percent got a positive response. In comparison, the figure was just 7.7 percent for Black male jobseekers with the same credentials. When the experiment was done on female community college graduates, the gap was even larger: 16.7 percent of White job applicants got a positive employer response versus 5.9 percent of Black applicants. These data substantiate that Blacks pay a steeper price than Whites for having spent time in prison. A criminal record blemishes their character to such a large degree that they become practically unemployable.[31]

Policing Drugs by Creating Scapegoats

4.4 Discuss how different drug scares throughout American history have been based on the scapegoating of minority groups.

Although thousands of crack offenders have served time in prison, the most common response to opioid abuse has been to treat it as a disease. The aim is to manage the harms associated with addiction rather than to punish drug abusers. They receive referrals to treatment centers and/or permission to participate in needle exchange programs, which allow heroin addicts to continue using with clean needles that prevent the spread of HIV and hepatitis.

It's tempting to see this shift as reflecting social progress since treatment seems more effective than punishment in reducing drug abuse and keeping people out of prison. After all, 75 percent of former inmates commit another crime.[32] However, by embracing the sociological perspective we can see that there's more to this story than meets the eye. Kimberlé Williams Crenshaw, an expert on racial inequality in the criminal justice system, noted that

> this new turn to a more compassionate view of those addicted to heroin is welcome. But, one cannot help notice that had this compassion existed for African-Americans caught up in addiction and the behaviors it produces, the devastating impact of mass incarceration upon entire communities would never have happened.[33]

Indeed, it's no coincidence that at a time of surging addiction to prescription opioids and heroin among middle-class Whites, there's been a backpedaling from the criminalization policy that began in the early 1970s when low-income Blacks were the public face of heroin addiction and continued for several decades as they became the face of crack addiction.[34]

In recent years, the families of opioid addicts have been able to use their class and race advantage to exert leverage on legislators to support medicalized responses to drug abuse. As a result, many states now have Good Samaritan laws that shield people who report an opioid overdose from being prosecuted. Most states also have statutes enabling family members to get a prescription for naloxone, a drug that counters the effects of a heroin overdose. Why do you think the families of addicts during the crack epidemic didn't have this same capacity to advocate for such laws?[35]

Even though about 90 percent of opioid addicts nowadays are White, the War on Drugs' legacy of producing and reinforcing racial inequality endures. We can see evidence in media coverage about the opioid epidemic. There are significant differences in reporting depending on the addict's race and class. News stories of suburban or rural, and typically White, addicts provide more details about why people abuse opioids than do accounts of urban, more typically minority addicts. These details depict suburban and rural addicts as victims of a devastating disease and therefore deserving of help. The absence of such details in news stories about urban minority addicts criminalizes these drug abusers and suggests they're undeserving of compassion.[36]

The fact that most heroin abusers nowadays are White is a key reason why opioid addiction is medicalized.
iStockphoto.com/LightFieldStudios

We can further see the effects of these media images in the type of medical treatment opioid addicts receive. A study found that Blacks and Latinos comprise 8 percent of those prescribed buprenorphine, yet they make up 47 percent of methadone patients. This difference is consequential given that buprenorphine can be taken in the privacy of one's home, whereas methadone must be administered at clinics by personnel from the federal Drug Enforcement Administration. There is an underlying message behind these different treatment options: White addicts are capable of monitoring themselves, but minority addicts require supervision. This message highlights that during the opioid epidemic, race remains central to the social construction of drug abuse.[37]

Indeed, responses to drugs often have a latent function beyond curbing abuse and addiction. "Drug scares are never about drugs per se," according to sociologist Craig Reinarman, "because drugs are inanimate objects without social consequences until they are ingested by humans. Rather, drug scares are about the use of a drug by particular groups of people who are, typically, already perceived by powerful groups as some kind of threat."[38] As we've seen, the War on Drugs was a disguised effort to disadvantage Black Americans. President Nixon made them into a **scapegoat**, which is someone or something that is unfairly given the blame for a social problem simply because they're an easy target. Nixon focused on Black drug abusers in order to appeal to the prejudices of Southern Whites and win their political support.

If this story sounds familiar, it's because history has recently repeated itself. In his ascent to the White House in 2016, Donald Trump tapped into voters' anti-immigrant sentiments. In a similar way to how prior presidential administrations scapegoated Black citizens for drug-related crimes, the target of attack more recently has been Brown people without legal status in the U.S. Here's what Trump said in the 2015 speech announcing his candidacy for president:

> When Mexico sends its people, they're not sending their best. They're not sending you. They're not sending you. They're sending people that have lots of problems, and they're bringing those problems with us. They're bringing drugs. They're bringing crime. They're rapists.[39]

The movement to prohibit the sale of alcohol that lasted from 1920 to 1933 was also based on scapegoating. This movement gained momentum by fueling the belief that European immigrant groups were chiefly responsible for drinking-related problems.
George Rinhart/Corbis via Getty Images

Scapegoating racial and ethnic minorities for social problems has been a recurrent theme throughout American history.
iStockphoto.com/Nalidsa Sukprasert

Whereas Trump's predecessors had a latent policy of mass incarceration, the manifest policy of the Trump administration has been mass deportation.

As has been the case time and again throughout American history, the recent scapegoating of minority groups hasn't been based on reality but on people's *perceptions* of reality. Many individuals believe undocumented immigrants deserve blame for drug-related crimes, but in truth these immigrants commit relatively few such crimes compared to native-born Americans. During the period from 1990 to 2013, when the foreign-born percentage of the population increased from 7.1 percent to 13.1 percent, the violent crime rate fell by nearly half. As with the War on Drugs, here the chief aim hasn't been to reduce crime but to subjugate people because they aren't native-born Whites.[40]

Comparing the War on Drugs to the current scapegoating of immigrants underscores that beneath the problem of drug abuse lies a story about racial inequality in American society. Knowing this hidden story adds depth to personal experiences of addiction you may hear about or be familiar with from people you know. While these stories vary from one individual to the next, they collectively reveal broader patterns in how our society responds to drug abuse—through policies that are colored by public perceptions about the racial groups most likely to use a particular drug.

Not as Simple as "Just Say No": The Social Forces Contributing to Drug Abuse

4.5 Identify the social forces that influence a person's decision to do drugs.

Several years ago when I first learned of Michelle Alexander's research about the racial inequalities produced by the War on Drugs, I had my doubts. I was interested in what she had to say but questioned her claim that there now was a new system of segregation similar to the Jim Crow system that existed in the South from the end of the Civil War until the 1950s. My skepticism stemmed from the comparison she was drawing. After all, whereas Jim Crow laws marginalized Blacks simply because of their race, it seemed to me that a person only went to prison for a drug crime if they'd made a conscious decision to break the law. I had grown up convinced that doing drugs was simply a matter of choice. I'd often seen First Lady Nancy Reagan's "Just Say No" anti-drug ads on TV, which many parents and teachers echoed. Hearing these three words again and again made it appear easy to resist drugs.

Once I began learning about Michelle Alexander's research, I started questioning this individual perspective. I began to recognize that the reasons many Black Americans have been convicted of drug offenses go well beyond personal choice. The view that drug abuse reflects a lack of willpower is short-sighted. I used to assume that by risking going to prison, a person was throwing away future opportunities for the immediate pleasure of getting high. You may hold this belief since there's a lot riding on your success in college. If you were arrested and convicted for drug possession, you might not finish school or find a good job. Therefore, it makes sense for you to choose not to do illegal drugs.

As Nancy Reagan was telling kids to just say no, "This Is Your Brain on Drugs" also appeared on TV. Depicting an egg sizzling in a pan, the ad's message was clear: Only a fool does drugs after knowing their dire effects.
iStockphoto.com/fermate

However, this individual perspective doesn't hold true for kids growing up in low-income Black neighborhoods. The reasons that many commit drug offenses go well beyond an inability to just say no. From a very young age, they learn to expect what is perhaps the most damning effect of the War on Drugs: that they, their families, and their neighbors will indefinitely be under surveillance. Even kids who avoid drugs and stay clear of other criminal activity grow up accustomed to being stopped and interrogated by the police and quite possibly also searched, beaten, or detained (see Chapter 3 for a fuller discussion). These humiliating experiences give kids the message that they're worthless and undeserving of a better life. Since they learn to see themselves as criminals regardless of how they behave, it may seem to them that using drugs carries little cost. Getting arrested for possessing drugs doesn't put low-income Blacks at risk of losing future opportunities if they never had those opportunities in the first place. It's convenient to believe that they shouldn't do

Many opioid addicts' troubles began not with a shady drug deal but with a doctor's prescription.
iStockphoto.com/PeopleImages

the crime if they don't want to do the time. But paying attention to the social forces influencing their behavior—which include police bias and a lack of job prospects—exposes the shortcomings of this belief.[41]

The sociological perspective offers an equally compelling explanation for why a White person from a middle-class family would choose to abuse a legal drug like OxyContin or other opioid painkillers. Most people who become addicts started using these drugs upon medical advice. Given that doctors are trusted professionals, why wouldn't a patient experiencing severe pain take a drug prescribed to provide relief? Yet what patients may not be aware of is that pharmaceutical companies have oversold the benefits of prescription painkillers to physicians and consumers alike. Their advertising has minimized opioids' addictiveness and dangerous side effects, while sometimes even promoting increased dosages when people become hooked. Once that happens, there's no easy option for turning back. While in hindsight it may seem that an addict made a bad choice by filling a prescription for opioid painkillers, it's a choice they made in response to significant social pressure: the professional credibility of doctors and the persuasive influence of pharmaceutical companies.[42]

This chapter's comparison of the different responses to the crack and opioid epidemics highlights the limitations of understanding drug abuse as a problem of individual irresponsibility. You can now see that a person's experiences with a dangerous drug have more to do with social forces outside their immediate control than with their own motivations for using the drug. These forces influence why many people do not just say no to drugs and explain why our society has shifted from a criminalized to a medicalized orientation toward addiction.

What Do You Know Now?

1. What does it mean to view drugs as social constructions? Why is it important to see drug abuse in this way?
2. While the manifest function of the War on Drugs was to curb abuse and addiction, how was its latent function to reinforce racial inequality? Answer by discussing actions taken by Richard Nixon and subsequent U.S. presidents.
3. How did the War on Drugs produce a new Jim Crow system?

4. In what subtle ways do remnants of criminalization linger amidst the medicalization of opioid addiction? Discuss media coverage of opioid addiction and the different types of treatment options that addicts receive.
5. From a sociological perspective, why might it be difficult for a person to say no to drugs? Answer by discussing the social forces that respectively produced the crack and opioid epidemics.

Key Terms

Drug addiction 59
Opioids 59
Epidemic 59
Crack 60
Criminalization 61
Medicalization 62
Social construction of drugs 63
War on Drugs 64
Racial discrimination 64
Jim Crow laws 64
Euphemism 65
Manifest function 66
Latent function 66
Mass incarceration 66
Scapegoat 71

Visit **edge.sagepub.com/silver** to help you accomplish your coursework goals in an easy-to-use learning environment.

Notes

1. Claire Hughes, "Outrunning Heroin Addiction: Ashley's Story," *Times Union*, October 14, 2015, http://www.timesunion.com/local/article/Outrunning-heroin-addiction-Ashley-s-story-6571339.php#photo-8766651.
2. American Addiction Center, "Quick Facts on Drug Addiction," https://americanaddictioncenters.org/rehab-guide/addiction-statistics/.
3. The estimate of Americans addicted to opioids is based on 2012 data and comes from Nora D. Volkow, "America's Addiction to Opioids: Heroin and Prescription Drug Abuse," National Institutes of Health, May 14, 2014, https://www.nih.gov/sites/default/files/institutes/olpa/20140514-senate-testimony-volkow.pdf.
4. Centers for Disease Control and Prevention, "Prescription Opioid Data," 2018, https://www.cdc.gov/drugoverdose/data/overdose.html.
5. The comparison with annual auto deaths and deaths from HIV/AIDS at the height of that epidemic come from Atul Gawande, "It's Time to Adopt Electronic Prescriptions for Opioids," *Annals of Surgery* 265, no. 4 (2017): 693–4.

6. As of 2016, sixteen states had passed laws placing limits on doctors' capacity to prescribe these addictive painkillers; see Christine Vestal, "The Days of Freely Prescribed Painkillers Are Ending. Here's What's Next," *Washington Post*, May 20, 2016, https://www.washingtonpost.com/national/health-science/the-days-of-freely-prescribed-painkillers-are-ending-heres-whats-next/2016/05/20/0081748c-15f8-11e6-9e16-2e5a123aac62_story.html?utm_term=.81a8df371c78.
7. German Lopez and Sarah Frostenson, "How the Opioid Epidemic Became America's Worst Drug Crisis Ever, in 15 Maps and Charts," *Vox*, March 29, 2017, https://www.vox.com/science-and-health/2017/3/23/14987892/opioid-heroin-epidemic-charts.
8. Jean Y. Ko et al., "Incidence of Neonatal Abstinence Syndrome—28 States, 1999–2013," Centers for Disease Control and Prevention, August 12, 2016, https://www.cdc.gov/mmwr/volumes/65/wr/mm6531a2.htm.
9. Seth Ferranti, "Crack Addict to Change Agent: Beverly Black's Redemption Story," *The Fix: Addiction and Recovery, Straight Up*, February 3, 2017, https://www.thefix.com/crack-addict-change-agent-beverly-black-redemption. Black recounts her period of addiction and recovery in her memoir *A Wretch Like Me: From Crack Addict to Change Agent* (Gumbo for the Soul Publications, 2016).
10. Jeremy Berke and Skye Gould, "This Map Shows Every U.S. State Where Pot Is Legal," *Business Insider*, January 4, 2019, https://www.businessinsider.com/legal-marijuana-states-2018-1; Jon Gettman, "Marijuana v. Cannabis: Pot-Related Terms to Use and Words We Should Lose," *High Times*, September 10, 2015, http://hightimes.com/culture/marijuana-vs-cannabis-pot-related-terms-to-use-and-words-we-should-lose/.
11. Schaffer Library of Drug Policy, "When and Why Were the Opiates and Cocaine Outlawed?" http://druglibrary.org/schaffer/library/opiates_outlawed.htm. These potency differences are based on research by Andy LaFrate reported in Alice G. Walton, "New Study Shows How Marijuana's Potency Has Changed over Time," *Forbes*, March 23, 2015, https://www.forbes.com/sites/alicegwalton/2015/03/23/pot-evolution-how-the-makeup-of-marijuana-has-changed-over-time/#3b967ada59e5.
12. Richard Nixon, "Remarks about an Intensified Program for Drug Abuse Prevention and Control," June 17, 1971, http://www.presidency.ucsb.edu/ws/index.php?pid=3047#axzz1PCJydjl5.
13. Data about funding for the War on Drugs come from a chart compiled by filmmaker Matt Groff based on data from the International Drug Policy Consortium. It can be found at http://www.mattgroff.com/questions-on-the-1315-project-chart/.
14. Quoted in Craig Reinarman and Harry G. Levine, "The Crack Attack: America's Latest Drug Scare, 1986–992," in *Images of Issues*, ed. Joel Best (New York: Transaction, 1995), 147–64.
15. Details about the Violent Crime Control and Law Enforcement Act of 1994 can be found in a fact sheet published by the U.S. Department of Justice at https://www.ncjrs.gov/txtfiles/billfs.txt.
16. The Johnson quote, as well as skepticism about whether he actually said it, comes from Steven J. Allen, "'We Have Lost the South for a Generation': What Lyndon Johnson Said, or Would Have Said if Only He Had Said It," Capital Research Center, October 7, 2014, https://capitalresearch.org/article/we-have-lost-the-south-for-a-generation-what-lyndon-johnson-said-or-would-have-said-if-only-he-had-said-it/.
17. For discussion of Nixon's focus on Black heroin addicts, see Maia Szalavitz, "These 'New Face of Heroin' Stories Are Just the Old Face of Racism," *Salon*, June 9, 2014, https://www.salon.com/2014/06/19/these_new_faces_of_heroin_stories_are_just_the_old_face_of_racism_partner/.
18. Quoted in Dan Baum, "Legalize It All: How to Win the War on Drugs," *Harper's*

Magazine, April 2016, https://harpers.org/archive/2016/04/legalize-it-all/.

19. Discussion of the distinction between the stated and hidden intent of the War on Drugs draws from Robert K. Merton, "Manifest and Hidden Functions," in *Social Theory Rewired: New Connections to Classical and Contemporary Perspectives*, eds. Wesley Longhofer and Daniel Winchester (New York: Routledge, 2016), 68–84.
20. In interviews, Michelle Alexander often recounts the story of the poster stapled to the telephone poll. This quote is from the April 21, 2016, interview she did for *On Being with Krista Tippett*, http://onbeing.org/programs/michelle-alexander-who-we-want-to-become-beyond-the-new-jim-crow/.
21. Alexander, *On Being with Krista Tippett.*
22. Data on the overrepresentation of Blacks in prison are from John Gramlich, "The Gap between the Number of Blacks and Whites in Prison Is Shrinking," Pew Research Center, January 12, 2018, http://www.pewresearch.org/fact-tank/2018/01/12/shrinking-gap-between-number-of-blacks-and-whites-in-prison/. The comparison of Blacks in the correctional system and those enslaved comes from Michelle Alexander, *The New Jim Crow: Mass Incarceration in the Age of Colorblindness* (New York: The New Press, 2010), 175, 185.
23. Cross-national comparisons of incarceration rates are from Adam Liptak, "U.S. Prison Population Dwarfs That of Other Nations," *New York Times*, April 23, 2008, https://mobile.nytimes.com/2008/04/23/world/americas/23iht-23prison.12253738.html. Data comparing the U.S. incarceration rate to the rest of the world come from World Prison Brief, http://prisonstudies.org/world-prison-brief-data.
24. Julie Netherland and Helena B. Hansen, "The War on Drugs That Wasn't: Wasted Whiteness, 'Dirty Doctors,' and Race in Media Coverage of Prescription Opioid Misuse," *Culture, Medicine, and Psychiatry* 40, no. 4 (2016): 664–86.
25. American Civil Liberties Union, "A Living Death: Life without Parole for Nonviolent Offenses," 2013, https://www.aclu.org/report/living-death-life-without-parole-nonviolent-offenses.
26. Kristen Gwynne, "4 Things You Probably Didn't Know About Crack, America's Most Vilified Drug," *AlterNet*, August 2, 2013, https://www.alternet.org/2013/08/4-things-you-probably-didnt-know-about-crack-americas-most-vilified-drug/.
27. Discussion of racial differences in use of crack and powder cocaine and the 100:1 disparity in sentencing between them comes from Joseph J. Palamar et al., "Powder Cocaine and Crack Use in the United States: An Examination of Risk for Arrest and Socioeconomic Disparities in Use," *Drug Alcohol Dependency* 149 (2015): 108–16. A 2010 law reduced the sentencing disparity to 18:1.
28. Data about the racial disparity in marijuana arrests are from 2010; see American Civil Liberties Union, "The War on Marijuana in Black and White: Billions of Dollars Wasted on Racially Biased Arrests," June 2013, https://www.aclu.org/sites/default/files/field_document/1114413-mj-report-rfs-rel1.pdf.
29. For the past several decades, marijuana has been the drug abused most often by college students; see Lloyd D. Johnston et al., "Monitoring the Future: National Survey Results on Drug Use, 1975–2016," The University of Michigan Institute for Social Research, January 11, 2017, http://www.monitoringthefuture.org//pubs/monographs/mtf-overview2016.pdf.
30. Alexander, *The New Jim Crow*, 192; Niraj Chokshi, "How Felon Voting Policies Restrict the Black Vote," *Washington Post*, February 12, 2014, https://www.washingtonpost.com/blogs/govbeat/wp/2014/02/12/how-felon-voting-policies-restrict-the-black-vote/?utm_term=.5b4d7d630bef.
31. Scott H. Decker et al., "Criminal Stigma, Race, and Ethnicity: The Consequences of

Imprisonment for Employment," *Journal of Criminal Justice* 43, no. 2 (2015): 108–21.

32. William A. Galston and Elizabeth McElvein, "Reducing Recidivism Is a Public Safety Imperative," The Brookings Institution, March 25, 2016, https://www.brookings.edu/blog/fixgov/2016/05/25/reducing-recidivism-is-a-public-safety-imperative/.
33. Katharine Q. Seelye, "In Heroin Crisis, White Families Seek Gentler War on Drugs," *New York Times*, October 30, 2015, https://www.nytimes.com/2015/10/31/us/heroin-war-on-drugs-parents.html.
34. Andrew Cohen, "How White Users Made Heroin a Public-Health Problem," *The Atlantic*, August 12, 2015, https://www.theatlantic.com/politics/archive/2015/08/crack-heroin-and-race/401015//.
35. Seelye, "In Heroin Crisis."
36. Netherland and Hansen, "The War on Drugs That Wasn't," 672–6; Maia Szalavitz, "These 'New Face of Heroin' Stories Are Just the Old Face of Racism," *Substance.com—The Stuff That Hooks Us*, June 8, 2014, http://www.substance.com/these-new-face-of-heroin-stories-are-just-the-old-face-of-racism/7555/.
37. Andrea Acevedo et al., "Performance Measures and Racial/Ethnic Disparities in the Treatment of Substance Use Disorders," *Journal of Studies on Alcohol and Drugs* 76, no. 1 (2015): 57–67; Helena B. Hansen et al., "Variation in Use of Buprenorphine and Methadone Treatment by Racial, Ethnic and Income Characteristics of Residential Social Areas in New York City," *Journal of Behavioral Health Services and Research* 40, no. 3 (2013): 367–77.
38. Craig Reinarman, "The Social Construction of Drug Scares," in *Constructions of Deviance: Social Power, Context, and Interaction*, eds. Patricia A. Adler and Peter Adler (Belmont, CA: Wadsworth, 1994), 92–104. Quote appears on pages 97–98.
39. Michelle Ye Hee Lee, "Donald Trump's False Comments Connecting Mexican Immigrants and Crime," *Washington Post*, July 8, 2015, https://www.washingtonpost.com/news/fact-checker/wp/2015/07/08/donald-trumps-false-comments-connecting-mexican-immigrants-and-crime/.
40. David Green, "The Trump Hypothesis: Testing Immigrant Populations as a Determinant of Violent and Drug-Related Crime in the United States," *Social Science Quarterly* 97, no. 3 (2016): 506–24; Walter Ewing, Daniel E. Martiniez, and Ruben G. Rumbaut, "The Criminalization of Immigration in the United States," American Immigration Council, July 13, 2015, https://www.americanimmigrationcouncil.org/research/criminalization-immigration-united-states.
41. For a general discussion of the significant police presence in Black kids' lives from a very young age, see Alexander, *The New Jim Crow*, 194–5. Based on a six-year study, sociologist Alice Goffman offers a gripping picture of the everyday experience of living amidst police surveillance in *On the Run: Fugitive Life in an American City* (New York: Farrar, Strauss, and Giroux, 2015).
42. Nicholas Kristof, "Drug Dealers in Lab Coats," *New York Times*, October 18, 2017, https://www.nytimes.com/2017/10/18/opinion/opioid-pharmaceutical-addiction-pain.html.

iStockphoto.com/wombatzaa

5

Slim Chances

Weight Anxiety in a Society That Prizes Thinness

Learning Objectives

1. Identify the importance of seeing weight as a social construction.
2. Describe how size discrimination affects individuals in the workplace, at the doctor's office, and in schools.
3. Explain why obesity rates are highest in low-income communities.
4. Recognize the social forces that can lead people of average weight to feel anxious about the size of their bodies.
5. Explain the importance of efforts by size acceptance activists to destigmatize the word *fat*.

Thin Is In: The Social Construction of Body Size

5.1 Identify the importance of seeing weight as a social construction.

During his nationally aired radio show on April 4, 2015, Mike Gallagher asked Fox News anchor Chris Wallace if he'd seen recent photos of Kelly Clarkson because, as Gallagher put it, "Holy cow, did she blow up." Wallace agreed, commenting that "[Clarkson] could stay off the deep dish pizza for a little while."[1] Their conversation came on the heels of similar criticism by British newspaper columnist and TV personality Katie Hopkins six weeks earlier when she tweeted, "Kelly Clarkson had a baby a year ago. That is no longer baby weight. That is carrot cake weight. Get over yourselves."[2] Hopkins subsequently told *Access Hollywood*, "Jesus, what happened to Kelly Clarkson? Did she eat all of her backing singers? Clarkson is a chunky monkey. . . . She's fat, she needs to get out, eat less, and move more."[3]

Notice that none of this discussion about Clarkson focused on the Grammy Award winner's powerful voice, which has propelled hit songs like "Because of You" and "Since U Been Gone." Instead, the main topic was her weight. The public humiliation she experienced is an example of **fat shaming**, which occurs when a person receives insults because they have a large body. In recent years, other celebrities, including Justin Bieber, Adele, Leonardo DiCaprio, and Kate Winslet, have also been fat shamed.[4] Although this negative publicity most often targets girls and women, boys experience fat shaming too.[5]

A couple of months before Clarkson's body became an object of ridicule, Toma Dobrosavljevic received media praise for winning season sixteen of *The Biggest Loser.* In the hit reality show, which aired from 2004 to 2016, overweight individuals competed to see who could slim down the most over thirty weeks. Dobrosavljevic triumphed by losing more than half his starting weight. He dropped 171 pounds, ending the competition at a slender 165 lbs.[6]

The title of "the biggest loser" doesn't only refer to the person who shed the most weight. It also conveys an opinion that viewers may have held toward all the contestants at the start of the competition: They were losers for being overweight. Media

Here is Kelly Clarkson in 2005, a few years after she won *American Idol* and several years before she experienced fat shaming.
AP Photo/Adam Rountree

Kelly Clarkson poses at the 2019 Academy of Country Music Awards, which took place a few years after she was shamed for how she looked after giving birth.
Jordan Strauss/Invision/AP

personalities cast similar aspersions on Kelly Clarkson after she became heavier. Whereas Toma Dobrosavljevic's victory upheld the idea that regimented dieting and vigorous exercise could produce dramatic weight loss, the fat shaming of Clarkson exposed the other side of the coin. Both of their stories reinforce the individual perspective, which characterizes each of us as personally to blame if we don't conform to societal norms about weight. Because of the widely held belief that a person is at fault for having a large body, being viewed by others or seeing oneself as "fat" is a source of **stigma**—a characteristic that marks a person as disreputable in the eyes of others. Because of this stigma, **weight anxiety**, or the unease many people have that their bodies are too big, is a pervasive social problem.

FIRST IMPRESSIONS?

1. Can you think of someone—either a celebrity or a person you know—who has experienced fat shaming? What was their experience like?
2. Why do you think media publicity of fat shaming more often focuses on instances where girls and women are the victims, rather than boys and men?
3. How do the negative media portrayals of Kelly Clarkson and the positive portrayals of Toma Dobrosavljevic similarly reflect the individual perspective toward body size?

Here's Toma Dobrosavljevic at the start of *The Biggest Loser*.
Trae Patton/NBC/NBCU Photo Bank via Getty Images

Toma Dobrosavljevic looked dramatically thinner after winning the competition.
Trae Patton/NBC/NBCU Photo Bank via Getty Images

When I was a kid, my older brother would sometimes tell me I was fat. I wasn't overweight according to the **body mass index** (BMI), which is a medical chart that labels a person's size relative to what doctors consider healthy. However, knowing this medical information didn't comfort me. Although my brother meant no harm, his teasing made me feel inadequate. After all, his opinions weren't isolated. Just about everywhere we turn—Instagram posts, movies, TV shows, and advertisements—there are images of thin people who are happy and successful and fat people who appear lazy and unkempt. These images fuel and reinforce weight anxiety.

Fat isn't merely a synonym for *overweight*, but a stigmatizing label. It's the f-word.
Cultura Creative (RF)/Alamy Stock Photo

Even people of average size are prone to experiencing weight anxiety because they believe they're fat.
Getty/Steve Niedorf Photography

Because of the psychological weight of living in a society where bodies reflect one's character, people whose BMIs indicate they are healthy may still be at risk of experiencing weight anxiety. In a poll of Americans of varying BMIs, 45 percent indicated they worried at least some of the time about being too big.[7] Approximately forty-five million people in the United States diet each year, spending about $33 billion on weight loss products. A massive amount of marketing is directed at these people.[8] Given the strong association in our society between feminine beauty and thinness, this marketing particularly targets girls and women. A common piece of advice directed at dieters is that they should get on the scale every *single* day.[9]

Weight anxiety is not only a condition that may afflict you or others you know; it also reflects how our society defines the ideal body. Start paying attention to the **social construction of weight**, or the idea that people assign bodies of varying sizes and shapes unequal amounts of social worth. In other words, the numbers on a scale or BMI chart do not speak for themselves; we give them meaning. For example, consider that for centuries in Fiji and other islands of the South Pacific, people equated plumpness with prosperity. The ideal body was fat, unlike in the contemporary U.S. Start taking notice of how prevalent the thinness ideal is in American culture. The popularity of *The Biggest Loser* is a prime illustration of this ideal. In fact, a survey done three years after the arrival of satellite TV in Fiji during the mid-1990s found that Fijian girls who watched three or more nights a week were 50 percent likelier to see themselves as fat and 30 percent likelier to diet than girls who watched infrequently. Indeed, across the world having a large body increasingly carries stigma.[10]

The internet has been exponentially more influential than TV in contributing to the global stigma of having a large body. Before reading this chapter, you may have believed thinness was inherently valuable. But there's more to this belief than meets the eye. The sociological perspective enables you to discover that because American culture defines thin bodies as beautiful and propagates this norm around the world, our *society* creates weight anxiety. Feelings of insecurity about being too big aren't simply a personal problem many of us experience; we've inflicted this problem upon ourselves.

This chapter highlights the array of social forces that jointly contribute to weight anxiety. Because these forces are often invisible, people experiencing anxiety are likely to believe they're personally responsible for feeling badly about themselves. Moreover, when we think about body size from the sociological perspective, we discover that the choices people make about how much to eat and exercise often reflect the economic, educational, and social opportunities available to them. This perspective, therefore, enables you to redefine *fat* as a word that describes—without judgment—people with BMIs beyond the range that doctors consider healthy. That's how I use the word throughout the chapter. Removing the stigma underscores that for those people who, medically speaking, are fat, as well as for others who feel fat, their condition isn't just a personal problem. It's a weight we *all* must bear.

Pierre-Auguste Renoir's paintings depicted the ideal body type for French women during the late nineteenth century. Notice that the woman depicted in this painting, titled *Odalisque*, is rounder than the body shape valued in Western women today.
The Print Collector/Alamy Stock Photo

Bias without Boundaries: The Pervasiveness of Size Discrimination in American Society

5.2 Describe how size discrimination affects individuals in the workplace, at the doctor's office, and in schools.

There are many stories online about fat people experiencing **size discrimination**, the unequal treatment people encounter simply for being overweight. Let's consider accounts by three women—Leah, Patti, and Serine. When Leah went shopping for a new bra the summer after her sophomore year of college, the sales associate said the store had nothing in 38 DD "because boobs aren't really supposed to be that big." Patti once interviewed for a job at an optometrist's office and, as the conversation was wrapping up, the interviewer told her, "You know, we have *very* small hallways here." When Serine was in high school and tried out for the cheering squad, the teacher evaluating her audition laughed at her large body. As Figure 5.1 indicates, these

FIGURE 5.1 • Unequal Treatment

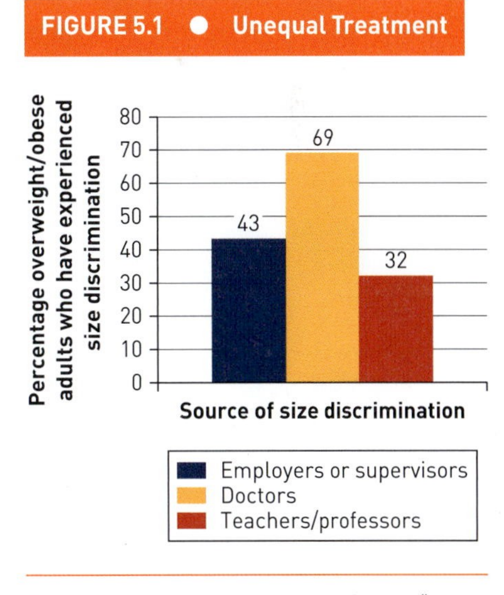

Source: Rudd Center for Food Policy & Obesity, "Weight Bias: The Need for Public Policy," 2008, https://www.naafaonline.com/dev2/about/Brochures/WeightBiasPolicyRuddReport.pdf.

Job recruiters may exhibit size discrimination toward fat applicants by seeing them as lacking ambition or self-discipline.
Christian Darkin/Science Source

stories are hardly isolated or exceptional. Size discrimination routinely occurs on the job, at school, and in the doctor's office.[11]

First, let's consider the workplace. Research indicates that for women, a gain of twenty-five pounds equates with an average salary loss of $14,000 a year.[12] Bosses may deny them a promotion or reassign them to a lower-paying job. Previously thin women are more prone to such discrimination than women who've already been overweight. Another study highlights the significant effects size discrimination has on men as well. Men of normal weight applied for jobs at retail stores. Then, while wearing overweight prosthetics, they applied for jobs at different stores and posed as customers in a third set of stores. In both of the contexts where they wore prosthetics, these men experienced size discrimination. Employees displayed more avoidance toward them and were less likely to nod or smile while interacting with them.[13]

Size discrimination also runs rampant in the classroom. Teachers may hold low expectations of fat students or judge them as lazy, untidy, and overly emotional. Physical education teachers often hold the strongest prejudices of all because they're professionally invested in seeing body size as a reflection of a person's physical health. Fat kids often start experiencing bias in the early grades, and it continues throughout their schooling. As a result, they tend to receive lower grades than their peers of average weight and are less likely to go to college.[14]

Weight is the number one reason kids are bullied. It's a more significant factor than religion, disability, race, intellectual ability, or sexual orientation (see Figure 5.2). Whereas one in four kids in the U.S. has experienced bullying, a survey of overweight high school students found that 58 percent of boys and 63 percent of girls had been bullied (see Chapter 12 for a fuller discussion of bullying).[15] This frequent victimization is a significant contributor, along with teacher bias, to why fat students tend to have lower academic achievement than average-weight students.[16]

Doctors and nurses sometimes make jokes or derogatory comments about overweight patients.
iStockphoto.com/jacoblund

Size discrimination is even common in health care settings where, one might expect, trained professionals would exhibit care in trying to alleviate weight anxiety. Doctors may regard fat patients as lazy or lacking in self-discipline and presume that any concerns

FIGURE 5.2 • Simply Mean

These are the most common reason kids are bullied.

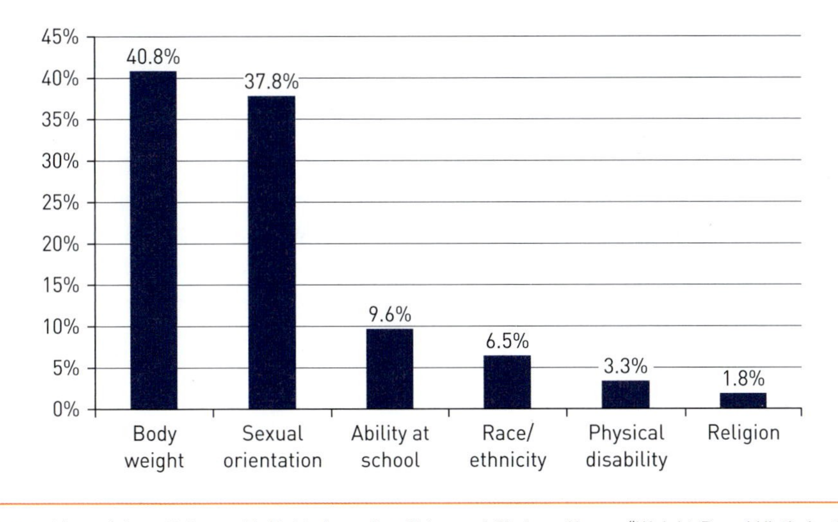

Source: Adapted from Rebecca M. Puhl, Joerg Luedicke, and Chelsea Heuer, "Weight-Based Victimization toward Overweight Adolescents: Observations and Reactions of Peers," *Journal of School Health* 81, no. 11 (2011): 696–703.

these patients express are related to their weight. As a result, doctors may not be sufficiently attentive to the patient's pain and neglect to run appropriate diagnostic tests.[17] A study of autopsy reports found that fat people were 1.65 times likelier than average and underweight people to have lived with undiagnosed conditions that may have compromised their quality of life.[18] The upshot of these data is that doctors sometimes allow health problems in fat people to fester and worsen instead of treating them.

There are legal steps a person can take to fight gender discrimination. However, fat people have few options for countering bias.
iStockphoto.com/CnOra

There is no federal legislation designating body size as a category deserving equal protection in the way that the law does for race, gender, and disability. The difference stems from a societal belief—the individual perspective—that since a person can change their body size, they're at fault for being overweight. A useful way to think about this belief is by comparing it to the public sentiment toward smokers. They are deemed undeserving of civil rights because they've consciously made the choice to do harm to their bodies. By the same token, since fat people purportedly

Michigan is the *only* state to prohibit size discrimination. Several cities do as well, including Madison, Wisconsin; Urbana, Illinois; Washington, D.C.; and San Francisco and Santa Cruz, California.
iStockphoto.com/Lady-Photo

Whereas Black people and women tend to have more favorable attitudes toward members of their own groups than does the population as a whole, this isn't true for fat people. They're more inclined to buy into societal biases against them.
iStockphoto.com/CreativaImages

can make healthier lifestyle decisions, they don't merit protection from weight bias.[19]

Think about why it matters whether a group has legal protection from discrimination. The response to the fat shaming of Kelly Clarkson would have been entirely different if she were Black and the ridicule pertained to her race. Such examples are common. For instance, DJ Don Imus called the all-Black Rutgers women's basketball team a bunch of "nappy-headed hos," and, as a result, CBS suspended his nationally syndicated show for several months. However, Chris Wallace—the Fox anchor who fat shamed Clarkson—got off with merely having to issue an apology. "I've sat in meetings with colleagues who wouldn't dream of disparaging anyone's color, sex, economic status or general attractiveness," commented Harriet Brown, a leading expert on body image issues. "Yet [these colleagues] feel free to comment witheringly on a person's weight."[20]

Because size discrimination is so widespread, fat Americans come to expect it and yet often aren't consciously even aware it's happening. Moreover, they may hold the very same prejudices that are common in the workplace, schools, and health care settings by, for example, regarding people like themselves as lazy or deserving unequal treatment. Fat people, therefore, play a key role in contributing to the acceptance of size discrimination across American society. They are certainly *not* to blame for this problem. The key point, rather, is that their internalization of societal prejudices against them is symptomatic of how pervasive size discrimination is in American society.[21]

Let's consider why size discrimination is so deep-seated in the U.S. Many people seem to view drinking a milkshake or eating a Big Mac as not only unhealthy but also as a moral failure. In their eyes, these behaviors express a preference for short-term pleasure over the enduring advantages one can accrue by delaying gratification. This focus on the value of future rewards is the essence of the **American Dream**, the idea that getting ahead hinges on self-discipline and hard work. From this perspective, fatness is a defiance of core values. Eating too much (gluttony) and exercising too little (sloth) are, after all, two of the seven deadly sins.[22]

Given that fat bodies are often seen as an indicator of immoral character, we can understand size discrimination as having a similar origin as violence against gays, lesbians, bisexuals, transgender, and queer people, who on average are killed about once a week. Overweight people and those in the LGBTQ population are both targeted for behaviors that carry stigma because others regard these behaviors—such as inhabiting a large body, being attracted to people of the same sex, or identifying outside the male/female gender binary—as fundamentally wrong. The policing of these behaviors aims to force people to become more "normal."[23]

At the 2016 Gay Pride parade in San Francisco, marchers stood in solidarity for the forty-nine people killed a week earlier at an LGBTQ nightclub in Orlando, Florida.
Justin Sullivan/Getty Images

The sociological perspective reveals that shaming a person for being fat can make them feel insecure and therefore prone to eating more, not less, to relieve their anxiety.
iStockphoto.com/Motortion

Discrimination and violence are *never* justified, period. Still, is it possible that shaming fat people can motivate them to embrace healthier lifestyles? It's an intriguing question. *The Biggest Loser* demonstrated, after all, that fat people can slim down within a short time frame if they have the competitive drive to do so. Of course, TV—even reality TV—is often unrealistic. Consider a study of fat women. One group was randomly assigned to watch a video with discriminatory content and a second group to watch a video with no such content. After offering food to all of them, the researchers observed that the first group consumed over three times as many calories. The anxiety the video produced activated the physiological urge to eat as a coping mechanism. Size discrimination, therefore, can create a **self-fulfilling prophecy**, a situation where a person behaves in ways that confirm how others label them. Because fat people may respond to size discrimination by overeating, in the eyes of those who mistreat them, such bias may seem justifiable.[24]

The Elephant in the Room: Explaining Why Obesity Is Most Prevalent in Low-Income Communities

5.3 Explain why obesity rates are highest in low-income communities.

The Biggest Loser was a hit TV show in part because of how often there is mention in the news and popular culture of **obesity**, a medical condition afflicting people with a BMI exceeding 30. Its many adverse effects include heart disease, type-2 diabetes, hypertension, and high blood pressure. Figure 5.3 reveals that over the past few decades, obesity in the U.S. has become an **epidemic**, or a harm that spreads rapidly across a larger and larger segment of people. During your lifetime, obesity has risen dramatically. In 1990, every state had an obesity rate below 15 percent. By 2000, the rate had risen to above 20 percent in about half the states, and by 2010 it exceeded 25 percent in the majority of states.[25] As of September 2018, the numbers were even more startling: Forty-eight states had an obesity rate exceeding 25 percent, twenty-nine states over 30 percent, and seven states over 35 percent.[26] One in three American adults is obese.[27] For children, it's one in six.[28] As these numbers indicate, obesity is producing a public health crisis.

The sociological perspective offers a powerful explanation for rising obesity rates. It reveals that a person's lifestyle choices—what and how much they eat, and when and how often they exercise—often reflect unequal opportunities. Figure 5.4 (on page 92) highlights a crucial story: Obesity is most prevalent among the least educated Americans and least prevalent among the most educated. Boys who drop out of high school are nearly twice as likely to be obese as boys who graduate from college: 21.1 percent versus 11.8 percent. Among girls, high school dropouts are nearly 2.5 times likelier to be obese: 20.4 percent versus 8.3 percent. What do you think explains these trends? What types of knowledge about nutrition and fitness are college graduates more likely to possess than high school dropouts?

Since college graduates typically earn much more than high school dropouts, this educational gap takes on added significance when we consider the high cost of eating well. Nutritious foods are typically expensive, which is why some people refer to Whole Foods as "Whole Paycheck." At the other extreme, the cheapest foods available in grocery stores are high in saturated fats and carbohydrates from sugar and starch. Therefore, lower-income people are inclined to make unhealthy choices. It's not their fault; they often can't afford fresh fruits and vegetables or lean meats. Although anyone can gain weight by eating poorly, some people—because of the earning power tied to their level of education—are at greater risk of becoming obese than others.[29]

Those who are the least able to afford nutritious foods are also the most likely to live in places where these foods aren't even available. Many poor communities in the United States are **food deserts**, neighborhoods where the only available foods are at convenience stores or fast food restaurants, both of which have few nutritious options.[30] Consider Chicago, for example. It reflects the U.S. as a whole in that Blacks and Latinos, who tend to be poorer than Whites, frequently live too far away from a grocery store to get fresh fruits and vegetables (see Figure 5.5 on page 93). Of the city's 610,000 residents living in food deserts, 480,000 (79 percent) are Black.[31] Because low-income people often don't own cars and can't easily get to supermarkets by public transportation, they're left with limited options for eating healthily.

Given that the cheapest and most accessible meals are often junk foods engineered with irresistible combinations of salt, sugar, and fat, low-income people are

FIGURE 5.3 • Evidence of an Epidemic

Notice the dramatic change in obesity rates in the U.S. over time.

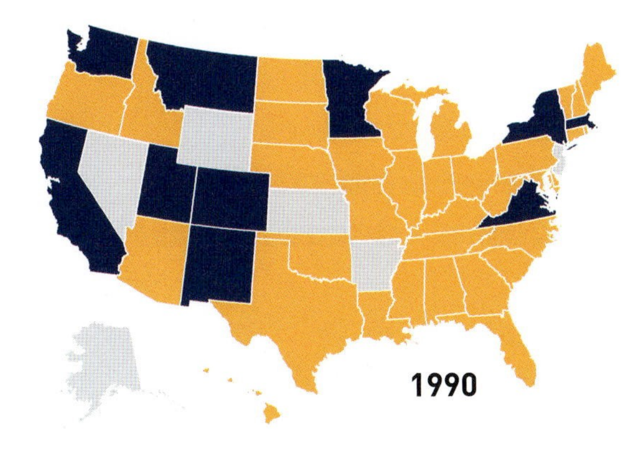

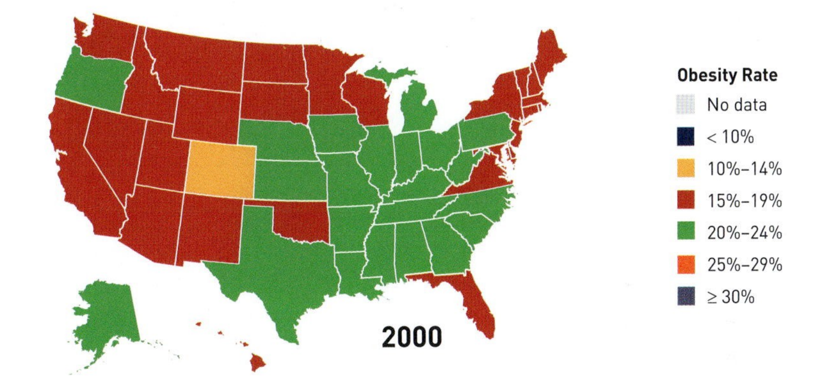

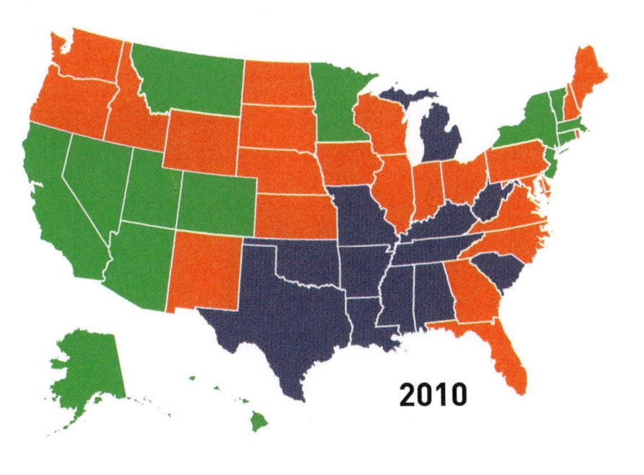

Source: Adapted from Harvard School for Public Health, "Obesity Prevention Source," https://www.hsph.harvard.edu/obesity-prevention-source/us-obesity-trends-map. Data source: Behavioral Risk Factor Surveillance System, Centers for Disease Control and Prevention.

FIGURE 5.4 ● Weight Reflects Education

People with lesser amounts of schooling are more likely to be obese.

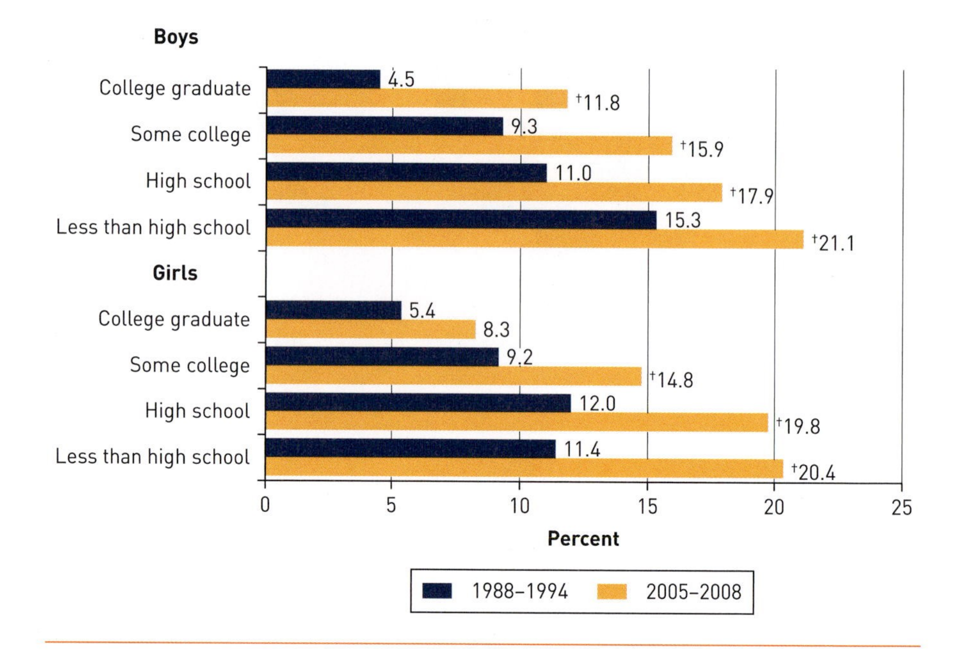

Source: Cynthia L. Ogden, Molly M. Lamb, Margaret D. Carroll, and Katherine M. Flegal, "Obesity and Socioeconomic Status in Children and Adolescents: United States, 2005–2008," National Center for Health Statistics Data Brief Number 51, Centers for Disease Control and Prevention, December 2010, https://www.cdc.gov/nchs/products/databriefs/db51.htm.

susceptible to developing cravings for chips, soda, hamburgers, French fries, and the like. Moreover, junk food marketing targets poor people. One study found that fast food chains advertised over 60 percent more frequently in low-income Black neighborhoods than in higher-income White neighborhoods, and that the advertising to Black consumers was more likely to target kids. Marketing junk foods to people who lack the resources to choose healthier options furthers the likelihood that they will consume these foods and be overweight.[32]

Limited fitness opportunities are yet another reason poor people are the most likely to become obese. Low-income neighborhoods often lack green spaces for exercising, and joining a gym is expensive. Since crime in these neighborhoods is typically high, fear further inhibits people from spending time outside. These income-related factors that foster a sedentary lifestyle are often invisible to people of greater means who may take for granted the availability of fitness opportunities. It's no wonder that many who enjoy these opportunities are inclined to see fat people as lazy.[33]

FIGURE 5.5 ● Food Desert

Chicago is emblematic of the U.S.more generally: Members of minority groups often live too far away from grocery stores to be able to access healthy food.

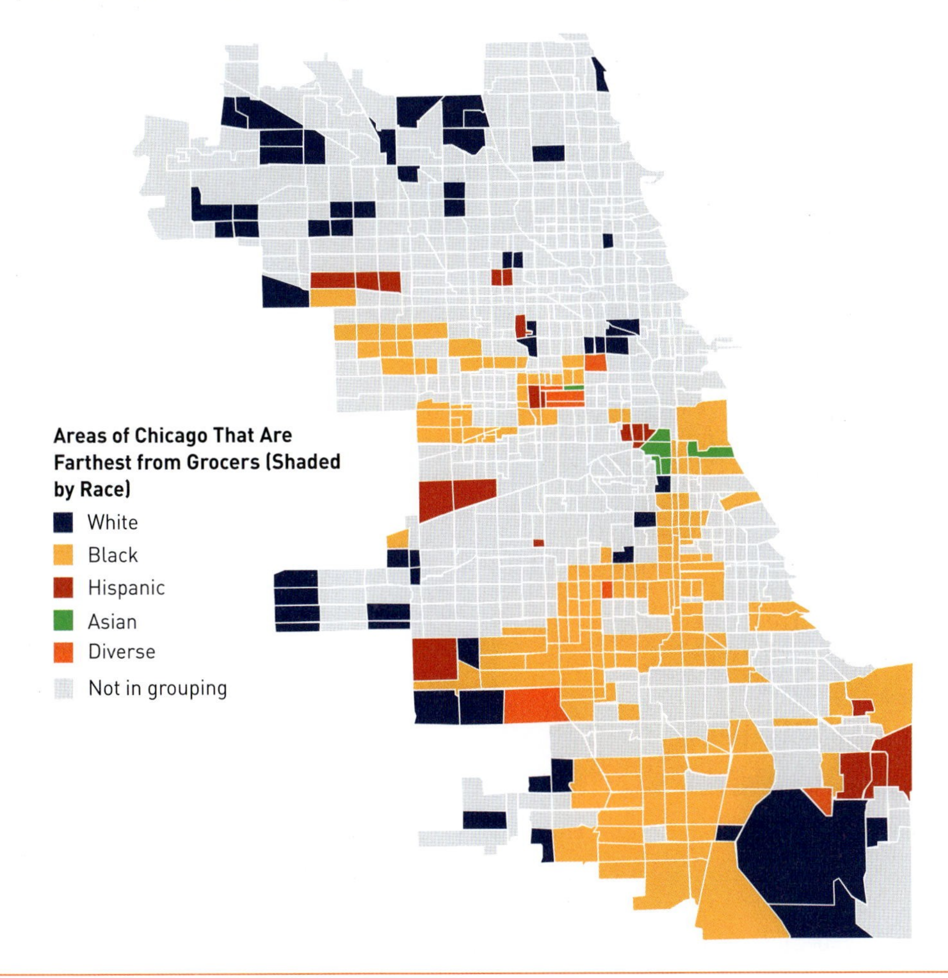

Source: Topher Gray, "From Farm to Food Desert," *Chicago Reader*, August 19, 2010, https://www.chicagoreader.com/chicago/chicago-food-deserts-hopkins-park-black-farmers/Content?oid=2272825. Adapted from Mari Gallagher Research & Consulting Group, "Examining the Impact of Food Deserts on Public Health in Chicago," July 18, 2006, https://www.marigallagher.com/2006/07/18/examining-the-impact-of-food-deserts-on-public-health-in-chicago-july-18-2006. Reprinted by permission of Mari Gallagher Research & Consulting Group.

In embracing the sociological perspective, we're seeing that there's so much more to the story of America's obesity epidemic than the lifestyle choices people make. This epidemic is a sign that at a time of unprecedented economic inequality, we've become a nation where prospects for low-income people to become upwardly

These are the food options typically available in low-income neighborhoods across the United States.
nik wheeler/Alamy Stock Photo

Since poor people often lack knowledge about nutrition, they are ripe targets of junk food marketing.
iStockphoto.com/skhoward

Feeling unsafe in their own neighborhoods is a barrier precluding low-income people from leading healthier lifestyles.
iStockphoto.com/TheaDesign

mobile—and enjoy the opportunity to lead healthier lives—have greatly diminished. An astounding 60 percent of American kids born into families in the bottom fifth of the income distribution are likely still to be in the bottom two-fifths at age forty. Over half of kids born into the bottom fifth who are Black or whose parents didn't finish high school remain in the bottom fifth as adults (see Chapter 2 for a discussion of this opportunity divide). These inequalities are the so-called **elephant in the room**—a major cause of a social problem that people often ignore—when it comes to explaining the obesity epidemic.[34]

Part of the reason you may be unfamiliar with the sociological perspective toward obesity is because media sources that provide information about it do not highlight this way of understanding the problem. It's common, instead, for journalists reporting about obesity to embrace the individual perspective. A study of news coverage over a ten-year period found that it consistently emphasized bad lifestyle choices as the leading cause of this epidemic. Consider how one article characterized a forty-something couple, Bruce and Lisa Smith:

> Chips, fried chicken, canned fruit, sodas—they ate as much as they wanted, whenever they wanted. Exercise? Pretty much nonexistent, unless you count working the TV remote or the computer mouse. "We were out of control," says Bruce, 42. And so was their son, Jarvae, who is 5 feet 4 and weighs 176 pounds.[35]

While a news report might mention sociological causes of obesity, it's likely to present these causes as secondary to bad personal choices.

News coverage about obesity often masks how poverty creates conditions that incline a person toward eating unhealthy foods and exercising infrequently.
iStockphoto.com/goir

Content on YouTube similarly tends to attribute responsibility for obesity largely to the individuals who suffer from it. A random sample of videos searched with the keywords *obesity* and *obese* revealed a common theme: Obese people eat poorly and lead a sedentary lifestyle. The videos could have delved into the social forces producing these behaviors, but they did not. Instead, many of them portrayed people being teased for excessive eating. By reinforcing individual explanations for obesity, YouTube videos legitimize the public shaming of fat people and the bias they often experience. As long as the inequalities at the root of the obesity epidemic remain hidden, there will be little public outcry about the mistreatment fat people frequently endure.[36]

Feeling the Weight: Why People of Average Size May Be Insecure about Their Bodies

5.4 Recognize the social forces that can lead people of average weight to feel anxious about the size of their bodies.

"I cannot tell you how many times throughout my life—especially during the 17 years I struggled with bulimia—I said the phrase, 'I feel fat,'" wrote Greta Gleissner, founder of an organization that helps people recover from eating disorders. "I could be having a perfectly pleasant day, only to suddenly find myself drowning in negative body image. Sometimes the trigger was situational, like putting on a pair of jeans that just came out of the dryer, feeling overly full after a meal or catching a glimpse of myself in a storefront window while shopping."[37]

Gleissner is hardly alone. Nearly half of Americans of varying BMIs worry all or some of the time about their weight—a 32 percent increase since 1990. This rise seems to be in response to the growing obesity epidemic over the past few decades. What's most telling is that from 1999 to 2014, weight anxiety rose only 6 percent among people who self-described as overweight, but 28 percent among those who believed their size was about right. This discrepancy underscores how susceptible people who are either of average weight or underweight are to experiencing emotional fallout from the public health crisis the obesity epidemic has produced. It's not just overweight people who feel the weight of this epidemic; healthy-sized people do too, to an even higher degree (see Figure 5.6 on page 96).

FIGURE 5.6 ● Feeling Fat

A growing segment of Americans, including those who believe their weight is about right, worries either some or all of the time about their weight.

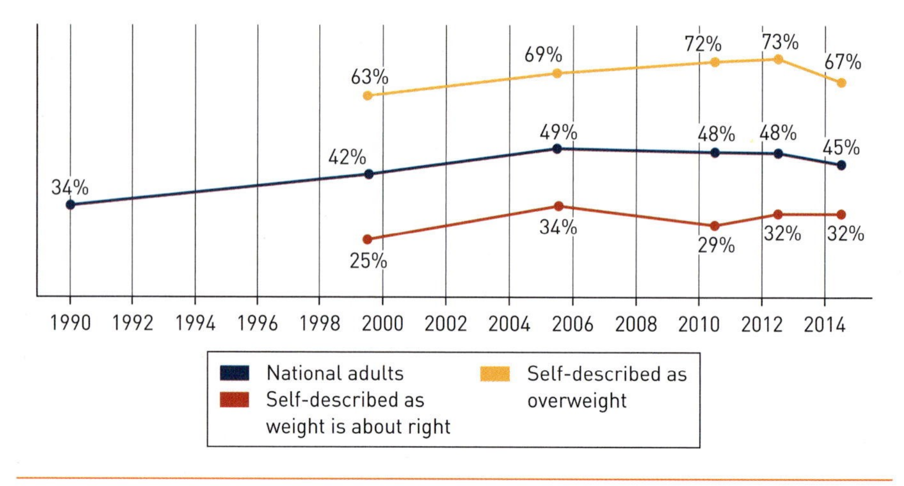

Source: Joy Wilke, "Nearly Half in U.S. Remain Worried About Their Weight," Gallup, July 25, 2014, http://news.gallup.com/poll/174089/nearly-half-remain-worried-weight.aspx.

Girls and women are particularly susceptible to feeling fat even when they aren't. Whereas 25 percent of college males reported in a study that they engaged in negative talk about their bodies, 90 percent of college women in a different study described themselves to others as fat, even though only 9 percent had an above-average BMI. To understand why, we need to pay attention to the many places where girls and women hear the message that they should evaluate themselves primarily based on their physical appearance.[38]

Whether it's from social media, TV shows, movies, advertising, or conversations with friends, girls come to believe from a young age that being thin will make them feel worthy. Shopping also reminds them that small bodies are the most valued. Consider the trendy "one-size-fits-all" clothes available at Brandy Melville, American Eagle, and elsewhere. It's misleading marketing since only the slim percentage of young women who are about 5′8″ and have a twenty-four-inch waist can wear them. The popularity of these clothes signals to the majority of girls and women who are differently shaped that their bodies do not fit the socially constructed norm. As a result, they may feel anxious about their size.[39]

For the many girls and women who feel fat, which may include you or people you know, it's valuable to know just how unattainable this norm is. Consider that whereas in 1975 models weighed 8 percent less than the average-sized American woman, nowadays they weigh 23 percent less. Therefore, the ideal female body has increasingly become just that—an image that exists on social media and television but that is essentially unattainable. How can knowing that the ideal-sized woman has gotten smaller over time lead girls and women who feel fat to become more accepting of their own bodies?[40]

Advertising contributes to why many girls and women who are not overweight feel insecure about their bodies.
Richard B. Levine/Newscom

It's no wonder that roughly two-thirds of women ages twenty-five to forty-five exhibit **disordered eating**, which is a condition characterized by an irregular meal schedule, limited food intake, binging, or self-induced vomiting after eating. Ten percent meet the diagnostic criteria for an eating disorder, such as anorexia or bulimia. Boys and men experience these disorders too, but to a lesser extent. That's because although they also derive status from their physical appearance, norms about body size are not as rigid for them as they are for girls and women. The greater prevalence of eating disorders among females reflects how common it is for cultural messages about body size to lead them to *feel* fat, regardless of their actual weight.[41]

While girls and women may experience this feeling as solely their own problem, the hidden story we've been exposing is that each of us contributes to the prevalence of weight anxiety in our society. In embracing the sociological perspective, we're seeing that several social forces produce this anxiety: the socially constructed norm of thinness, the prevalent belief that body size is a choice, the absence of legal protections against size discrimination, and the messages directed at girls and women that they should define their self-worth based on their physical appearance. This perspective highlights that although how people feel about their bodies may seem strictly personal, this feeling reflects ideas and beliefs that permeate throughout our society.

Debunking Conventional Wisdom about Body Size

5.5 Explain the importance of efforts by size acceptance activists to destigmatize the word *fat*.

Contestants on *The Biggest Loser* succeeded in slimming down because they were under a strict regimen. Trained professionals told them when, what, and how much to eat; when and how long to exercise; and how long to sleep. Given that the goal of weight loss is to keep the pounds off, consider the revealing findings from a six-year study of former show contestants. The study tracked the weight of fourteen people from the 2009 season and found that, on average, they regained 70 percent of what they'd lost. Dropping a significant number of pounds in a short amount of time slowed their metabolism while diminishing their levels of

"Dad bod" memes, which depict guys who appear attractive despite having a gut, underscore that men have more latitude to conform with body norms than do women.
Erik Isakson/Getty Images

the hormone that tells the body it's satisfied. The former contestants felt the need to increase their calories at the same time that they were burning them more slowly—a perfect storm for significant weight gain.[42]

Many people believe anyone can lose weight by exerting self-discipline and that weight loss is a way to feel better about oneself. These beliefs are, unfortunately, only half-truths. The reality underscored by the study of former contestants on *The Biggest Loser* is that diets often don't have lasting effects. This isn't only the case for obese people trying to shed a significant number of pounds quickly; it's also true for people of any weight who aim to become thinner over a gradual period of time. To understand why, you just need to know some basics about the brain. It has a survival instinct that works to regulate a person's weight and keep it within a certain range. If it dips below that range, the body's response is exactly what *The Biggest Loser* contestants experienced: a greater urge to eat combined with slower metabolism. Research shows, incredibly, that dieters are *more* likely to become obese than nondieters—a finding that holds true through middle age, across ethnic groups, and for men and women alike. This effect is actually strongest among people who had been at a normal weight before they started dieting.[43]

Because dieters often do not maintain weight loss, their inability to do so may *further* their anxiety about body size. Dieting can fuel a vicious circle of failure and self-blame. Therefore, our society not only produces weight anxiety but propels people onto a treadmill where they relentlessly feel they must push forward in order to feel better about themselves. I'm not only referring to the type of treadmill found at the gym; there's also a second type. It's not a machine but a state of mind whereby people keep striving for—yet may continually struggle to attain—a greater acceptance of their own bodies.

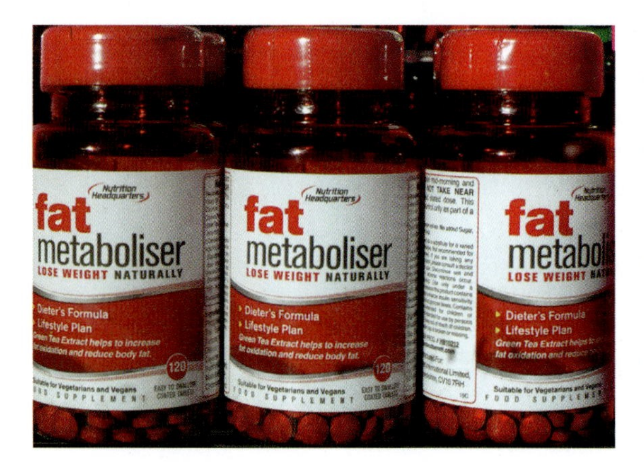

In playing up how easy it is to slim down, diet ads don't mention how much harder it is to maintain weight loss.
Jeffrey Blackler/Alamy Stock Photo

Whereas dieting is a visible yet elusive path toward greater bodily acceptance, there's another path that may be more promising. However, it may not be familiar to you because it doesn't have the slick marketing of the multibillion-dollar diet industry. Its proponents are outspoken but don't have a lot of clout. They are **size acceptance activists**, a group of people who aim to diminish the importance of weight to a person's feelings of self-worth.[44]

These activists challenge the conventional wisdom that fat people necessarily lead unhealthy lives. Indeed, research indicates that a person can be both overweight and healthy. In one study, a third of the obese people sampled had normal heart disease risks based on factors like cholesterol level and blood pressure. Other research reveals that the key determinant of health isn't a person's weight but their level of exercise. While people who are both obese and physically fit account for just 2 to 3 percent of deaths in the United States, inactivity across the BMI spectrum is responsible for 16 to 17 percent.[45] These findings lend support to "health at every size," a slogan size acceptance activists use to distinguish constructive concerns about health from destructive angst about large bodies.

Size acceptance activists believe that removing stigma from the word *fat* is necessary to alleviate weight anxiety experienced by people of all shapes and sizes.
Frances Roberts/Alamy Stock Photo

Promoting this slogan is part of these activists' goal to redefine the term *fat* so that it no longer carries stigma and instead becomes synonymous with *overweight*. While this may be a lofty goal, it underscores the significant payoff of understanding weight anxiety from a sociological perspective. As we've seen throughout this chapter, this perspective examines the social forces that create such anxiety. Exposing the tremendous influence these forces have on individuals across the size spectrum reveals how foolish we'd be if we were to continue seeing weight anxiety strictly as a problem of personal irresponsibility; doing so would make us the biggest losers of all.

What Do You Know Now?

1. What does it mean to see weight as a social construction? How might recognizing this concept enable people who experience weight anxiety to feel better about their bodies?
2. How does the individual perspective toward body size legitimize size discrimination?
3. Why is it harder for low-income people to make healthy choices when it comes to eating and fitness than it is for people from higher-income families and communities? In what sense, then, is obesity a social problem for which all of us, regardless of our weight, bear responsibility?

4. Why are people of average weight, and even those who are underweight, at risk of feeling anxious about being fat?
5. How successful do you believe fat acceptance activists can be at removing stigma from the word *fat*? To what degree do you think it's possible for our society to address the health problems associated with being overweight without casting a negative judgment on fat people?

Key Terms

Fat shaming 81
Stigma 82
Weight anxiety 82
Body mass index 83
Social construction of weight 84
Size discrimination 85
American Dream 88
Self-fulfilling prophecy 89
Obesity 89
Epidemic 90
Food deserts 90
Elephant in the room 94
Disordered eating 97
Size acceptance activists 98

Visit **edge.sagepub.com/silver** to help you accomplish your coursework goals in an easy-to-use learning environment.

Notes

1. Stephanie Webber, "Chris Wallace Fat-Shames Kelly Clarkson: 'She Could Stay off the Deep Dish Pizza,'" *Us*, April 4, 2015, http://www.usmagazine.com/celebrity-news/news/chris-wallace-fat-shames-kelly-clarkson-she-could-stay-off-the-pizza-201544.
2. Medeline Boardman, "Katie Hopkins Takes Kelly Clarkson Fat-Shaming Comments Even Further, Calls Singer a 'Chunky Monkey,'" *Us*, March 6, 2015, http://www.usmagazine.com/celebrity-body/news/katie-hopkins-takes-kelly-clarkson-fat-shaming-comments-even-further-201563.
3. Ibid.
4. For a list of celebrities who've been targets of fat shaming, see http://www.nickiswift.com/2061/stars-fat-shamed/.
5. For discussion of evidence that boys and men are increasingly experiencing fat shaming, see Michael Andor Brodeur, "Why Male Body Shaming Is on the Rise in the Media," *Boston Globe*, March 9, 2015, https://www.bostonglobe.com/lifestyle/2015/03/08/manshaming/F4IOidjmYSzlbTvMGua0sJ/story.html.
6. Patrick Gomez, "Biggest Loser Winner Toma Dobrosavljevic: I Know My Father Would Be Proud of Me," *People*, January 30, 2015,

https://people.com/tv/biggest-loser-winner-toma-dobrosavljevic-my-father-would-be-proud-of-me/.

7. Joy Wilke, "Nearly Half in U.S. Remain Worried about Their Weight," Gallup, July 25, 2014, http://news.gallup.com/poll/174089/nearly-half-remain-worried-weight.aspx.
8. Figures on dieting and on money spent on weight loss comes from Boston Medical Center, "Weight Loss," https://www.bmc.org/nutrition-and-weight-management/weight-management.
9. "New Advice for Weight Loss: Get on the Scale Every Day," *USA Today*, January 3, 2016, https://www.usatoday.com/story/life/2016/01/03/weight-loss-scales-daily/77584478/.
10. Ann E. Becker, "Television, Disordered Eating, and Young Women in Fiji: Negotiating Body Image and Identity during Rapid Social Change," *Culture, Medicine and Psychiatry* 28, no. 4 (2004): 533–59; Alexandra Brewis et al., "Body Norms and Fat Stigma in Global Perspective," *Current Anthropology* 52 (2011): 2.
11. These stories are part of a collection chronicled in "Fat Discrimination: 14 Women Open Up about Their Experiences with Size Prejudice," *The Huffington Post*, May 21, 2013, http://www.huffingtonpost.com/2013/05/20/fat-discrimination-women-experiences-size-prejudice_n_3308012.html.
12. Timothy A. Judge and Daniel M. Cable, "When It Comes to Pay, Do the Thin Win? The Effect of Weight on Pay for Men and Women," *Journal of Applied Psychology* 96, no. 1 (2011): 95–112.
13. Enrica Ruggs, Michelle R. Hebl, and Amber Williams, "Weight Isn't Selling: The Insidious Effects of Weight Stigmatization in Retail Settings," *Journal of Applied Psychology* 100, no. 5 (2015): 1483–96.
14. Roberta R. Friedman and Rebecca M. Puhl, "Weight Bias: A Social Justice Issue," Rudd Center for Food Policy & Obesity, 2012, http://www.uconnruddcenter.org/files/Pdfs/Rudd_Policy_Brief_Weight_Bias.pdf; Carolyn MacCann and R. D. Roberts, "Just as Smart but Not as Successful: Obese Students Obtain Lower School Grades but Equivalent Test Scores to Nonobese Students," *International Journal of Obesity* 37, no. 1 (2013): 40–6; K. S. O'Brien, J. A. Hunter, and M. Banks, "Implicit Anti-fat Bias in Physical Educators: Physical Attributes, Ideology and Socialization," *International Journal of Obesity* 31 (2007): 308–314; Rebecca M. Puhl and Chelsea A. Heuer, "The Stigma of Obesity: A Review and Update," *Obesity* 17, no. 5 (2012), https://onlinelibrary.wiley.com/doi/full/10.1038/oby.2008.636.
15. The one-in-four statistic comes from http://www.bullyingstatistics.org/content/school-bullying-statistics.html. For data about the prevalence of weight-based bullying among high school students, see JoAnn Stevelos, "Bullying, Bullycide, and Childhood Obesity," Obesity Action Coalition, https://4617c1smqldcqsat27z78x17-wpengine.netdna-ssl.com/wp-content/uploads/Bullying-and-Bullycide.pdf.
16. Rebecca A. Krukowski et al., "Overweight Children, Weight-Based Teasing and Academic Performance," *International Journal of Pediatric Obesity* 4, no. 4 (2009): 274–80.
17. For a discussion of derogatory comments made by doctors and nurses, see Friedman and Puhl, "Weight Bias." Concerning health care providers' susceptibility to misdiagnosing illness in overweight people, see "Fat Shaming in the Doctor's Office Can Be Mentally and Physically Harmful," *Science Daily*, August 3, 2017, https://www.sciencedaily.com/releases/2017/08/170803092015.htm.
18. Simon Gabriel, Edward J. Gracely, and Billie S. Fyfe, "Impact of BMI on Clinically Significant Unsuspected Findings as Determined at Postmortem Examination," *American Journal of Clinical Pathology* 125, no. 1 (2006): 127–31.
19. Samantha Kwan and Mary Nell Trautner, "Weighty Concerns," *Contexts* 10, no. 2 (2011): 52–7. Friedman and Puhl, "Weight Bias."
20. Harriet Brown, "For Obese People, Prejudice in Plain Sight," *New York Times*, March 15, 2010,

https://www.nytimes.com/2010/03/16/health/16essa.html.

21. S. S. Wang, K. D. Brownell, and T. A. Wadden, "The Influence of the Stigma of Obesity on Overweight Individuals," *International Journal of Obesity and Related Metabolic Disorders* 28 (2004): 1333–7.
22. Louise Townend, "The Moralizing of Obesity: A New Name for an Old Sin?" *Critical Social Policy* 29, no. 2 (2009): 171–90.
23. LGBTQ homicide data are for 2017 and reported in National Coalition of Anti-Violence Programs, "A Crisis of Hate: A Report of Lesbian, Gay, Bisexual, Transgender, and Queer Hate Violence Homicide in 2017," http://avp.org/wp-content/uploads/2018/01/a-crisis-of-hate-january-release-12218.pdf.
24. Natasha Schvey, Rebecca Puhl, and Kelly Brownell, "The Impact of Weight Stigma on Caloric Consumption," *Obesity* 19, no. 10 (2011): 1957–62; Janet Tomiyama, "Weight Stigma Is Stressful. A Review of Evidence for the Cyclic Obesity/Weight-Based Stigma Model," *Appetite* 82 (2014): 8–15.
25. For evidence linking obesity to adverse health outcomes, see Y. Claire Wang et al., "Health and Economic Burden of the Projected Obesity Trends in the USA and the UK," *The Lancet* 378, no. 9793 (2011): 815–25.
26. "Adult Obesity in the United States," *The State of Obesity*, 2018, https://stateofobesity.org/adult-obesity/.
27. National Institute of Diabetes and Digestive Kidney Diseases, "Overweight & Obesity Statistics," 2017, http://www.niddk.nih.gov/health-information/health-statistics/Pages/overweight-obesity-statistics.aspx.
28. Centers for Disease Control and Prevention, "Childhood Overweight and Obesity," 2018, http://www.cdc.gov/obesity/childhood/.
29. Data on education and income disparities come from the Bureau of Labor Statistics, http://www.bls.gov/emp/ep_chart_001.htm.
30. General discussion of food deserts is based on Alison Hope Alkon et al., "Foodways of the Urban Poor," *Geoforum* 48 (2013): 126–35 and Cynthia Gordon et al., "Measuring Food Deserts in New York City's Low-Income Neighborhoods," *Health and Place* 17, no. 2 (2011): 696–700.
31. Data about Chicago's Black population living in food deserts comes from Topher Gray, "From Farm to Food Desert," *Chicago Reader*, August 19, 2010, https://www.chicagoreader.com/chicago/chicago-food-deserts-hopkins-park-black-farmers/Content?oid=2272825.
32. Michael Moss, "The Extraordinary Science of Addictive Junk Food," *New York Times Magazine*, February 20, 2013, https://www.nytimes.com/2013/02/24/magazine/the-extraordinary-science-of-junk-food.html; Punam Ohri-Vachaspati et al., "Child-Directed Marketing Inside and on the Exterior of Fast Food Restaurants," *American Journal of Preventive Medicine* 48, no. 1 (2015): 22–30.
33. Amy M. Burdette and Terrence D. Hill, "An Examination of Processes Linking Perceived Neighborhood Disorder and Obesity," *Social Science and Medicine* 67, no. 1 (2008): 38–46.
34. Russell Sage Foundation, *Opportunity, Mobility, and Increased Inequality* (New York: Author, 2016).
35. Abigail C. Saguy and Kjerstin Gruys, "Morality and Health: News Media Constructions of Overweight and Eating Disorders," *Social Problems* 57, no. 2 (2010): 231–50. The article quoted in their study is Karen Springen, "Health: Battle of the Binge," *Newsweek*, February 19, 2007.
36. Jina H. Yoo and Junghyun Kim, "Obesity in the New Media: A Content Analysis of Obesity Videos on YouTube," *Health Communication* 27, no. 1 (2012): 86–97; Kimberly J. McClure, Rebecca M. Puhl, and Chelsea A. Heuer, "Obesity in the News: Do Photographic Images of Obese Persons Influence Antifat Attitudes?" *Journal of Health Communication: International Perspectives* 16, no. 4 (2011): 359–71.
37. Greta Gleissner, "'I Feel Fat': How to Feel Instantly Better in Your Body," *Huffington Post*, February 3, 2013, https://www.huffing

tonpost.com/greta-gleissner/body-image_b_2238291.html.

38. Friedman and Puhl, "Weight Bias"; Rachel H. Salk and Renee Engeln-Madoxx, "'If You're Fat, Then I'm Humongous!' Frequency, Content, and Impact of Fat Talk among College Women," *Psychology of Women Quarterly* 35, no. 1 (2011): 18–28; Renee Engeln, Michael R. Sladek, and Heather Waldron, "Body Talk among College Men: Content, Correlates, and Effects," *Body Image* 10, no. 3 (2013): 300–8.
39. "This Is What 'One Size Fits All' Actually Looks Like on All Body Types," *BuzzFeed*, December 3, 2014, https://www.buzzfeed.com/candacelowry/heres-what-one-size-fits-all-looks-like-on-all?utm_term=.plOzg2jOQ#.gnepNjyP7.
40. Carolyn Coker Ross, "Why Do Women Hate Their Bodies?" *PsychCentral*, July 8, 2018, https://psychcentral.com/blog/why-do-women-hate-their-bodies/.
41. Lauren Reba-Harrelson et al., "Patterns and Prevalence of Disordered Eating and Weight Control Behaviors in Women Ages 25–45," *Eating and Weight Disorders* 14, no. 4 (2009): 190–8; "Eating Disorders: Differences in Young Men and Women," American Psychological Association, May 14, 2016, https://www.psychiatry.org/newsroom/news-releases/eating-disorders-differences-in-young-men-and-women.
42. Erin Fothergill et al., "Persistent Metabolic Adaptation 6 Years after 'The Biggest Loser' Competition," *Obesity* 24, no. 8 (2016): 1612–9.
43. Sandra Aamodt, "Why You Can't Lose Weight on a Diet," *New York Times*, May 6, 2016, https://www.nytimes.com/2016/05/08/opinion/sunday/why-you-cant-lose-weight-on-a-diet.html.
44. The handbook for this activism is Linda Bacon, *Health at Every Size: The Surprising Truth about Your Weight* (Dallas: Benbella Books, 2010).
45. Rachel P. Wildman et al., "The Obese without Cardiometabolic Risk Factor Clustering and the Normal Weight with Cardiometabolic Risk Factor Clustering," *JAMA Internal Medicine* 168, no. 15 (2008): 1617–24; Steven N. Blair, "Physical Inactivity: The Biggest Public Health Problem of the 21st Century," *British Journal of Sports Medicine* 43 (2009): 1–2.

6

What's Sex Got to Do with It?

Uncovering the Roots of Teen Pregnancy

Learning Objectives

1. Define the sexualization of childhood and discuss the fears it generates in American parents.
2. Describe trends in rates of premarital sex, contraceptive use, and pregnancy over the past several decades.
3. Explain why kids from low-income families are the likeliest to become teen parents.
4. Recognize how popular ideas about teenage pregnancy reinforce the sexual double standard.
5. Identify the characteristics of early childhood education that can lower teen pregnancy rates.
6. Compare attitudes toward teen sex and pregnancy in the U.S. and the Netherlands.

Parental Anxiety Gone Wild: Raising Kids in a Society Where Sexualized Images Are Everywhere

6.1 Define the sexualization of childhood and discuss the fears it generates in American parents.

In November 2015, police in Cañon City, Colorado, discovered that over one hundred high school students had been **sexting**—using mobile devices to exchange sexually explicit photos and videos. The students had been sharing provocative images via vault apps, which are social media sites that veil inappropriate content behind an innocent-looking façade like a calculator.[1] At the time, sexting was punishable under Colorado's and other states' child pornography laws.[2] The size of the bust gave parents across the country reason to quiver. Might their own teenagers also be sexting? My son was fifteen and my daughter was thirteen, so the thought certainly crossed my mind. I'd recently read a survey in which over a quarter of teens had reported having sexted. As Figure 6.1 (on page 106) illustrates, the percentage actually varies by age and increases as kids get older.

Some parents see sexting as an indicator that kids coming of age nowadays face more menacing dangers than in the past. It taps parents' concerns about the **sexualization of childhood**, or kids' continuous exposure to provocative images as they're growing up. The fear is that these images entice kids to have sex before they're emotionally mature enough to understand its dangers. Compared to other sexualized images, photos and videos transmitted via mobile devices fuel unique parental worries because the people exchanging them know one another. While teens may recognize the damaging messages in graphic song lyrics or ads depicting

Antonio Guillem Fernández/Alamy Stock Photo

FIGURE 6.1 ● Sharing Nude Images

The percentage of kids who sext increases throughout middle and high school.

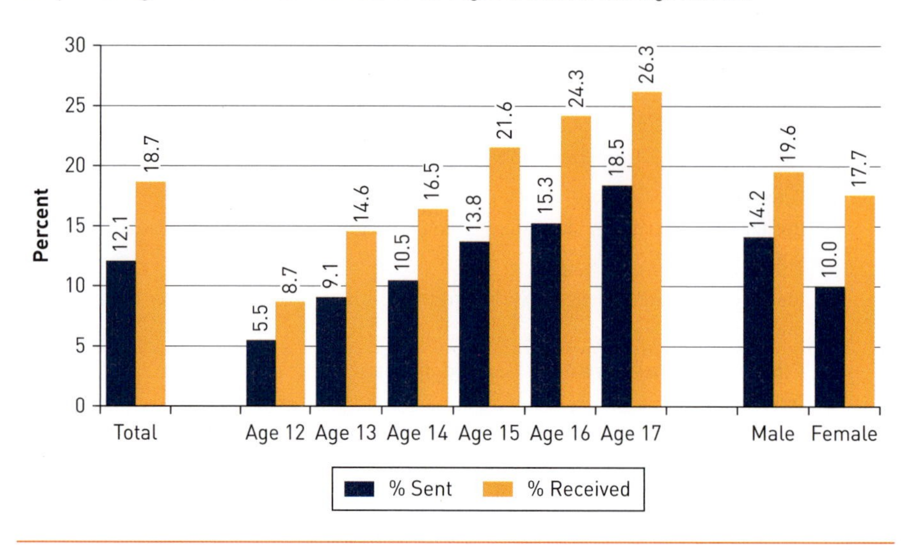

Source: Justin W. Patchin, "New Teen Sexting Data," Cyberbullying Research Center, February 24, 2017, https://cyberbullying.org/new-teen-sexting-data.

Sexting has become a normal part of many teens' lives.
iStockphoto.com/ bymuratdeniz

scantily clad girls, they may be less inclined to acknowledge that the sexualized photos and videos they share amongst themselves can be exploitative too.[3]

Among parents' top concerns over sexting and other forms of sexualization is that they make kids susceptible to intimate partner violence, sexually transmitted infections, and pregnancy. The risk of having a baby at such a young age is the focus of this chapter. Abortion is unthinkable for many pregnant teens and inaccessible to others. Giving birth at a stage of life when people are not emotionally or financially ready for the responsibilities of raising kids often has adverse effects. Relative to their peers from similar backgrounds who postpone childbearing, teen mothers are less likely to complete high school and be in a position to obtain the postsecondary skills needed to land a decent-paying job. Babies of teen mothers are more likely to suffer health complications from premature birth, which elevates their chances of performing poorly in school and dropping out. Like their parents, kids born to teens often get stuck in poverty and, not surprisingly, are also inclined to become parents before they're ready. Given the many social problems stemming from teen pregnancy, it's an issue that deserves our investigation.[4]

FIRST IMPRESSIONS?

1. In your view, how does the potential danger of sexted photos and videos compare with that of other sexualized images?
2. What types of concerns do your parents have about teen sex?
3. What have you heard about teen pregnancy? From which sources?

Many parents worry about how the sexualization of childhood particularly harms girls. There's a long-standing belief in the U.S. that because girls are at risk of pregnancy, they need to be protected from vice. It's no wonder some parents become alarmed when they see female celebrities wearing low-cut and tight-fitting clothing, which may appear to give their daughters permission to dress promiscuously. Parents similarly fear that sexting will increase their heterosexual daughters' likelihood of having sex at a young age. This fear is legitimate insofar that the reason girls sext is often to capture male attention. Boys frequently coerce girls to demonstrate their attractiveness and romantic interest through graphic images. In a society where girls derive considerable self-worth from their attractiveness, parents certainly have legitimate concerns about the sexualization of childhood.[5]

Britney Spears rose to fame as a teen pop star in part because of her sex appeal. Many girls idolized her.
dpa picture alliance archive/ Alamy Stock Photo

These concerns are often misplaced, however. Whereas middle- and upper-class parents often have lots of angst about the risk of their teens getting pregnant, these teens actually face a low chance of doing so relative to kids from lower-income families. Roughly 80 percent of teen mothers were poor before they became pregnant. In the relatively rare instances when more economically advantaged teens have babies, parenting usually doesn't adversely affect the trajectory of their lives. Because they're likely to have grown up in stable families—often with the support of two parents—having children rarely causes them to drop out of school or get stuck in low-wage jobs.[6]

What are the social forces that lead low-income teens to be the likeliest among their age group to become pregnant? By using the sociological perspective to uncover these forces, we'll discover how this social problem reflects kids' unequal opportunities to get ahead in life. This chapter explains why teen pregnancy is *not* simply or primarily about sex.[7]

Who's Doing What and When? Shifting Views about Sex, Birth Control, and Pregnancy

6.2 Describe trends in rates of premarital sex, contraceptive use, and pregnancy over the past several decades.

"Your generation didn't invent sex," my dad would sometimes tell me with a smirk when I was in my twenties. I took these words as his way of trying to bridge our forty-one-year age difference. It was a monumental gap, given the dramatic changes that occurred during the years between when he came of age in the 1940s and I did in the 1980s. Beginning in the 1960s, public attitudes toward sex loosened significantly—a transformation known as the **sexual revolution**.

Just 18 percent of adults in a 1963 survey indicated it was right for an engaged heterosexual couple to have sexual intercourse. There was even less acceptance (15 percent) if the couple was in love but not engaged, and substantially less (9 percent) if they felt little affection for one another. These views dramatically shifted over the ensuing decades. The birth control pill (often referred to simply as "the pill") came on the market in the early 1960s, and its usage became widespread after 1965 when the Supreme Court gave married couples the right

FIGURE 6.2 ● Sex before Marriage

Since the complete legalization of birth control in 1972, the belief that premarital sex is wrong has continually declined.

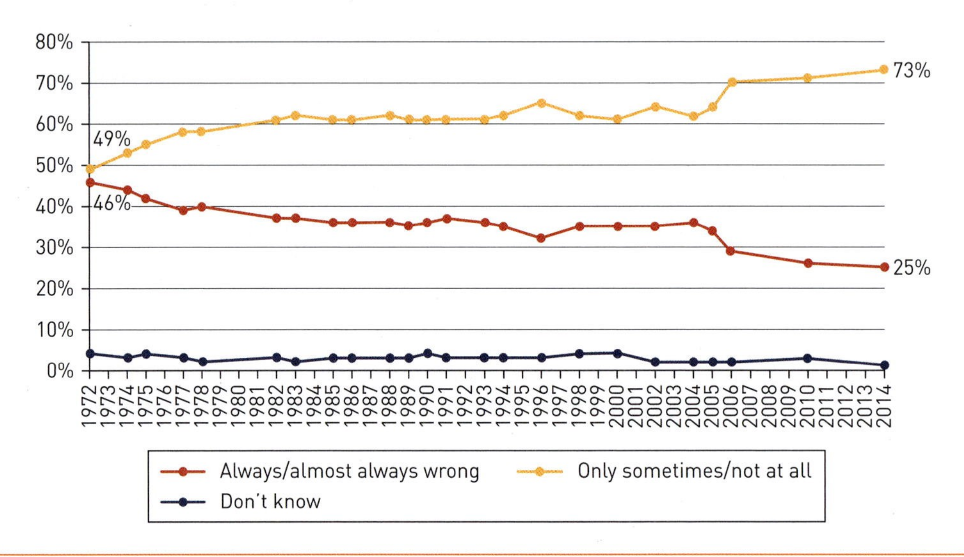

Source: Graph—Public Beliefs about Whether Premarital Sex is Wrong from "Going All the Way: Public Opinion and Premarital Sex," Roper Center for Public Opinion Research, Cornell University, July 7, 2017, https://ropercenter.cornell.edu/blog/going-all-way-public-opinion-and-premarital-sex.

to access contraception. As the feminist movement was concurrently asserting that women should have the freedom to express their sexual desires, the pill was making it possible for them to do so while controlling their fertility. As Figure 6.2 indicates, by 1972 *fewer* people (46 percent) thought it was immoral to have sex before marriage than thought it was acceptable (49 percent).[8] After the Supreme Court that same year legalized contraceptives for anyone to use, views toward premarital sex continued to loosen.[9] By 2014, just 25 percent of Americans believed it was wrong.[10]

Given this transformation, parents have reason to believe that their kids are having sex at a younger age than in prior generations. After all, the sexualization of childhood occurred alongside these attitudinal changes, and nowadays kids see racy images practically everywhere. Despite parental fears, however, there has *not* actually been a dramatic change in the average age when heterosexuals lose their virginity. It was eighteen in the 1940s and 1950s, and currently it's seventeen. Moreover, parents nowadays seemingly have less reason to be concerned about teenage pregnancy. Whereas in 1957 there were 96.3 births for every one thousand girls ages fifteen to nineteen, the teen pregnancy rate plummeted nearly fourfold over the next several decades. In 2014, there were only 24.2 births for every one thousand girls ages fifteen to nineteen (see Figure 6.3).[11]

FIGURE 6.3 • Fewer Kids Pushing Strollers

Public angst about teens getting pregnant overshadows that birth rates within this age group have steadily declined since the years immediately following World War II.

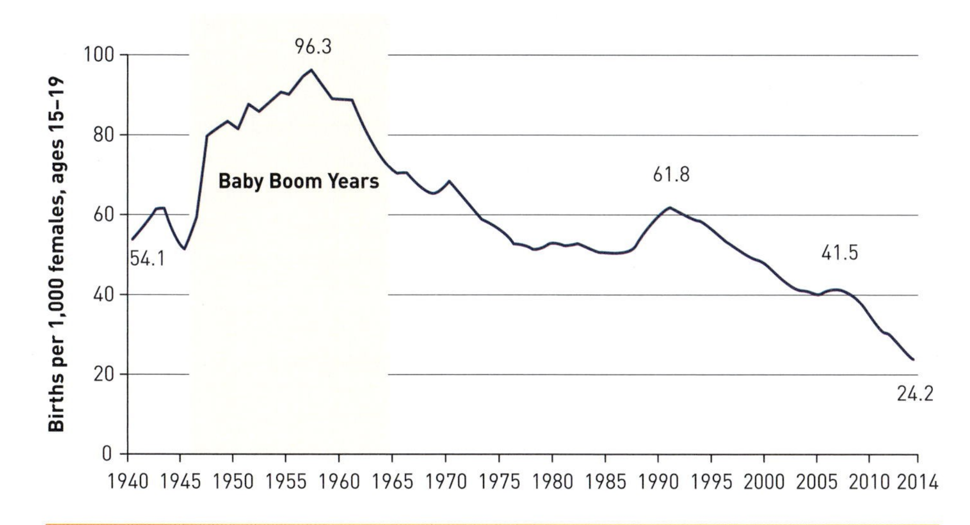

Source: Pew Research Center, "U.S. Teen Birth Rate Has Fallen Dramatically Over Time," April 29, 2016, http://www.pewresearch.org/ft_16-04-29_teenbirths_longterm_640.

During the Baby Boom era from 1946 to 1964, adults didn't generally regard the high birth rate to teen parents as a social problem since most of these teens (roughly 85 percent) were married. Of course, many had a **shotgun marriage**, which occurs when a young heterosexual couple quickly decides to marry after the girl becomes pregnant as a way to avoid embarrassment over her having had premarital sex.[12] In contrast, nowadays about 89 percent of pregnant teens are unmarried. This reality worries many of their parents, who fear that if their teens have children but lack the stability and support that marriage can provide, there will be detrimental lifelong consequences. The concern is that child-rearing will impair these teens' ability to stay in school, land a good job, and/or provide their own children the financial and emotional supports needed to become productive adults.[13]

The popularity of TV shows about teen pregnancy may feed this fear since they create the impression that this social problem is more pervasive than it actually is. The irony is that these shows are partially responsible for declining teen birth rates. A study of internet use during the twenty-four hours after a new episode of *16 and Pregnant* found that tweets with the words *birth control* increased by 23 percent and Google searches for how to get birth control pills similarly spiked. The researchers estimated that the show was responsible for a 4.3 percent drop in teen births across the United States. Who says reality television can't be educational![14]

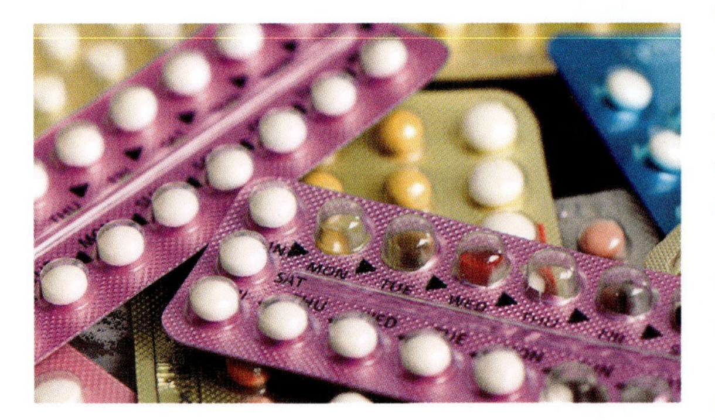

Watching TV shows about the everyday challenges of teen parenting motivates some viewers to seek out birth control.
iStockphoto.com/areeya_ann

Let's take a closer look at the decline in teen birth rates over the past few decades. Figure 6.4 tracks the period from 1991 to 2016, focusing on race. Even though birth rates plummeted across all racial groups, there were differences among them. In 2016, Black teens had over twice as many babies as White teens, and Hispanic teens even more. Whereas 8 percent of Whites live in poverty, the poverty rate is 16 percent for Hispanics and 20 percent for Blacks. Therefore, these racial gaps reflect social class differences too; in other words, teen pregnancy is concentrated among the poor. It's curious that while many parents target their concerns about this problem at the sexualization of childhood, the concentration of teen birth rates among low-income kids indicates that the source is entirely different. Teen pregnancy chiefly reflects inequalities in access to economic opportunity.[15]

Dim Hopes for the Future: Highlighting the Social Forces That Contribute to Teens Having Babies

6.3 Explain why kids from low-income families are the likeliest to become teen parents.

"In the opinion of many well-meaning middle-class people," wrote sociologist Kristin Luker, "the trouble with poor and pregnant teenagers is that they do not do

FIGURE 6.4 • Teen Pregnancy and Race

Although teen birth rates have fallen steadily during your lifetime, long-standing racial differences in teens' likelihood of becoming parents persist.

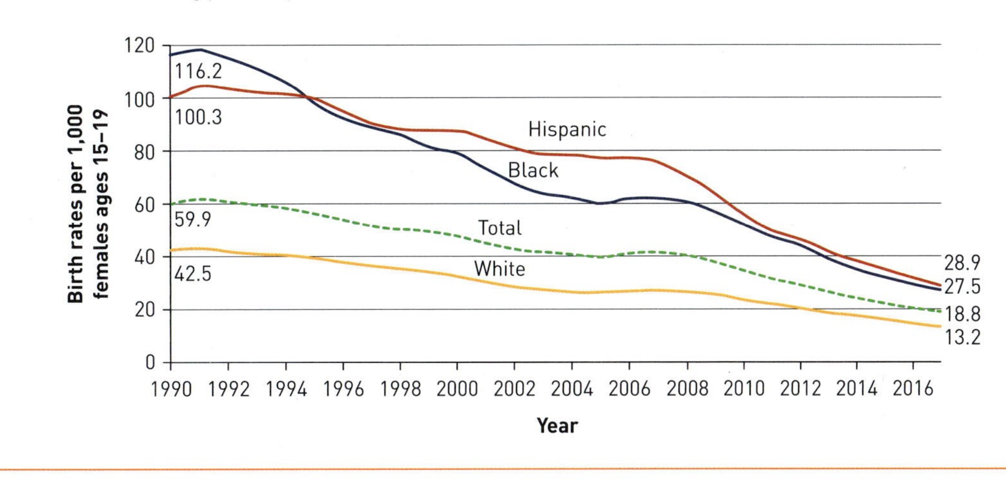

Source: Joyce A. Martin, Brady E. Hamilton, Michelle J. K. Osterman, and Anne K. Driscoll, "Births: Final Data for 2016," National Vital Statistics Reports, January 31, 2018, U.S. Department of Health and Human Services, https://www.cdc.gov/nchs/data/nvsr/nvsr67/nvsr67_01.pdf.

what middle-class people do: invest in an education, establish themselves in a job, marry a sensible and hardworking person, and only then begin to think about having a baby."[16] Luker is characterizing the individual perspective, which asserts that teens become pregnant because they don't exercise prudence and restraint. From this perspective, pregnancy is the irresponsible result of not tempering immediate sexual desires and focusing instead on long-term academic and career goals.

Pay attention to the phase "well-meaning middle-class people." Who do you think she means? If you look around campus, the answer will be staring at you. Though you and your peers come from diverse backgrounds, something you all share in common is having a **middle-class mindset**, which is an understanding that there are certain things a person needs to do to achieve what our society defines as success. Your understanding likely resembles the sequence Luker describes:

1. Finish school.
2. Attain a good job.
3. Marry a person with similar values.
4. Have children together.[17]

This sequence illustrates the long-range thinking that defines the **American Dream**, the belief that any person can get ahead if they work hard and exercise self-discipline. For the many people who subscribe to this idea, it makes sense to see teen parents

Although college students come from a variety of backgrounds, you and your peers similarly share the aspiration to work toward a brighter future.
iStockphoto.com/FatCamera

as irresponsible. By having a child so young, these teens have reduced the likelihood that they will complete school, land a good job, or get married.

However, the individual perspective overlooks the full story of what's going on in the lives of low-income kids, who are the most likely to become teen parents. To tell this story, let's widen the lens and explore social forces that influence the choices teens of different race and class backgrounds make. In order to understand these choices, we must look beyond their sexual behavior and recognize variation in how they see their place in the world. To do so, try to put yourself in the shoes of someone who grew up in a low-income family. Maybe this is your story. If not, think about how you might look at the world differently if your parents struggled to make ends meet and there were few examples in your life of people achieving the American Dream.

Kids from these backgrounds often attend underfunded schools ill-suited for learning and live in neighborhoods with a concentration of adults who have experienced similar disadvantages. These adults either work in low-wage jobs with few opportunities for advancement, are unemployed, or are engaged in criminal activities. Therefore, it makes sense why low-income kids might not care much about doing well in school. Since few of their friends or family members are role models of academic success, they may see little point in exerting effort to achieve a brighter future. For teens with leveled aspirations, therefore, childbirth doesn't carry the costs that it does for those who grow up with a middle-class mindset.[18] "The idea that young people would be better off if they worked harder, were more patient, and postponed their childbearing is simply not true," noted sociologist Kristin Luker, "and is unlikely to become true in the foreseeable future for a great many people at the bottom of the income scale."[19]

Low-income kids often attend overcrowded schools that may also be dirty, violence ridden, and unable to attract high-quality teachers.
Marmaduke St. John/Alamy Stock Photo

It's no wonder that teen birth rates are concentrated in low-income communities and are over three times higher in states that have the largest inequalities between the poor and middle class, such as Mississippi and Arkansas, than in states that have the smallest, such as Massachusetts and New Hampshire. Growing up in a place where economic inequalities seem insurmountable leads kids to believe it's impossible to improve their lives, so they see no point in even trying. For these kids, the costs of teen parenting are small.[20]

Not only do a large percentage of low-income teens give birth relative to their higher-income peers but many of these poor teens also subsequently drop out of high school (see Figure 6.5). However, parenthood isn't the underlying reason. "If a teenager has a baby because her life chances seem so limited that her life will not be any better if she delays childbearing," wrote economists Melissa S. Kearney and Philip B. Levine, "then teen childbearing is unlikely to be causing much of a detrimental effect."[21] These words underscore that becoming a parent isn't fundamentally why poor kids are at the greatest risk of quitting school. Rather, teenage pregnancy and dropping out both reflect a lack of access to educational opportunity. Because many low-income kids view the future as bleak, they see no point of investing in academics.

In addition to being a catalyst for teen childbearing, leveled aspiration for upward mobility can also inhibit poor kids from getting married. Since the future feels unstable, they often can't imagine sharing it with another person.
iStockphoto.com/Pollyana Ventura

FIGURE 6.5 ● High School Dropouts

Although the rate of kids leaving school without a diploma has fallen dramatically, there continue to be notable racial differences in kids' likelihood of finishing high school.

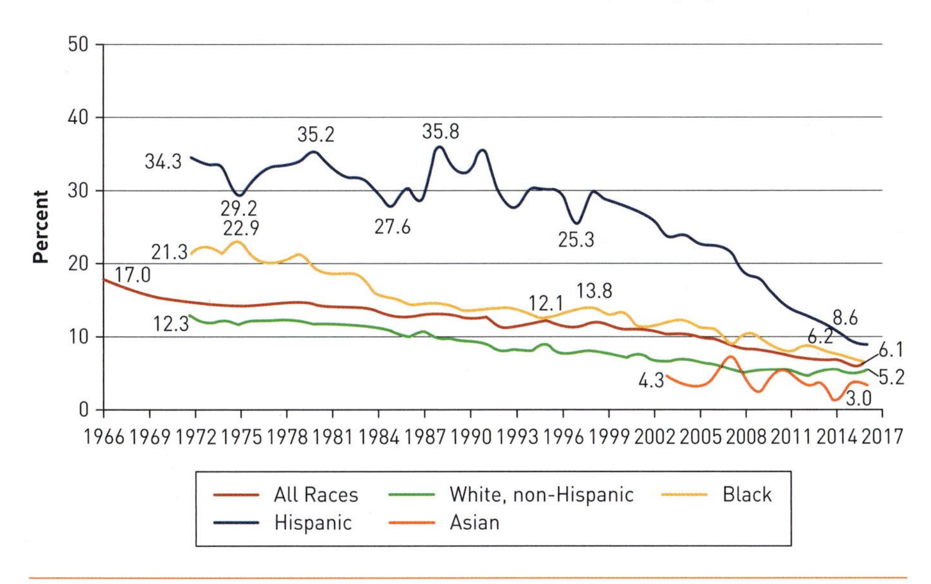

Source: Child Trends, "High School Dropout Rates," 2018, https://www.childtrends.org/indicators/high-school-dropout-rates.

Low-income kids who receive high-quality mentoring are inclined to work toward long-term goals and stay out of trouble instead of being resigned to menial work and possibly crime.
iStockphoto.com/StockRocket

Given many low-income kids' pessimism toward the future, it's unsurprising that relatively few of them go to college. Among all people who receive bachelor's degrees, only about 10 percent are students from families earning under $35,000 a year. The reasons for this go beyond cost (see the discussion of cultural capital in Chapter 2). Often, it stems from a lack of sustained guidance. I've learned through conversations over the years with low-income students I've taught that they've often benefitted from having a strong mentor who provided ongoing encouragement and support. This person motivated them to work hard in school and emphasized that despite obstacles in their lives, they could have a brighter future. The mentor may have been a family member, a teacher, or a volunteer at a nonprofit organization like Big Brothers Big Sisters. Research I've conducted reveals that these organizations enable low-income kids to build relationships with accomplished adults who motivate them to set defined academic goals and work to pursue those goals.[22]

The fact that many Black college students balance parenting with school reflects the same constraints that lead a much higher percentage of Black kids than White kids to become teen parents.
iStockphoto.com/Steve Debenport

Poor kids who go to college are much likelier than their higher-income, typically White or Asian peers to be among the nearly one in four undergraduates with children of their own (see Figure 6.6). Over a third of low-income and first-generation students juggle parenting with school, and often with a paid job too. The takeaway from these data mirrors what we've seen in looking at teen pregnancy from a sociological perspective: This social problem stems from inequality of opportunity.[23]

Given that you're committed to earning a college degree, before reading this chapter you may have viewed a sixteen-year-old girl as irresponsible for having had a baby at such a young age. But now that you've closely considered the sort of girl who's typical of teen mothers, you can see why it's shortsighted to regard her as having made an irresponsible decision. She likely grew up in a low-income family and, because of a variety of factors, learned not to care about academics. She probably lived in a neighborhood where most adults worked in jobs with few prospects for upward mobility. Teachers at the public school she attended may not have given students individualized attention. Maybe she never had a mentor to encourage her to work toward a better life. These social forces shaped her decision to behave in a way that someone like you, with your middle-class mindset, might consider impulsive and careless.

FIGURE 6.6 • Earning a Degree while Raising Kids on Their Own

College students who are single mothers are most likely to be women of color.

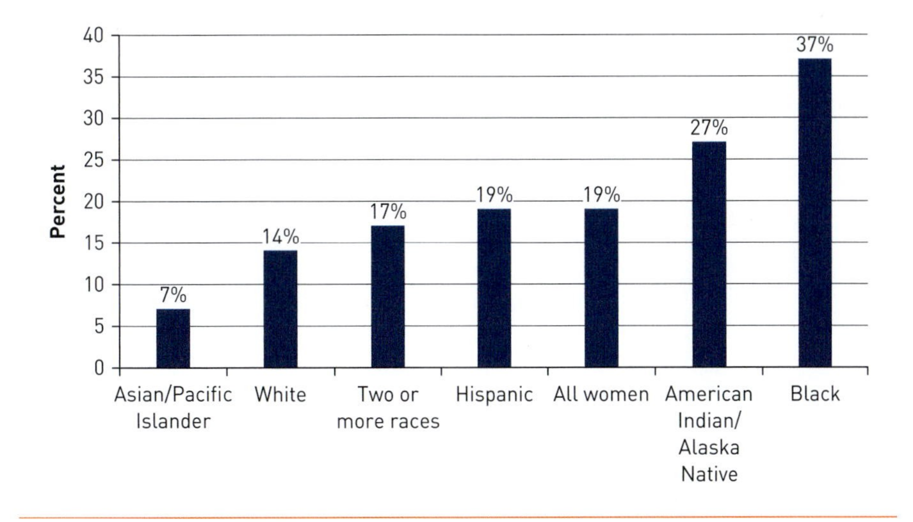

Source: Melanie Kruvelis, Lindsey Reichlin Cruse, and Barbara Gault, "Single Mothers in College: Growing Enrollment Financial Challenges, and the Benefits of Attainment," Institute for Women's Policy Research, September 20, 2017, https://iwpr.org/publications/single-mothers-college-growing-enrollment-financial-challenges-benefits-attainment.

Put differently, here's a plausible story for why you may have just seen a girl walking out of Walmart carrying a pack of diapers: After barely getting by in ninth grade, she failed out of tenth grade, so she decided to drop out of school. That summer, she and her boyfriend had unprotected sex, but not because they couldn't afford birth control or didn't know where to get it. Rather, the girl thought she might be better off spending her days nurturing a baby who needed her and whom she could love instead of continuing to be bored sitting in an overcrowded classroom without direction about where her life was heading. Indeed, there's much more to her life story than that she prioritized present sexual gratification over future educational and career goals and is now paying the price for her irresponsible behavior. Because she grew up without these goals, she came to see her life as going nowhere regardless of the decisions she made. Therefore, having a baby seemed like no big deal.[24]

The sociological perspective highlights the limitations of using a middle-class mindset to understand teen pregnancy.
iStockphoto.com/ArtisticCaptures

"Acting Like Sluts": How the Individual Perspective toward Teen Pregnancy Reinforces Gender Inequality

6.4 Recognize how popular ideas about teenage pregnancy reinforce the sexual double standard.

Do your own web search and see for yourself: people commonly associate only girls with teenage pregnancy even though, of course, it takes two.
Ian Hooton/Science Source

What comes to mind when you think of teen parents? I did a Google search and scrolled through the first fifty photos that appeared on the screen. It wasn't until the thirtieth that I saw a father and his baby, and it was the only one of its kind among the fifty. Of the others, twenty-one were of both mother and father and twenty-eight of just the mother. While it's true that much of the time the children of teen parents live with their mothers, fathers still often visit and provide financial support. Not only are our ideas about this social problem skewed but they also characterize girls in disparaging ways. We saw in the last section that the individual perspective obscures the class and race inequalities at the root of teen childbearing. We'll now uncover how this perspective perpetuates gender inequality too.[25]

Seeing teen pregnancy as stemming from females' impulsive desires reinforces the **sexual double standard**, which is the idea that girls and women deserve criticism for sexual behavior that is praiseworthy if carried out by boys or men. Although we saw earlier that the sexual revolution made it acceptable for women to have sex before marriage and to derive pleasure from sex, this double standard persists. Heterosexual guys may regard a woman as a slut for hooking up with a friend of theirs last night, yet give the friend a high-five for his triumph. Even though such guys may see committed relationships as involving mutual pleasure, they may also be inclined to view hookups as only for their own benefit. Given the implicit message behind their beliefs—that women who hook up are sluts and thus don't deserve pleasure—it's no wonder that research indicates college women are less likely to have orgasms during hookups than when having sex with a long-term male partner. Where else have you seen evidence of the sexual double standard?[26]

Other research among college students indicates that girls and women also sometimes reinforce the sexual double standard. One study found that women from higher-income backgrounds used the term *slut* as a way to criticize lower-income women on campus who hooked up with guys. This label perpetrates not only gender inequality but class inequality too. Recall our earlier discussion that teenage pregnancy is concentrated in poor communities. A derogatory label like *slut*—which characterizes the individual perspective—obscures this reality by blaming girls for sexual behavior that is symptomatic of these teens' lack of access to economic opportunity.[27]

The individual perspective also perpetuates racial inequality. Because minority teens are, on average, more sexually active than White teens (see Figure 6.7), it's convenient to believe minority teens should keep their clothes on and focus their priorities more on doing well in school. Seeing teen mothers, who are disproportionately Black, as needing to exercise sexual restraint doesn't just obscure the class disadvantages that may make childbearing attractive to them. It also reinforces the long-standing view that Black females have uncontrollable sexual desires—a view that justified the raping of girls during the slave period and in the decades that followed it.[28]

The most significant way the individual perspective toward teen pregnancy reinforces gender inequality is by obscuring how frequently girls who become mothers have endured violence. Possibly as many as 20 percent of teen mothers give birth to a child conceived during forced sex. An even higher portion—between 30 and 44 percent—are the survivors of rape or attempted rape that occurred at some point in their young lives.[29] Indeed, 60 percent of girls who've had sex before age fifteen, and 74 percent before age fourteen, report having been forced to do so.[30] Seeing

FIGURE 6.7 ● Sexually Active

There are stark racial differences among teens in the percentage who report having had sexual intercourse.

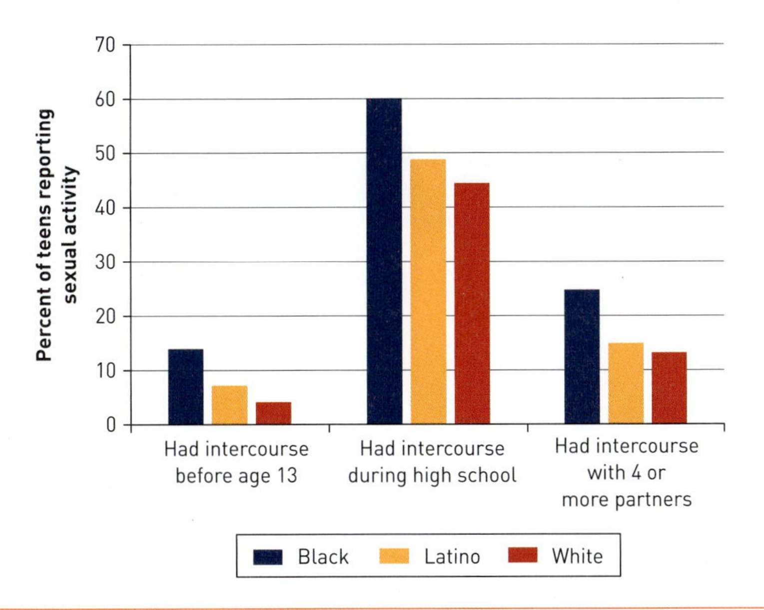

Source: Adapted from Joel Best and Kathleen A. Bogle, *Kids Gone Wild* (New York: New York University Press, 2014), 142.

teen pregnancy as stemming from slutty behavior, therefore, adds insult to injury. Moreover, this individual perspective legitimizes the sense of entitlement that boys and men often have toward girls' and women's bodies (see Chapter 9 for a fuller discussion of the link between male entitlement and gender violence).

Reducing Teen Pregnancy: Early Educational Opportunity Is Key

6.5 Identify the characteristics of early childhood education that can lower teen pregnancy rates.

In doing the research for this chapter, one study particularly caught my eye. It found that 24 percent of people in their twenties who had no more than a high school degree (and who therefore tended to be lower income) believed unprotected sex didn't put them at much risk of getting pregnant. Only 9 percent of people the same age with a college degree held this belief. Twenty-five percent of the former group thought they were unlikely to get pregnant if they had sex standing up, versus 7 percent of their college-educated peers. These data point to glaring differences in knowledge about conception among people of different social class backgrounds.[31]

These differences stem from unequal exposure to **comprehensive sex education**, or school curricula that promote birth control as a way to avoid pregnancy and sexually transmitted infections. Table 6.1 reveals that Whites, who typically come from higher-income families than minorities, are more likely to have received comprehensive sex education than Blacks or Hispanics, and likelier to have done so in both

TABLE 6.1 ● Getting Educated about Sex

There are significant racial gaps in kids' exposure to comprehensive sex education.

	All Millennials	White	Black	Hispanic	API
Middle school only	17	18	16	13	16
High school only	19	17	18	25	18
Both middle and high school	39	42	32	34	41
Did not take sex education class	23	21	29	27	22
Refused	2	2	5	1	3
Total	100	100	100	100	100

Source: Robert P. Jones and Daniel Cox, "How Race and Religion Shape Millennial Attitudes on Sexuality and Reproductive Health," Public Religion Research Institute, March 27, 2015, https://www.prri.org/wp-content/uploads/2015/03/PRRI-Millennials-Web-FINAL.pdf.

middle and high school. Common sense tells us that lacking sound knowledge about contraception elevates the risks of getting pregnant. Therefore, race- and class-based disparities in kids' knowledge about birth control contribute toward explaining which teens are likeliest to have babies.

The reasons poor and minority teens have the highest risk of getting pregnant go beyond the availability and accessibility of birth control.
iStockphoto.com/pederk

However, providing low-income kids with better access to comprehensive sex education or teaching them to abstain from sex entirely aren't enough to reduce social class and racial inequalities in teen pregnancy rates. That's because gaps in knowledge about how to minimize the risks of conception compound the significant differences in educational opportunity that we explored earlier. Since the kids who are the least educated about sex are also the most disposed toward seeing their futures as bleak, they have the highest chance of becoming teen parents. Therefore, better education about sex, by itself, doesn't make much of a difference. It's not the right type of education since the reasons teens become pregnant go beyond sex. Instead, low-income kids must be encouraged to believe from a very young age that, like kids from higher-class backgrounds, they too can achieve a better life.[32]

Research conducted in Michigan beginning in the mid-1960s demonstrates that giving preschoolers access to high-quality educational experiences can motivate them to invest in long-term goals. However, a lack of early childhood educational enrichment is the reality for many poor kids since preschool is not federally mandated, and therefore not free. The High/Scope Perry Preschool Study randomly assigned fifty-eight low-income Black children to attend its preschool, while sixty-five other low-income Black kids did not attend any preschool at all. Researchers have followed these two groups ever since, and Figure 6.8 (on page 120) highlights the differences between them forty years later. Those who attended the High/Scope Perry Preschool were more likely to have achieved indicators of educational, occupational, and financial success. These data reveal that having access to a high-quality preschool can powerfully influence children's beliefs about the future, motivating them to strive for academic success and avoid pregnancy. Low-income kids learn what middle- and upper-class kids have grown up taking for granted: that becoming a teen parent should be avoided because it will jeopardize their futures.[33]

When kids believe from a young age that education can better their lives, they tend to avoid becoming teen parents.
iStockphoto.com/FatCamera

FIGURE 6.8 ● Building Blocks for Life

An educationally rich preschool experience can have significant positive effects on low-income kids.

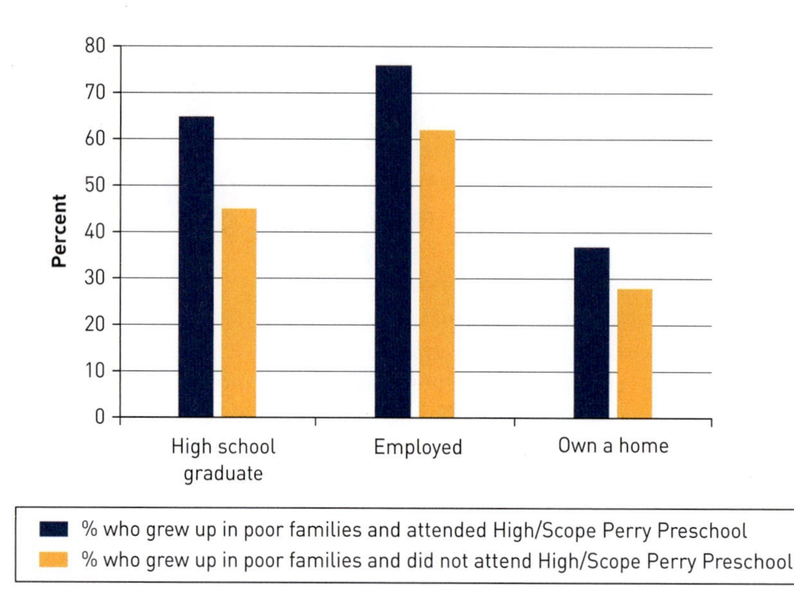

Source: Adapted from Lawrence J. Schweinhart, Jeanne Montie, Zongping Xiang, W. Steven Barnett, Clive R. Belfield, and Milagros Nores, "The High/Scope Perry Preschool Study through Age 40," http://nieer.org/wp-content/uploads/2014/09/specialsummary_rev2011_02_2.pdf.

Looking Cross-Culturally; Exposing an Alternative Way to Understand Teens and Sex

6.6 Compare attitudes toward teen sex and pregnancy in the U.S. and the Netherlands.

As the father of two teens, I'm keenly aware that my son could impregnate a girl and that my daughter could become pregnant. However, as a sociologist I recognize that these are unlikely possibilities. My kids have access to one of the best forms of birth control: the belief that they have bright futures ahead of them. While they've had the financial and emotional comforts to nourish this belief, other teens who attend their diverse high school outside Boston have grown up with any of a number of disadvantages. These include a lack of sufficient income to meet basic needs, parents whose jobs offer limited opportunities for advancement, and a dearth of academic role models. Because of such constraints, these teens are more inclined than kids like mine to believe that a brighter future is unavailable to them and that there are comparatively few costs to becoming parents. Differences among teens in the likelihood that they are pushing strollers stem from unequal childhoods, not the sexualization of childhood.

Still, parents have legitimate reasons for not wanting their kids exposed to sexualized images in ads, movies, or TV shows, or on their phones. These images contribute to girls' evaluating themselves in relation to the seemingly perfect bodies they see and often feeling insecure as a result. Moreover, boys and men may judge girls' and women's worth based on how closely their bodies compare to these images. And, because sexualized images reinforce some males' view that females exist to satisfy their sexual pleasure, such boys and men may feel justified in acting violently toward females. These are all very serious concerns. Nonetheless, sexualized images present a smokescreen for us in our efforts to get to the root of teen pregnancy.[34]

Class, race, and gender inequalities—not sexualized images—are the driving social forces behind teen pregnancy.
In Pictures Ltd./Corbis via Getty Images

Since the crux of this social problem is leveled aspirations, racially and economically disadvantaged kids need to be exposed to opportunities for seeing their lives differently. These opportunities can arise during a high school class or at an after-school job. Both are settings where a teen might establish relationships with a supportive mentor. However, fostering aspirations in a person who hasn't ever thought much about the future is challenging. That's why the most likely prospect for reducing social inequalities in teen pregnancy rates lies in providing poor children access to high-quality preschool curricula. Recognizing the need to address these inequalities long before kids are sexually active is an insight we gain from the sociological perspective.

This perspective also enables us to uncover a way that American parents might respond to teen sexuality other than through fear. To see how, consider research by Amy Schalet. She was born in the United States, grew up in the Netherlands, and returned to the U.S. for college and graduate school. For her Ph.D. dissertation, she studied the different meanings of sex within the two countries. Schalet learned during interviews with White, middle-class American parents of teenagers that they tend to view their kids in the following ways:

- Too immature to fully recognize the consequences of their actions
- Consumed by raging hormones that lead them to act impulsively
- Untrustworthy to make important decisions

These views underlie why the sexualization of childhood fuels anxieties in many American parents. These parents see sex as threatening because kids are unable to defend themselves against its vices.[35]

Schalet found that Dutch parents' views starkly differ. These parents see teens as mature and responsible, and they feel that once kids learn about the dangers of sex (comprehensive sex education is much more common in the Netherlands than in the U.S.), it's up to them to determine their readiness for becoming sexually active. Moreover, Dutch parents believe that if their child is in a committed and loving relationship, it is acceptable for the couple to spend time alone in the bedroom.[36]

Not only do Dutch teens grow up in more sexually permissive families, they also experience fewer of the harms of sex. Even though earlier in this chapter we saw that a massive drop in the U.S. teen birth rate has occurred over the past sixty years, this rate is still excessive compared to many other countries and is nearly five times higher than in the Netherlands (see Figure 6.9). Compared to their Dutch counterparts, American teens are also many times more likely to contract a sexually transmitted infection: 2.7 more times for syphilis, 19 for chlamydia, and 33 for gonorrhea (see Table 6.2).

FIGURE 6.9 ● A Particularly American Problem

The teen pregnancy rate in the U.S. is higher than in most other industrialized countries.

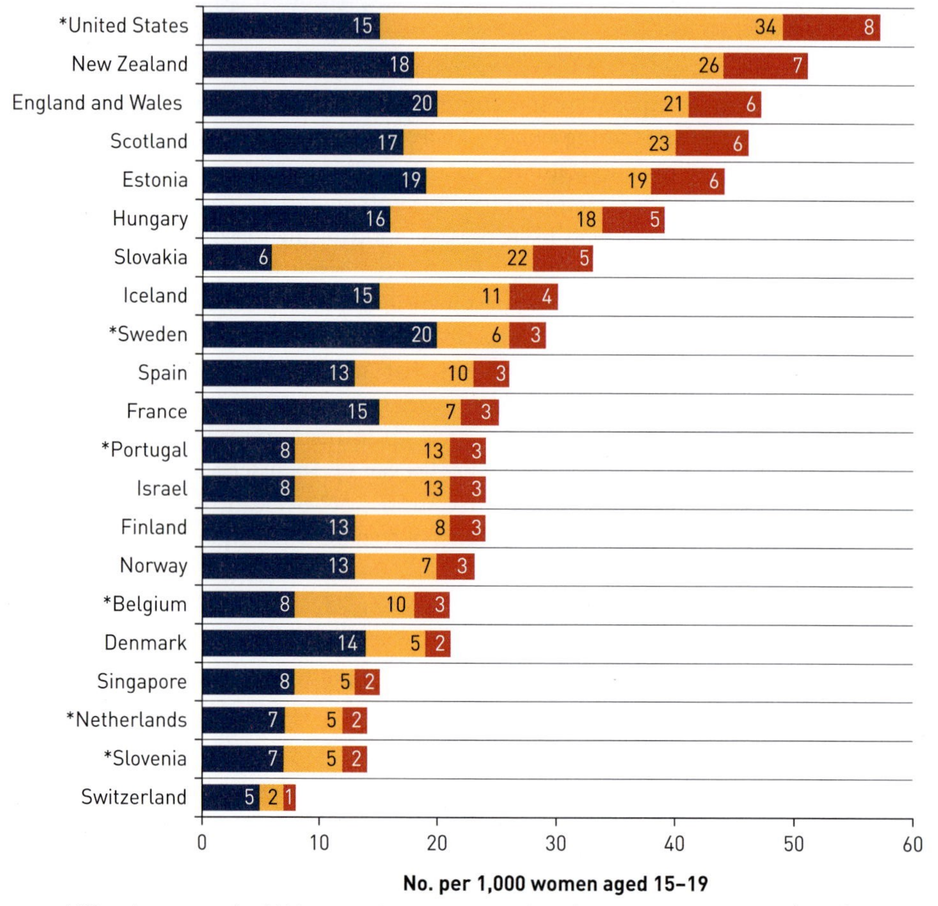

*All estimates are for 2011 except the Netherlands (2008), Belgium and Slovenia (2009), and Sweden and the United States (2010).

Source: Guttmacher Institute, "Adolescent Pregnancy and Its Outcomes across Countries," Fact Sheet, New York: Guttmacher Institute, 2015, https://www.guttmacher.org/sites/default/files/factsheet/fb-adolescent-pregnancy-outcomes-across-countries.pdf.

TABLE 6.2 • A Tale of Two Countries

Problems associated with teen sex occur much more frequently in the U.S. than in the Netherlands.

Country	Birth Rate: Per 1,000 girls ages 15–19	Syphilis Rate: Reported cases per 100,000 teens ages 15–19	Chlamydia Rate: Reported cases per 100,000 teens ages 15–19	Gonorrhea Rate: Reported cases per 100,000 teens ages 15–19
United States	24.2	2.7	2862.7	458.8
Netherlands	3.7	1.0	150.4	13.92

Source: Data about teen birth rates come from Centers for Disease Control and Prevention, "Reproductive Health: Teen Pregnancy," March 2019, https://www.cdc.gov/teenpregnancy/about/ and "Live Births Women Aged 'Under 18' and 'Under 20,' (per 1,000 Women Aged 15 to 17 and 15 to 19) in EU28 Countries, 2004, 2013, and 2014," Office for National Statistics, March 11, 2016, https://www.ons.gov.uk/peoplepopulationandcommunity/birthsdeathsandmarriages/livebirths/adhocs/005466livebirthswomenagedunder18andunder20per1000womenaged15to17and15to19ineu28countries20042013and2014. Data about teen abortions, HIV/AIDS rates, and rates of other sexually transmitted diseases come from Advocates for Youth, "Adolescent Sexual Health in Europe and the U.S.—Why the Difference?" September 2009, http://www.advocatesforyouth.org/storage/advfy/documents/fsest.pdf.

These data about Dutch society point to a way of understanding teen sexuality that's largely absent in the United States. Most American kids grow up learning that sex is dangerous. Comprehensive sex education programs promote birth control as a way to prevent harms like pregnancy, sexually transmitted infections, and violence. Programs that alternatively teach kids to abstain from sex until marriage portray premarital sex as immoral. Although Dutch kids also learn that sex can be dangerous, parents don't focus on only its risks. Instead, they emphasize that teens should regard sex as a training ground for exploration and discovery.[37]

A teen couple spending time alone may make American parents anxious, but there's a different way to understand teen sexuality than through the prism of fear.
MBI/Alamy Stock Photo

Although this view is entirely outside the American mainstream, it need not be. Parents already regard the teenage years as a critical period for social and emotional development. Viewing adolescence as also a key time for sexual growth doesn't condone promiscuity but conversely asserts that healthy sex involves commitment. Partners learn to recognize what they desire and need as well as how to communicate with one another in ways that produce mutual pleasure and closeness. Learning as teens about the importance of sex in building fulfilling relationships can enable people to experience greater sexual, romantic, and emotional satisfaction as adults. Acquiring these skills as teens can also counteract damaging messages in popular culture about sex being frivolous and devoid of consequences.[38]

This alternative way of understanding teen sexuality is sure to spark debate given the uncertainties and anxieties many Americans have about teens coming of age. Such debate is constructive because it considers whether there might be value in viewing teen sex through a prism besides danger. Just as we've seen in this chapter that race, class, and gender inequalities give rise to teenage pregnancy, seeing adolescence as a healthy sexual training ground also exposes a fascinating hidden story behind parental anxieties about sex.

What Do You Know Now?

1. Based on evidence discussed in the chapter, how is teenage pregnancy fundamentally not about sex but instead reflective of the inequality of opportunity in American society?
2. Elaborate on your answer to Question 1 by thinking about your own life. Discuss whichever of the following best describes you:
   - If you are or were a teen parent, what opportunities did or didn't you have growing up? How did your access to opportunity contribute to your having a child at a young age?
   - If you come from a financially secure family, how might your outlook on life differ if you hadn't grown up with the comforts and supports you've enjoyed?
   - If you come from a low-income family, why do you think other kids you knew growing up didn't similarly go to college?
3. How does the individual perspective toward teen pregnancy reinforce gender inequality?
4. Why is exposing kids to a high-quality preschool a more effective way to prevent teens from having babies than giving them free condoms?
5. How does Amy Schalet's data about Dutch society highlight a way that Americans could think about teens and sex that *isn't* focused on fear and danger?

Key Terms

Sexting 105
Sexualization of childhood 105
Sexual revolution 108
Shotgun marriage 110
Middle-class mindset 111
American Dream 111
Sexual double standard 116
Comprehensive sex education 118

Visit **edge.sagepub.com/silver** to help you accomplish your coursework goals in an easy-to-use learning environment.

Notes

1. Scott Calvert, "'Sexting' Case Rocks Colorado Town," *Wall Street Journal*, November 9, 2015, http://www.wsj.com/articles/sexting-case-rocks-colorado-town-1447029748; Katie Rogers, "The Vault Apps That Keep Sexts a Secret," *New York Times*, November 6, 2015, https://mobile.nytimes.com/2015/11/07/us/the-vault-apps-that-keep-sexts-a-secret.html.
2. The intent of child pornography laws—enacted long before it was possible to sext—was to punish adults who possessed or distributed nude images of minors. However, many child pornography cases have been brought against teens caught sexting. One study estimated that minors comprised 23 percent of all indictments on such charges. See Janis Wolak, David Finkelhor, and Kimberly J. Mitchell, "Trends in Arrests for Child Pornography Production: The Third National Juvenile Online Victimization Study (NJOV-3)," Crimes against Children Research Center, 2012, http://www.unh.edu/ccrc/pdf/CV270_Child%20Porn%20Production%20Bulletin_4-13-12.pdf and Heidi Strohmaier, Megan Murphy, and David DeMatteo, "Youth Sexting: Prevalence Rates, Driving Motivations, and the Deterrent Effect of Legal Consequences," *Sexuality Research and Social Policy* 11, no. 3 (2014): 245–55.
3. Joel Best and Kathleen A. Bogle, *Kids Gone Wild* (New York: New York University Press, 2014), 1.
4. Data about social problems stemming from teens having babies come from Schuyler Center for Analysis and Advocacy, "Teenage Births: Outcomes for Young Parents and Their Children," December 2008, http://www.scaany.org/documents/teen_pregnancy_dec08.pdf.
5. Amy Adele Hasinoff, "Blaming Sexualization for Sexting," *Girlhood Studies* 7, no. 1 (2014): 102–20; Sara E. Thomas, "'What Should I Do?' Young Women's Reported Dilemmas with Nude Photographs," *Sexuality Research and Social Policy* 15, no. 2 (2018): 192–207.
6. Mary Patrice Erdmans, "On Becoming a Teen Mother: Life before Pregnancy," University of California Press Blog, August 17, 2015, https://www.ucpress.edu/blog/18683/on-becoming-a-teen-mom-life-before-pregnancy/; Best and Bogle, *Kids Gone Wild*, 143–4.
7. Kristin Luker, *Dubious Conceptions: The Politics of Teenage Pregnancy* (Cambridge, MA: Harvard University Press, 1996), 107.
8. The 1963 data are from a survey conducted by the National Opinion Research Center. Since attitudes varied slightly when respondents were asked about the permissibility of men having premarital sex versus women, I've averaged the different percentages. These data are reported by the Roper Center for Public Opinion Research in "Going All the Way: Public Opinion and Premarital Sex," July 7, 2017, https://ropercenter.cornell.edu/blog/going-all-way-public-opinion-and-premarital-sex.
9. Kristen M. J. Thompson, "A Brief History of Birth Control in the U.S.," *Our Bodies, Ourselves*, December 14, 2013, https://www.ourbodiesourselves.org/book-excerpts/health-article/a-brief-history-of-birth-control/.
10. Roper Center, "Going All the Way."
11. "Researchers Measure Increasing Sexualization of Images in Magazines," PBS, December 21, 2013, https://www.pbs.org/newshour/nation/social_issues-july-dec13-sexualization_12-21; Best and Bogle, *Kids Gone Wild*, 125–7.
12. The percentage of married teen mothers during the Baby Boom years comes from Heather D. Boonstra, "Teen Pregnancy: Trends and Lessons Learned," *Guttmacher Policy Review* 5, no. 1 (2002), https://www.guttmacher.org/gpr/2002/02/teen-pregnancy-trends-and-lessons-learned. The estimate that many of them had shotgun marriages is based on data from the 1930s, when about half of girls who became pregnant

quickly married prior to having their baby; see Duke University, "Is Shotgun Marriage Dead?" November 1, 2016, https://phys.org/news/2016-11-shotgun-marriage-dead.html.
13. Data about the percentage of unmarried teen mothers nowadays are based on 2016 figures and is reported in Child Trends, "Key Facts About Teen Births," https://www.childtrends.org/indicators/teen-births.
14. Melissa S. Kearney and Philip B. Levine, "Media Influences on Social Outcomes: The Impact of MTV's *16 and Pregnant* on Teen Childbearing," *American Economic Review* 105, no. 12 (2015): 3597–632.
15. Henry J. Kaiser Family Foundation, "Poverty Rate by Race/Ethnicity" 2017, https://www.kff.org/other/state-indicator/poverty-rate-by-raceethnicity/?currentTimeframe=0&sortModel=%7B%22colId%22:%22Location%22,%22sort%22:%22asc%22%7D#note-1.
16. Luker, *Dubious Conceptions*, 107.
17. Luker, *Dubious Conceptions*.
18. Jonathan Kozol, "Still Separate, Still Unequal," *Harper's Magazine* 311, no. 1864 (2005): 41–54, 44.
19. Luker, *Dubious Conceptions*, 107.
20. Philip B. Levine and Melissa S. Kearney, "Forget Plan B: To Fight Teen Pregnancy, Focus on Economic Opportunity," *The Atlantic*, May 8, 2013, https://www.theatlantic.com/business/archive/2013/05/forget-plan-b-to-fight-teen-pregnancy-focus-on-economic-opportunity/275623/.
21. Melissa S. Kearney and Philip B. Levine, "Why Is the Teen Birth Rate in the United State So High and Why Does It Matter?" *Journal of Economic Perspectives* 26, no. 2 (2012): 141–66. Quote appears on page 163.
22. The Pell Institute for the Study of Equity in Higher Education, "Indicators of Higher Education Equity in the United States," 2016 Historical Trend Report, http://www.pellinstitute.org/downloads/publications-Indicators_of_Higher_Education_Equity_in_the_US_2016_Historical_Trend_Report.pdf; Ira Silver, *Giving Hope: How You Can Restore the American Dream* (Scotts Valley, CA: CreateSpace, 2013), 93–121.
23. Institute for Women's Policy Research, "College Students with Children Are Common and Face Many Challenges in Completing Higher Education," March 2013, https://files.eric.ed.gov/fulltext/ED556715.pdf.
24. Luker, *Dubious Conceptions*, 107–8; Judith Herman, "Adolescent Perceptions of Teen Births," *Journal of Obstetric, Gynecologic, and Neonatal Nursing* 37, no. 1 (2008): 42–50.
25. Sandra K. Danziger and Norma Radin, "Absent Does Not Equal Uninvolved: Predictors of Fathering in Teen Mother Families," *Journal of Marriage and Family* 52, no. 3 (1990): 636–42; Best and Bogle, *Kids Gone Wild*, viii.
26. Elizabeth A. Armstrong, Paula England, and Alison C. K. Fogarty, "Accounting for Women's Orgasm and Sexual Enjoyment in College Hookups and Relationships," *American Sociological Review* 77, no. 3 (2012): 435–62.
27. Elizabeth A. Armstrong et al., "'Good Girls': Gender, Social Class, and Slut Discourse on Campus," *Social Psychology Quarterly* 77 (2014): 100–22.
28. Equal Justice Institute, "History of Racial Injustice: Sexual Exploitation of Black Women," https://eji.org/history-racial-injustice-sexual-exploitation-black-women.
29. Data about teen births from forced sex and on the likelihood of teen girls having experienced either rape or attempted rape come from Malika Saada Saar, "A Missing Piece of the Prevention Puzzle," Center for American Progress, August 6, 2008, https://www.americanprogress.org/issues/women/news/2008/08/06/4768/a-missing-piece-of-the-prevention-puzzle/.
30. Alan Guttmacher Institute, "Sex and America's Teenagers," 1994, https://www.guttmacher.org/sites/default/files/pdfs/pubs/archive/SaAT.pdf.
31. Catherine Rampell, "The Sex Ed Gap," *Washington Post*, October 17, 2014, https://www.washingtonpost.com/news/rampage/

wp/2014/10/17/the-sex-ed-gap/?utm_term =.3d41c5d234a5.

32. Erdmans, "On Becoming a Teen Mother."
33. Lawrence J. Schweinhart et al., "The High/Scope Perry Preschool Study through Age 40," 2011, file:///C:/Users/isilver/AppData/Local/Microsoft/Windows/Temporary%20Internet%20Files/Content.Outlook/D4R34F38/specialsummary_rev2011_02_2.pdf.
34. Maddie Oatman, "Miss Representation Shows the Ugly Side of Women in Media," *Mother Jones*, October 20, 2011, https://www.motherjones.com/media/2011/10/miss-representation-doc-shows-ugly-side-women-media/.
35. Amy T. Schalet, *Not under My Roof: Parents, Teens, and the Culture of Sex* (Chicago: University of Chicago Press, 2011).
36. The discussion here is based on an earlier analysis Amy Schalet did that draws on the same research: "Must We Fear Adolescent Sexuality?" *Medscape General Medicine* 6, no. 4 (2004): 44.
37. Kristin Luker, *When Sex Goes to School: Warring Views on Sex—and Sex Education—since the Sixties* (New York: W. W. Norton, 2006), 91–118.
38. Amy T. Schalet, "Beyond Abstinence and Risk: A New Paradigm for Adolescent Sexual Health," *Women's Health Issues* 21, no. 3 (2011): S5–S7; Hasinoff, "Blaming Sexualization for Sexting," 457–8.

7

"Everybody's Doing It"

Getting Ahead by Cheating

Learning Objectives

1. Recognize how Lance Armstrong's story confirms the conventional wisdom that people who cheat are morally weak.
2. Describe the prevalence of doping in sports.
3. Explain how the *winner-take-all society* concept offers a sociological perspective on cheating across different settings.
4. Discuss how the use of various common drugs resembles the use of performance-enhancing drugs by athletes but is not considered cheating.
5. Explain how the medicalization of certain conditions supports the winner-take-all mindset.
6. Discuss how cheating can be understood as an affirmation of mainstream American values.

From Champ to Chump: How Lance Armstrong Tarnished His Reputation Riding the Road to Victory

7.1 Recognize how Lance Armstrong's story confirms the conventional wisdom that people who cheat are morally weak.

"In all seven of your Tour de France victories, did you ever take banned substances or blood dope?" Oprah Winfrey knows how to cut to the chase. Looking battered and defeated as he listened to her matter-of-fact question, Lance Armstrong softly muttered, "Yes." "In your opinion was it humanly possible to win the Tour de France without doping seven times in a row?" "Not in my opinion," Armstrong said with his head hung low.[1] In the remainder of the interview, Armstrong detailed how he repeatedly ingested illegal substances during the years he won the world's most prestigious cycling competition. (This chapter interchangeably refers to these illegal substances as *performance-enhancing drugs* and as *doping*.)

Lance Armstrong won the Tour de France every year from 1999 to 2005.
iStockphoto.com/pejft

The drugs Armstrong took included strength enhancers like testosterone, cortisone, and human growth hormone. He also took erythropoietin, which increases endurance by elevating oxygen levels in the bloodstream. Armstrong's revelation was shocking not only because he admitted to having cheated but also because

iStockphoto.com/LuckyBusiness

he fessed up to having left a long trail of lies. For over a decade, he'd adamantly denied doping allegations.

For years before these allegations surfaced, Armstrong was an international hero. Over an unprecedented seven consecutive years, he was the fastest person to cycle the Tour de France. The 2,200-mile race takes place in twenty-one segments over twenty-three days, much of it through the steep and rugged Alps. Armstrong built his streak against unimaginable odds. In 1996, he was diagnosed with metastatic testicular cancer, a life-threatening disease. After two surgeries and four rounds of chemotherapy, he returned to competitive training. Just a year later, Armstrong triumphed in Paris for the first time. He reflected on this period of glory before the ensuing storm of controversy: "You overcome the disease, you win the Tour de France seven times, you have a happy marriage, you have children, I mean, it's just this mythic perfect story."[2]

Armstrong became a leading figure in the fight against cancer when he established the Livestrong Foundation. It partnered with Nike in marketing a bright yellow wristband with "LIVESTRONG" emblazoned across it. The gimmick became an overnight sensation. The wristbands were trendy to wear since people felt good in knowing that some of the proceeds went to charity. I remember each guest receiving one at my cousin's bat mitzvah. People teared up when she announced that she'd be donating a portion of her gifts to the Livestrong Foundation.

Few could have imagined that within a few years of his last Tour de France victory, Armstrong's reputation would tank. Amidst the repeated doping allegations, some people crossed out the "V" on their wristband so that it now said "LIESTRONG." By the time Oprah interviewed him in 2013, Armstrong had become a public disgrace. Their conversation was so chilling because it vividly confirmed what many people had suspected yet had a hard time fathoming could actually be so: This one-time hero had not only cheated but repeatedly lied about having done so. It's no wonder the interview was must-see TV for millions of people. It exposed the truth behind Armstrong's antics and highlighted the justifiable reasons his reputation shattered.[3]

Americans look upon performance-enhancing drugs as the most serious among a variety of current threats to the integrity of sports (see Figure 7.1). Given this sentiment, it's no surprise that Lance Armstrong and other star athletes proven to have doped have been demonized. The vilification of people who at one time were heroes highlights the individual perspective toward cheaters: Their choice not to play by the rules demonstrates their moral weakness. This chapter develops a compatible sociological perspective that expands the focus beyond sports by exploring many different contexts in which people take shortcuts to success. Seeing cheating from this vantage point reveals that this behavior reflects the intense competition to get ahead in American society. This competition motivates individuals to do whatever it takes to win, even if that means sidestepping what's right. So instead of viewing cheaters as people who deviate from societal norms, our aim is to see how they *conform* to the expectations such competition fosters.[4]

The wristband that the Livestrong Foundation created as an emblem of Lance Armstrong's heroism became a symbol of his deception.
iStockphoto.com/nailzchap

FIGURE 7.1 ● Tarnishing the Game

Americans view performance-enhancing drugs as the most serious threat to the integrity of sports.

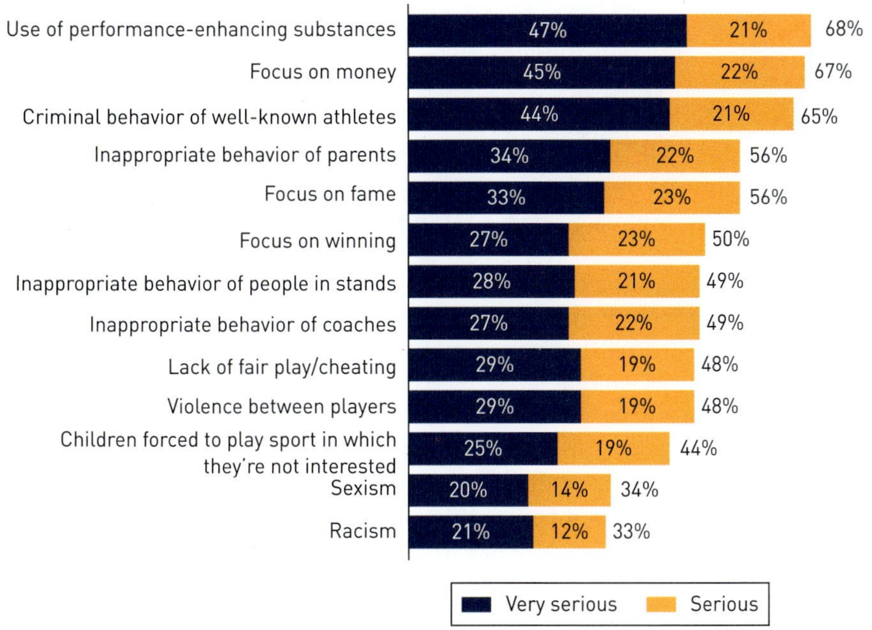

	Very serious	Serious	Total
Use of performance-enhancing substances	47%	21%	68%
Focus on money	45%	22%	67%
Criminal behavior of well-known athletes	44%	21%	65%
Inappropriate behavior of parents	34%	22%	56%
Focus on fame	33%	23%	56%
Focus on winning	27%	23%	50%
Inappropriate behavior of people in stands	28%	21%	49%
Inappropriate behavior of coaches	27%	22%	49%
Lack of fair play/cheating	29%	19%	48%
Violence between players	29%	19%	48%
Children forced to play sport in which they're not interested	25%	19%	44%
Sexism	20%	14%	34%
Racism	21%	12%	33%

Very serious Serious

Source: National Collegiate Athletic Association, "NCAA Doping, Drug Education and Drug Testing Task Force," 2013, http://www.ncaa.org/sport-science-institute/ncaa-doping-drug-education-and-drug-testing-task-force. Reprinted by permission of US ADA.

FIRST IMPRESSIONS?

1. What are your reactions to learning the details of Lance Armstrong's story?
2. Why do you think athletes cheat?
3. What other stories are familiar to you about celebrities whose misdeeds led to the tarnishing of their reputations?

A Sketchy Path to the Top: Doping in Sports

7.2 Describe the prevalence of doping in sports.

Lance Armstrong's story added a shocking twist to a scandal that had been occurring for many years, not just in professional sports like competitive cycling but

If you've ever played competitive sports, you know there are lots of acceptable ways to train for success. These include eating nutritiously, getting plenty of sleep, and exercising regularly.
iStockphoto.com/skynesher

among athletes at all different levels. It's hard to know the exact prevalence of doping in school sports; many student athletes who cheat in this way don't get caught because systematic monitoring of their behavior is rare. And of course these athletes have a motive not to disclose their behavior to researchers when asked. That said, studies based on self-reports indicate that over the past year, perhaps as many as 6.6 percent of high school and college athletes have taken **anabolic steroids**, which are drugs that promote muscle growth by mimicking the effects of testosterone. These steroids are just one type of performance-enhancing drug; others include human growth hormone, creatine, androstenedione, and erythropoietin. Doping is, therefore, likely to be much more extensive among student athletes than the reported numbers indicate.[5]

Data about Olympians are more conclusive since several regulatory organizations, including the U.S. Anti-Doping Agency and the World Anti-Doping Agency, randomly conduct tests for banned substances. Over the past fifty years, there's been a steady rise in the number of athletes disqualified from competing because they've tested positive (see Figure 7.2). These numbers only begin to tell the story of the pervasiveness of doping among Olympians. More than 1 percent of those who competed at the 2016 Summer Games in Rio (120 athletes in all) had either been previously suspended or had prior medals revoked because of doping violations.[6] Based on evidence of government-sponsored doping that occurred when Russia hosted the 2014 Winter Olympics in Sochi, the International Olympic Committee barred Russia from sending a team to the 2018 Winter Olympics in Pyeongchang, South Korea. While a select group of 169 athletes who hadn't failed drug tests were allowed to compete under the designation "Olympic Athlete from Russia," their presence hardly distracted attention from the Russian doping scandal as two of these competitors tested positive for banned substances during the games.[7]

While Russia has been in the spotlight concerning Olympians' use of performance-enhancing drugs, American athletes cheat with comparable frequency. A 2014 composite of professional athletes around the world found that the U.S. actually had more proven cases of doping than Russia (see Figure 7.3 on page 134). These numbers, however, did not reflect the massive doping by Russian athletes at the Sochi games that same year. Nonetheless, the U.S. and Russia far exceeded the country with the third most cases, the Dominican Republic. Whereas cheating occurred across many different sports, 65 percent of the cases occurred in just three: weightlifting, baseball, and athletics (which in England refers to track and field events). Athletics had the most by far, with ninety-five cases.

FIGURE 7.2 ● Olympic Doping on the Rise

Since 1968, the number of athletes who've tested positive for performance-enhancing drugs at the Summer Olympics has steadily increased.

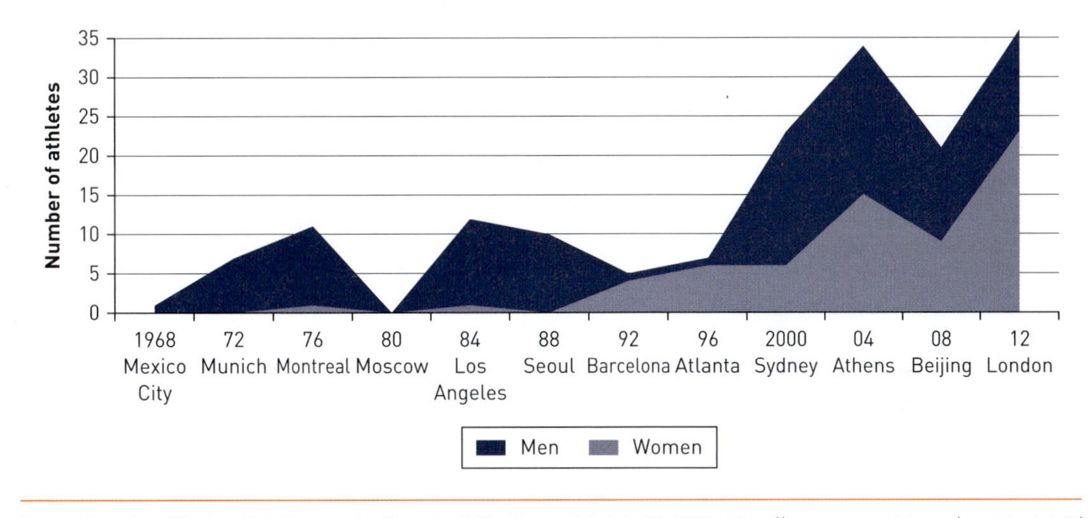

Source: Based on "Doping Violations at the Olympics," *The Economist*, July 25, 2016, https://www.economist.com/graphic-detail/2016/07/25/doping-violations-at-the-olympics.

It's telling that baseball was second in number of cases. At the time of Lance Armstrong's transgressions, nowhere in the U.S was the scandal surrounding performance-enhancing drugs in sports bigger than in Major League Baseball. Twice during the 2000s, Congress conducted investigations. Some of the most accomplished players of their generation, including Roger Clemens, Mark McGwire, and Alex Rodriguez, were either proven to have participated in, or heavily suspected of, doping. In the history of baseball through 2018, a player hit fifty or more home runs in a single season forty-four times. Over half of those feats (twenty-four) occurred during the so-called steroids era from the late 1980s until the late 2000s. Barry Bonds is a poster child of this cheating scandal. His strength went off the charts after he allegedly started doping, and he hit seventy-three home runs in 2001 to shatter the single season record by twelve. Bonds then went on to break the career home run record.[8]

The prevalence of doping in Major League Baseball has tarnished the game's image as America's pastime.
George Mattei/Science Source

If we only pay attention to Bonds and other star players who took performance-enhancing drugs, it seems obvious that the drugs gave them a leg up on the competition. However, many of the players who doped during Major League Baseball's steroids era did *not* actually perform better. For example, the season after Baltimore

FIGURE 7.3 ● Athletes Who Cheat

Here's a global snapshot from 2014 of performance-enhancing drug usage across different sports.

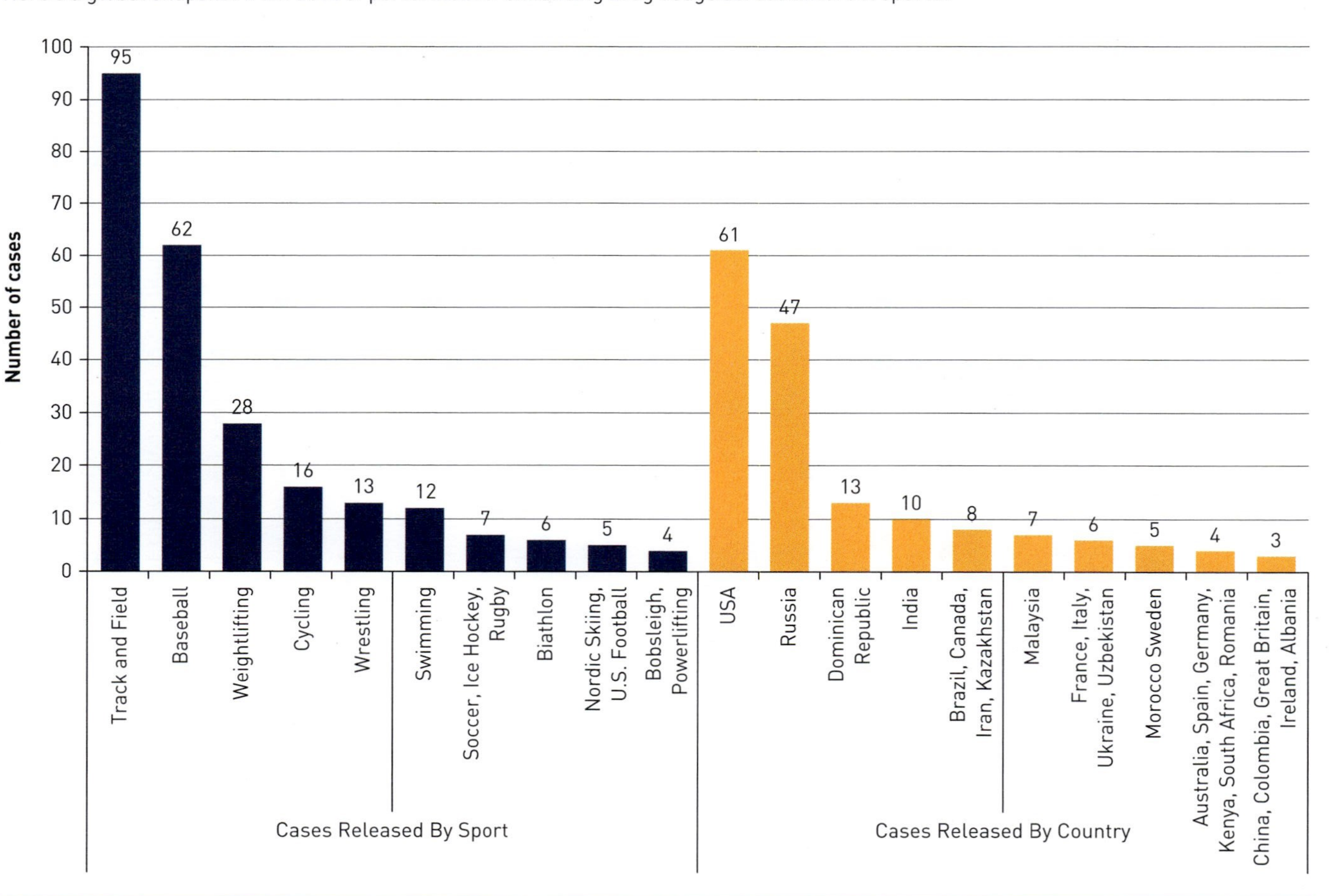

Source: David Owen, "Figures Indicate Athletics Was Sport with Most Doping Cases in 2014," *Inside the Games: The Inside Track on World Sport*, January 8, 2015, https://www.insidethegames.biz/articles/1024740/figures-indicate-athletics-was-sport-with-most-doping-cases-in-2014.

Orioles outfielder Jay Gibbons is alleged to have used human growth hormone, his home run total dropped from twenty-three to ten.[9] Of course, cheating is still wrong even if the cheater doesn't perform better as a result. Teachers sometimes have to remind students that someone who copies incorrect answers from a classmate's exam or lifts a poorly written essay from the internet has still broken the rules.

Like Lance Armstrong, many people used to hold Bill Cosby in high regard as an all-American family man. But after many sexual assault allegations and a 2018 conviction, Cosby is now a source of public shame.
Richard Levine/Alamy Stock Photo

Professional athletes proven to have used performance-enhancing drugs and achieved success offer some of the clearest evidence we have of people who get ahead by cheating. Since these athletes supposedly serve as role models to kids, vilifying them sends the message that cheating is an unacceptable way to win. Their public condemnation, more generally, reflects Americans' twisted fascination with athletes, actors, entertainers, and other celebrities who fall from grace, plummeting from the penthouse to the doghouse.[10]

Striving to Be #1: Competing in a Winner-Take-All Society

7.3 Explain how the *winner-take-all society* concept offers a sociological perspective on cheating across different settings.

The stakes in the Tour de France, World Series, and Super Bowl are so high because they only crown a champion once a year. For Olympic sports, it's only once every four years. The winner gets all of the glory, while few people remember who came in second. If we think for a moment about the structure of athletic competitions, we can begin to see that they're a microcosm of the United States as a whole. Ours is a **winner-take-all society** in that the educational system and many occupations follow the same competitive structure that's so apparent in sports. In these other settings, people have an incentive to beat out the competition given that being the very best yields most of the rewards. People who perform the next best get fewer rewards, and the rest practically none.[11] As a result, **income inequality**, or the disparity in earnings between the richest and poorest people, is the highest in the U.S. that it's been since 1928 (see Figure 7.4 on page 136).[12]

Your parents may remember a time when it didn't used to be this way. From the end of World War II until the late 1970s, the earnings of poor people grew at a higher rate than those of the rich. Even though in sports and certain other fields—namely entertainment and the arts—there was a substantial difference in compensation between winners and the next best, the overall income gap in American society was diminishing. Moreover, there were substantially greater opportunities for low-income people to increase their earnings than exist today. (For a fuller discussion of this, see Chapter 2.)[13]

Nowadays there's evidence all around you of the disproportionate rewards given to people who perform exceptionally versus those who merely perform well:

FIGURE 7.4 ● The Winner Takes It All

Comparing these two trends highlights that the rewards structure across American society has increasingly come to resemble the glaring income inequalities among athletes.

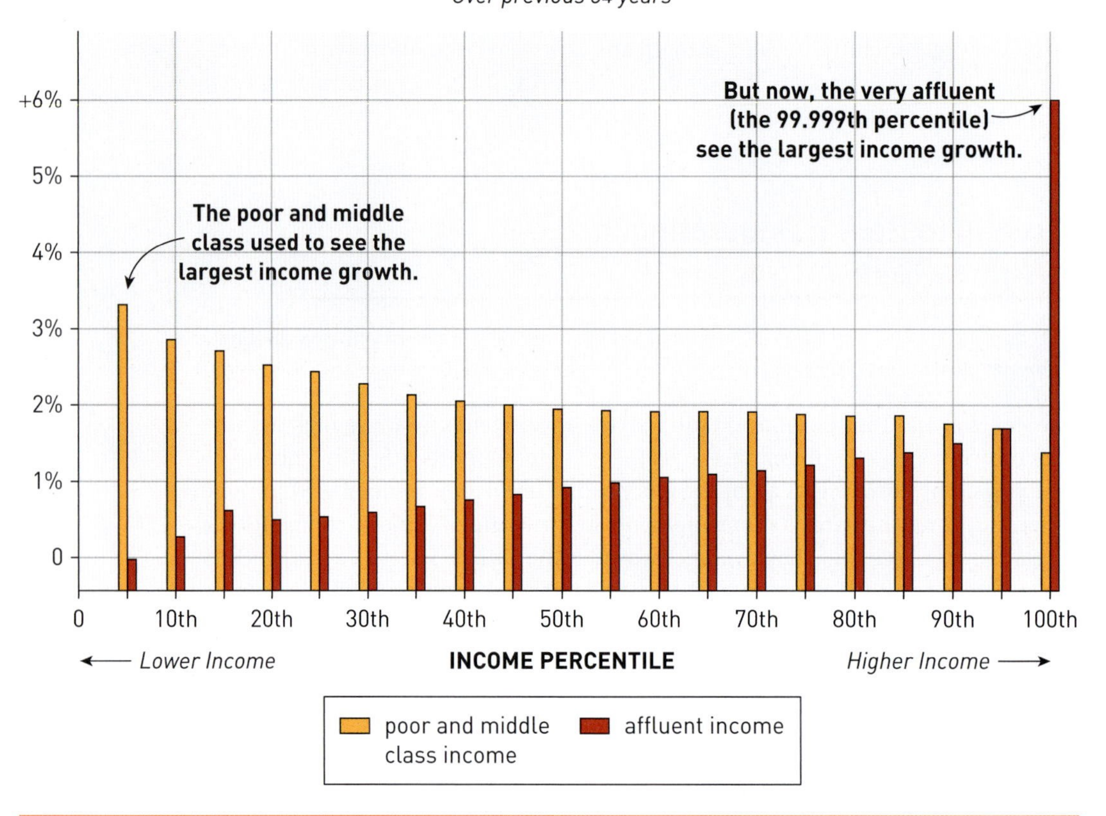

Source: Based on David Leonhardt, "Our Broken Economy, in One Simple Chart," *New York Times*, August 7, 2017, https://www.nytimes.com/interactive/2017/08/07/opinion/leonhardt-income-inequality.html.

- Pop singers at the top of the iTunes chart who sell out stadium concerts versus obscure musicians who play gigs at local bars
- The heart surgeon whose compensation is many times greater than the primary care physician
- The inventor whose patent earns millions in royalties versus the salaried scientist doing important laboratory experiments
- The tenured professor who receives a decent salary and benefits versus the adjunct lecturer who struggles to make ends meet and lacks health insurance

J. K. Rowling, author of the *Harry Potter* series, may be the best example of all. These books have been an international sensation since the late 1990s, grossing billions of dollars in sales and related merchandise. While Rowling has basked in huge success, many talented writers around the world persistently struggle to publish their work. Can you think of other examples of the winner-take-all society?[14]

This lopsided reward structure gives people an incentive to be the absolute best they can be. That's obviously a good thing as long as they pursue greatness in legitimate ways, such as by working hard and making sacrifices. The problem arises when the desire to be number one fuels the impetus to take shortcuts. Ken Caminiti, who played third base for the San Diego Padres in the 1990s and took performance-enhancing drugs en route to winning the Most Valuable Player award, told *Sports Illustrated* that he "looked around and everybody was doing it."[15] Caminiti elaborated:

> If a young player were to ask me what to do, I'm not going to tell him it's bad. Look at all the money in the game. You have a chance to set your family up, to get your daughter into a better school. . . . So I can't say "Don't do it" when the guy next to you is as big as a house and he's going to take your job and make the money.[16]

Popular music is a clear example of a context where winners get most of the rewards. Only a handful are megastars like Adele; thousands of other talented people struggle to build careers.
Gareth Cattermole/Getty Images for September Management

In a setting where winning yields vastly greater rewards than being second best, the ends can seem to justify the means.

This scenario isn't unique to baseball or to sports more generally. You know quite well that the educational system also rewards winning much more than being the next best. Think about what it means to get an "A" versus a "B." Indeed, school is where—from a very young age—you became introduced to the winner-take-all society. Although it's the place you likely hear the most consistent message that cheating is wrong, it's also

Teachers may be reluctant to talk about the prevalence of cheating and hesitant to acknowledge the role they play in fueling the competitive pressure that gives rise to it.
Roman Kosolapov/Alamy Stock Photo

This scene is particularly common at top-ranked high schools and colleges since the competitive stakes to be the very best are so high.
iStockphoto.com/mediaphotos

where people are apt to cheat for the first time and to do so most often. About 60 percent of high school students report having cheated in the past year. Cheating is also pronounced in college (see Table 7.1).[17]

As a teacher, I see my role as encouraging students to regard their personal and intellectual growth as more important than getting top grades. After all, grades don't necessarily measure what a person has learned or the impact the content has had on them. Yet as a sociologist I recognize why students may enter my classes focused on getting the best grade and be unwilling to change their mindset no matter what I say or do. After all, our relationship rests on my evaluation of their performance. Moreover, it can make a huge difference in their future opportunities and earnings whether they get good grades or top grades. Given the high stakes of education, students cheat on papers and exams in many ways, such as using SparkNotes, scribbling formulas on their calculators, discreetly checking their phones, and copying from the person sitting next to them. I doubt any of this is news to you.

Cheating is especially common in top-ranked schools where the rewards attached to winning are the greatest. In such schools, students are apt to reinforce to one another the message that a person's self-worth hinges on their academic success. Expensive private high schools are a case in point. At Horace Mann in New York City, students feel intense pressure to remain part of the elite. Just a few miles away, low-income kids who score well enough on entrance exams to attend prestigious Stuyvesant High School experience similar pressures. These kids carry parental expectations that academic success will lead to upward mobility. Students at both schools are keenly aware that a small change in their grade point average can have huge lifelong consequences. It can mean the difference between being accepted at very good colleges versus top-tier schools that offer a vaster array of opportunities.[18]

In 2019, federal prosecutors arrested fifty people who'd been involved in an eight-year cheating scheme to get kids from wealthy families accepted at several

TABLE 7.1 ● Students Taking Shortcuts

Here's a snapshot of the prevalence of cheating among undergraduates. Do you think these numbers accurately reflect how often cheating occurs at your school? Why or why not?

Written Assignment	Percentage Who Cheated	Test or Exam	Percentage Who Cheated	Other Assignment	Percentage Who Cheated
Working with others on an assignment when asked for individual work	42	Learning what is on a test from someone who has already taken it	33	Fabricating or falsifying lab data	19
Paraphrasing/copying few sentences from written source without footnoting it	38	Using false excuse to delay taking test	16	Copying someone else's program in a course requiring computer work	11
Paraphrasing/copying few sentences from internet source without footnoting it	36	Copying from another student on a test/exam without their knowledge	11	Fabricating or falsifying research data	8
Receiving unpermitted help from someone on an assignment	24	Helping someone else cheat on test	10		
Fabricating/falsifying a bibliography	14	Copying from another student on a test/exam with their knowledge	9		
Turning in work copied from another	8	Using unauthorized crib/cheat notes	8		
Copying material almost word for word from a written source without citation	7	Using an electronic/digital device as an unauthorized aid during a test/exam	5		
Turning in work done by another	7				
Obtaining paper from term paper mill	3				

Source: Adapted from Donald L. McCabe, "Cheating among College and University Students: A North American Perspective," *International Journal for Educational Integrity* 1, no. 1 (2005).

Note: Based on a survey of over seventy thousand undergraduates across eighty-three campuses in the U.S. and Canada.

Actresses Felicity Huffman (left) and Lori Loughlin (right) were among the parents charged in the college admissions cheating scandal, nicknamed Operation Varsity Blues.
LISA O'CONNOR,TOMMASO BODDI/AFP/Getty Images

top-tier schools: Yale, Stanford, the University of Southern California, Georgetown, Wake Forest, and the University of Texas at Austin. Their parents paid thousands of dollars or more—in one instance, $1.2 million—to Edge College & Career Network. The California-based college-prep company used the money to bribe test administrators to falsify SAT and ACT scores and sports coaches to guarantee spots on their teams. The kids involved were not arrested since they were supposedly unaware of the extensive efforts their parents had undertaken to cheat on their behalf. The biggest victims of these scams were the other, hard-working students denied admission at top schools because their spots had been bought.[19]

Although cheating is most prominent in school, it hardly ends when people have finished their formal education. Many continue to cheat as they enter other winner-take-all settings. For example, people have an impetus to misrepresent credentials in the search for employment and to take shortcuts once on the job. Mechanics may do unnecessary car repairs in order to meet quotas for the amount of revenue their work is expected to generate. Lawyers may exaggerate their billable hours. These behaviors are responses to situations that incentivize people to do whatever they can to get ahead given the huge rewards that come with being the best.[20]

Consider what Lance Armstrong told Oprah in their interview: "The definition of cheat is to gain an advantage on a rival or foe—you know—that they don't have. I didn't view [what I was doing] that way. I viewed it as a level playing field."[21] When I heard him say those words, I nearly fell over. I wondered how he could have possibly

It's not uncommon for jobseekers to lie on their résumés in an effort to stand out from the competition.
iStockphoto.com/peepo

People are motivated to cheat when they're figuring out how much they owe in taxes for the year. Less money paid means keeping more of one's earnings. . .unless, of course, the government catches wind of it!
iStockphoto.com/Bill Oxford

believed that doping was a form of fair play. Many viewers probably had a similar reaction. After all, there are limits to what athletes may acceptably do to win. Because drugs that increase strength and endurance offer shortcuts to success, taking them is cheating.

A couple of years after Oprah's interview of Lance Armstrong, I started reading about the winner-take-all society and noticed my perspective toward him begin to change. I came to see why it was shortsighted to view him as lacking the character or integrity to play by the rules. Since in training for a high-stakes race like the Tour de France there's an incentive to take shortcuts, it was reasonable for Armstrong to think his rivals would be cheating too. So if he didn't try to one-up the competition, he risked being one down.

From a sociological perspective, being a competitive cycler—where only the very best earn riches from sponsorships and endorsements—blurred Lance Armstrong's understanding of the difference between doing what was best and what was right.
iStockphoto.com/HasseChr

Indeed, what looks like an immoral act from afar can seem acceptable from the standpoint of a person taking part in a winner-take-all competition. In his study of cheating in the United States, David Callahan wrote that "the pervasiveness of this rationalization [that 'everybody is doing it'] shows how easily cheating can create a downward spiral: The more cheating there is, the more it becomes a routine part of life. The more it's normalized, the less it becomes a conscious choice driven by any meaningful motive at all."[22]

The truth is that if Armstrong hadn't cheated, he would have been at a competitive disadvantage. This doesn't justify his actions, but it acknowledges the high competitive stakes that people experience in winner-take-all situations.[23]

Challenging the Conventional Wisdom about Who's a Cheater

7.4 Discuss how the use of various common drugs resembles the use of performance-enhancing drugs by athletes but is not considered cheating.

"The first time I took Adderall, I was a sophomore at Brown University, lamenting to a friend the possibility of my plight: a five-page paper due the next afternoon on a book I had only just begun reading," wrote Casey Schwartz.[24] She became "locked in a passionate embrace with the book I was reading and the thoughts I was having about it, which tumbled out of nowhere and built into what seemed an amazing pile of riches."[25] Schwartz frequently took the drug while pulling all-nighters to complete course assignments. After graduating from college, she continued using it while working toward a master's degree from University College London.

Adderall is an amphetamine stimulant that, when used as directed, alleviates the symptoms of attention deficit hyperactivity disorder (ADHD) by improving a

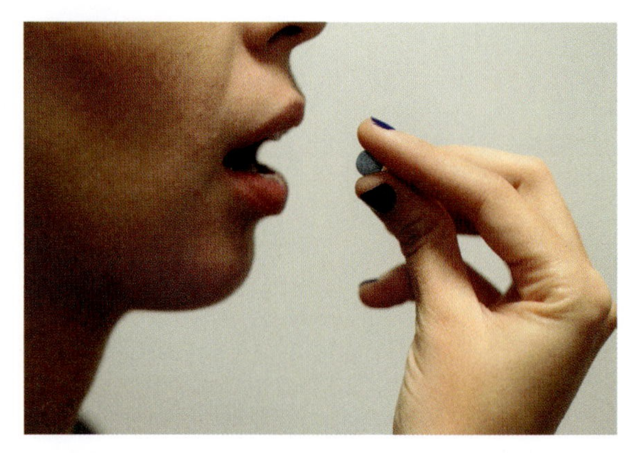

Sometimes called the magic blue pill, Adderall enabled Casey Schwartz to hone her writing skills and develop a strong portfolio in a field of shrinking opportunities.
GDA via AP Images

person's focus and concentration. Schwartz didn't have ADHD symptoms, but she acted as if she did in order to obtain a prescription. She also bought the drug from other students willing to sell their pills. Her story is unremarkable in that, as you may know, the nonmedical use of Adderall, Ritalin, and other amphetamine stimulants is common on campuses across the nation (see Figure 7.5). These drugs are second only to marijuana in popularity among college students.[26]

When she began using Adderall, Schwartz knew becoming a successful professional hinged on having excellent credentials. As she discovered her aptitude for writing, taking this drug made sense given the winner-take-all stakes in journalism. Many traditional media outlets—newspapers, magazines, TV, and radio—had downsized or gone out of business. It had gotten even harder to make a living as a writer due to the norm in the digital world that content should be free. Taking Adderall, therefore, became game changing for Schwartz. It increased her attention and focus so she could keep developing her talent for writing. Currently, she's on the staff at *The Daily Beast* and often freelances too. A top press published her first book.[27]

FIGURE 7.5 ● Taking Drugs to Pull All-Nighters

College students are the age group that's most likely to use Adderall for nonmedical purposes.

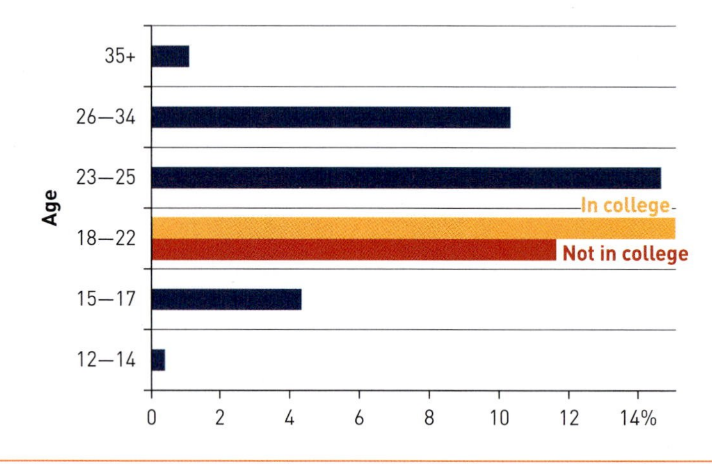

Source: Based on Emma Pierson, "College Students Aren't the Only Ones Abusing Adderall," *FiveThirtyEight*, November 5, 2015, https://fivethirtyeight.com/features/college-students-arent-the-only-ones-abusing-adderall.

Amphetamine stimulants are one of several types of drugs that people outside of sports may take to help them achieve success. Consider two others:

- **Inderal.** Most people know what it's like to experience jitters in front of an audience. For some, the feeling can be paralyzing. The first time I taught a college class, my heart was racing and my palms were sweating. I felt a pit in my stomach and kept stuttering. People who encounter stage fright may be tempted to use Inderal. It's considered a beta blocker, which, when used as directed, treats high blood pressure. Nonmedically, it calms people so they can perform effectively in front of an audience.[28]
- **Caffeine.** Where I live in Massachusetts, there's a Dunkin' every half a mile or so. Emblazoned on its cups is the slogan "America Runs on Dunkin'." Not only is this slogan catchy, it's also an indicator of how indispensable caffeine is for millions of people every day. Imagine how listless and unproductive many people would be without their morning jolt.[29]

Thinking about these drugs alongside amphetamine stimulants invites us to consider how they all resemble the performance-enhancing drugs athletes use.

In what sense is my drinking a cup of dark roast while I teach akin to Lance Armstrong's doping as he trained for races? Why haven't the universities that awarded Casey Schwartz her degrees revoked them, just as the United States Anti-Doping Agency stripped Armstrong of his seven Tour de France victories? How ethical is it for a college student to use an amphetamine stimulant without a prescription in order to stay focused in studying all night for a test the next day?[30]

These questions suggest that an athlete who dopes and the person sitting next to you in your Social Problems class who uses Adderall nonmedically may not be all that different. Yet my guess is that you've never given much thought to their similarities. Here are two reasons why:

Since there's often an asterisk next to official written mention of Barry Bonds' home run records, why doesn't Casey Schwartz's name similarly carry public criticism?
Francis Specker/Alamy Stock Photo

1. Since winning is everything in sports, our society punishes athletes who take performance-enhancing drugs to gain an unfair advantage over the competition.
2. Since professional athletes are in the public eye for supposedly being role models to kids, they deserve to be singled out for unethical behavior.

Since many students feel anxious about speaking out in class, in what sense are those who take a beta blocker getting an unfair advantage over their classmates?
iStockphoto.com/monkeybusinessimages

But there's more to this picture. School is also a winner-take-all context, even though learning is supposed to be more important than getting the best grades. Moreover, while coffee may not make me more skilled in the classroom, it enhances my classroom performance by removing fatigue, grogginess, and other barriers to effective teaching. Because so many people drink coffee, take amphetamine stimulants, or ingest beta blockers nonmedically, they're personally invested in believing that their drug use completely differs from athletes doping. But is this a meaningful difference?[31]

When Second Best Feels Like Losing: The Medicalization of Imperfection

7.5 Explain how the medicalization of certain conditions supports the winner-take-all mindset.

Think of an advertisement you've seen recently that has stuck in your mind. Some ads seem unforgettable, which is exactly what advertisers intend. I can remember the first time I saw a commercial where Queen's iconic "We Are the Champions" provided the soundtrack. Several different guys with grins on their faces were jumping up and down throughout a suburban neighborhood. There was no dialogue, leaving me in suspense for a minute about the product being sold. Finally, in bold letters "VIAGRA" appeared on the screen. Ever since hitting the market in 1998, Viagra has been one of the most successfully marketed drugs in terms of both sales and brand familiarity. It's also been the brunt of innumerable jokes on late-night TV and has caused students of mine over the years to turn bright red when I make mention of it in class.[32]

The marketing of Viagra and other erectile dysfunction drugs has led many men to believe taking them is essential for their masculinity.
iStockphoto.com/RapidEye

Before the development of Viagra, Cialis, and other drugs to help men get and sustain an erection, being unable to do so was considered a natural part of the male aging process. Now this inability is a treatable, named condition: erectile dysfunction. The same is true for infertility. Throughout human history, some adults have always been unable to conceive children. It's only because of advances over the past several decades in the development of hormone therapy and fertility drugs that doctors and patients have come to see infertility as a disorder. Erectile dysfunction and infertility are examples of **medicalization**, which is the process by which a group of people experiencing the same health problem come to be seen as having a treatable disease. Because of medicalization, a person may regard natural human imperfections as conditions they *must* try to fix.[33]

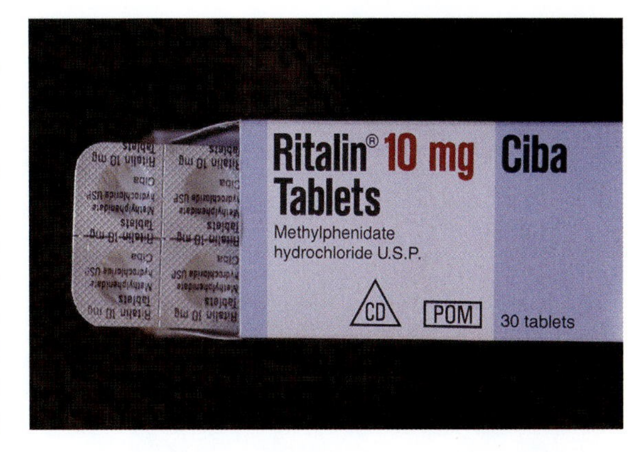

The development and marketing of Ritalin changed how people think about the symptoms that constitute ADHD. These symptoms went from being a natural and normal part of life to a deficit requiring pharmacological intervention.
Jenny Matthews/Alamy Stock Photo

The acceptability of drugs as remedies for human imperfections extends well beyond sex and reproduction. Consider ADHD treatments. The Swiss pharmaceutical company Ciba-Geigy, which manufactures Ritalin—the first drug of its kind—played a critical role in the creation of ADHD. Saying ADHD was "created" may seem like an odd way of putting it since people diagnosed with this disorder certainly suffer in serious ways. This choice of words is in no way meant to downplay the significant difficulties focusing and concentrating that someone who has ADHD experiences. My aim is instead to highlight that Ciba-Geigy, with support from many doctors, transformed public understanding of these difficulties and they became a treatable medical disorder.[34]

Another aspect of the medicalization of ADHD was its listing within the *Diagnostic and Statistical Manual of Mental Disorders* (*DSM*). Therapists consult this book to determine if a person has symptoms that fit the criteria for a particular psychiatric disorder. Originally called hyperkinetic impulse disorder, ADHD first appeared in the second edition of the *DSM* in 1968. In the 1980 edition it became known as attention deficit disorder, and then ADHD in 1987. Therapists diagnose the disorder and prescribe Adderall, Ritalin, or another amphetamine stimulant when they observe patients whose difficulties focusing and concentrating impair their ability to perform well at school or work, or in completing tasks of daily living (see Chapter 13 for a fuller discussion of the role of the *DSM* in the creation of ADHD).[35]

Not only do millions who suffer from ADHD obtain a prescription to take these drugs but, as we've seen, many people also do so nonmedically—particularly college students. Perhaps you or someone you know who doesn't have ADHD has still used amphetamine stimulants, maybe to pull an all-nighter and complete an upcoming assignment. The popularity of these drugs highlights the blurry line between a person using drugs to compensate for a deficit (which is acceptable because it produces greater equity) and using drugs to enhance performance (which is unacceptable because it offers an unfair advantage). Given that in a winner-take-all society being the very best carries far greater rewards than simply being a high achiever, people may see second best as akin to not performing well—aka failure. And since there are many drugs to fix this problem, some people may believe that taking them without a prescription is a way to compensate for a deficit rather than a form of cheating. It seems that a person's inability to be number one is increasingly becoming medicalized in American society.

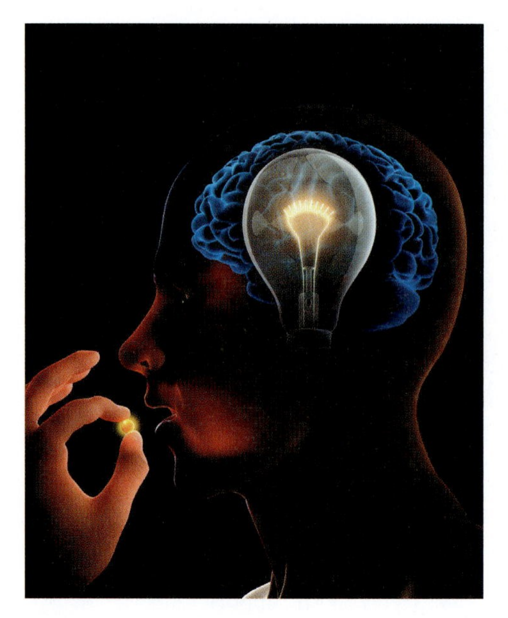

As the development of drugs that increase focus and attention becomes more advanced and their usage more common, in what sense is taking these drugs nonmedically becoming a *necessary* part of life?
JOSE ANTONIO PENAS/Science Source

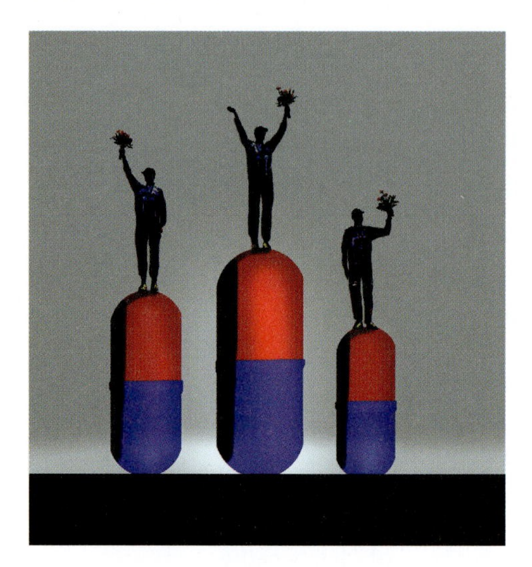

There is intense public outcry when athletes cheat because their behavior violates the ethos in sports that people should follow the rules of the game. But what does "fair play" mean when the incentive to cheat is built into the very rewards structure of athletic competition?
Gary Waters/Science Source

Game On: Seeing Cheating as a Way to Conform with Mainstream Values

7.6 Discuss how cheating can be understood as an affirmation of mainstream American values.

In the riveting documentary *Bigger, Stronger, Faster*, director Christopher Bell encourages audiences to think about the meaning of the **American Dream**—the belief that hard work and self-discipline enable people to get ahead. This belief continues to unite people even at a time of political polarization when so little else does. Bell tells the story of his two brothers' quest to become as bulked up as the bodybuilders they viewed as heroes while growing up: Arnold Schwarzenegger, Sylvester Stallone, and Hulk Hogan. All of these men doped, and so too did Bell's brothers. The film makes reference to the doping scandal in Major League Baseball and suggests that the drugs many nonathletes take to get ahead in academic and occupational contexts may similarly be forms of cheating.

The film's subtitle, *The Side Effects of Being American*, fittingly captures the central message in this chapter: Striving to be number one in a winner-take-all society can blur the distinction between doing what's right and doing what's best. When people appear to be sidestepping the rules, they're not being un-American but following a script in our culture for how to get ahead. Therefore, cheaters share something fundamental in common with those who abide by the old rules and play fairly. Indeed, because of the imperative to be the best in winner-take-all contexts, otherwise moral people may take shortcuts to success.

This discussion highlights that the sociological perspective enables us to see cheating differently than how we typically do. Becoming outraged about specific individuals who gain unfair advantages overlooks the many contexts in our society where people hear the message that such behavior is acceptable. You may be most familiar with the prevalence of this message in professional sports since doping by athletes has become a topic of broad interest over the past several years, even among those who are not sports fans. Yet cheating also occurs in many other contexts within our society. Embracing the sociological perspective enables us to recognize the similarities among all cheaters, regardless of how or where they take shortcuts to success.

Of course, not *everyone* is doing it. An example from the sports world is Ken Griffey Jr., who had one of the most illustrious careers in Major League Baseball history and was elected into the Hall of Fame in 2016. He twice hit fifty home runs in a season, but he never took performance-enhancing drugs. His story is particularly noteworthy given that he played from 1999 to 2010, during the peak of the steroids era. I'm sure you can think of others—perhaps people you know at school or at your job—who similarly play by the rules. Moreover, *you* may be committed to earning success via hard work and sacrifice rather than by cheating. I hope so!

However, the key takeaway is that just because many people like Ken Griffey Jr. attain extraordinary success by playing fairly doesn't mean those who don't are deviant. There are innumerable examples of ordinary people in varied walks of life who take shortcuts to success. Though we most often hear about baseball players and other athletes who use performance-enhancing drugs, we've seen in this chapter that cheating has become the national pastime for people across American society.

What Do You Know Now?

1. How is cheating a reflection of our winner-take-all society? Explain this concept and illustrate it by comparing the winner-take-all stakes within professional sports and in the educational system.
2. What do you think about people who take Adderall or other amphetamine stimulants in order to perform at a higher level instead of as these drugs are intended—to compensate for a medical deficit? Are they cheaters? Why or why not?
3. Consider two other situations where people take drugs: (1) to conquer stage fright and (2) to become more awake and alert. Discuss why you either view or don't view these instances of drug use as cheating.
4. Why is there public outrage toward athletes who take performance-enhancing drugs but not toward people who take drugs to get an edge at school or at work?
5. What evidence have you seen to indicate that at least some people view *not* performing well in a similar way as they view problems like erectile dysfunction and infertility—as problems they must medically fix?

Key Terms

Anabolic steroids 132
Winner-take-all society 135
Income inequality 135
Medicalization 144
American Dream 146

Visit **edge.sagepub.com/silver** to help you accomplish your coursework goals in an easy-to-use learning environment.

Notes

1. A complete transcript of Oprah Winfrey's January 2013 interview with Lance Armstrong can be found at https://armchairspectator.wordpress.com/2013/01/23/full-transcript-lance-armstrong-on-oprah/.
2. Lance Armstrong shared the details of his battle against cancer and comeback to competitive cycling in a summer 2006 article in *NIH MedicinePlus* available at https://medlineplus.gov/magazine/issues/summer06/articles/summer06pg6-9.html. The quote comes from the interview with Oprah Winfrey.
3. Connor Simpson, "Lance Armstrong Killed the Livestrong Bracelet," *The Atlantic*, May 28, 2013, http://www.theatlantic.com/business/archive/2013/05/nike-livestrong-lance-armstrong/314850/.
4. David Callahan, *The Cheating Culture: Why More Americans Are Doing Wrong to Get Ahead* (New York: Harcourt, 2007), 20.
5. Claudia L. Reardon and Shane Creado, "Drug Abuse in Athletes," *Substance Abuse Rehabilitation* 5 (2014): 95–105.
6. Gegor Aisch and K. K. Rebecca Lai, "Rio 2016," *New York Times*, August 18, 2016, https://www.nytimes.com/interactive/2016/08/18/sports/olympics/athletes-at-the-rio-olympics-who-were-previously-suspended-for-doping-.html.
7. BBC, "Winter Olympics 2018: 169 Russians Invited to Compete as Neutrals," January 27, 2018, https://www.bbc.com/sport/winter-olympics/42848595; Umair Irfan, "A Second Russian Olympian Has Failed a Doping Test," *Vox*, February 23, 2018, https://www.vox.com/2018/2/23/17044718/russia-bobsled-olympics-doping.
8. "50 Homeruns in a Season," *Baseball Reference*, 2014, http://www.baseball-reference.com/bullpen/50_Home_Runs_in_a_Season.
9. Allan Schwarz, "Baseball's Devil May Not Be in the Details," *New York Times*, February 10, 2008. Only slightly more than half of the eighty-nine players mentioned in a 2007 report about alleged use of performance-enhancing drugs improved their hitting or pitching after they began doping (see http://files.mlb.com/mitchrpt.pdf).
10. For reporting about athletes and performance-enhancing drugs, see Kate Zernike, "The Difference between Steroids and Ritalin Is. . .," *New York Times*, March 20, 2005, http://www.nytimes.com/2005/03/20/weekinreview/the-difference-between-steroids-and-ritalin-is.html. For a discussion of the public curiosity with celebrities who damage their reputations, see Orin Starn, *The Passion of Tiger Woods: An Anthropologist Reports on Golf, Race, and Celebrity Scandal* (Durham, NC: Duke University Press, 2011).
11. Philip J. Cook and Robert H. Frank, *The Winner-Take-All Society: How More and More Americans Compete for Ever Fewer and Bigger Prizes, Encouraging Economic Waste, Income Inequality, and an Impoverished Cultural Life* (New York: The Free Press, 1995).
12. Drew DeSilver, "U.S. Income Inequality, on Rise for Decades, Is Now Highest since 1928," Pew Research Center, December 5, 2013, http://www.pewresearch.org/fact-tank/2013/12/05/u-s-income-inequality-on-rise-for-decades-is-now-highest-since-1928/.
13. Chad Stone et al., "A Guide to Statistics on Historical Trends in Income Inequality," Center on Budget and Policy Priorities, May 15, 2018, https://www.cbpp.org/research/poverty-and-inequality/a-guide-to-statistics-on-historical-trends-in-income-inequality.
14. Callahan mentions these and other examples of the winner-take-all society throughout his book *The Cheating Culture*.

15. Quoted in Tom Verducci, "Totally Juiced," *Sports Illustrated*, June 3, 2002, http://www.si.com/mlb/2014/09/09/totally-juiced-tom-verducci-ken-caminiti-si-60. In honor of the magazine's sixtieth anniversary in 2014, it reposted Verducci's article as one of the sixty best over the magazine's history.
16. Quoted in Verducci, "Totally Juiced."
17. Richard Perez-Pena, "Studies Find More Students Cheating, with High Achievers No Exception," *New York Times*, September 7, 2012, https://www.nytimes.com/2012/09/08/education/studies-show-more-students-cheat-even-high-achievers.html.
18. Discussion of the two New York City schools comes from Callahan, *The Cheating Culture*, 196–203, 214–5.
19. Jennifer Medina, Katie Benner, and Kate Taylor, "Actresses, Business Leaders and Other Wealthy Parents Charged in U.S. College Entry Fraud," *New York Times*, March 12, 2019, https://www.nytimes.com/2019/03/12/us/college-admissions-cheating-scandal.html.
20. Callahan mentions these and other examples of cheating throughout *The Cheating Culture*.
21. Quoted in interview with Oprah Winfrey.
22. Callahan, *The Cheating Culture*, 179.
23. Donald Palmer and Christopher B. Yenkey, "Drugs, Sweat, and Gears: An Organizational Analysis of Performance-Enhancing Drug Use in the 2010 Tour de France," *Social Forces* 94, no. 2 (2015): 891–922.
24. Schwartz, "Generation Adderall," *New York Times Magazine*, October 12, 2016.
25. Ibid.
26. Ibid.
27. For a discussion of the winner-take-all stakes in journalism, see Callahan, *The Cheating Culture*, 87–8, 214–5.
28. Blair Tindall, "Better Playing through Chemistry," *New York Times*, October 17, 2004, https://www.nytimes.com/2004/10/17/arts/music/better-playing-through-chemistry.html.
29. Zernike, "The Difference between Steroids and Ritalin Is. . ."
30. Benedict Carey, "Brain Enhancement Is Wrong, Right?" *New York Times*, March 9, 2008, https://www.nytimes.com/2008/03/09/weekinreview/09carey.html; Alan Schwarz, "Workers Seeking Productivity in a Pill Are Abusing A.D.H.D. Drugs," *New York Times*, April 18, 2015, https://www.nytimes.com/2015/04/19/us/workers-seeking-productivity-in-a-pill-are-abusing-adhd-drugs.html.
31. Many sources praise people who take amphetamine stimulants nonmedically. For example, see Vicky Castro, "The 'Brain-Enhancing' Drugs That Are Sweeping Silicon Valley," *Inc.*, March 5, 2015, http://www.inc.com/vicky-castro/the-brain-enhancing-drug-thats-sweeping-silicon-valley.html; Melinda Wenner Moyer, "A Safe Drug to Boost Brainpower," *Scientific American*, March 1, 2016, https://www.scientificamerican.com/article/a-safe-drug-to-boost-brainpower/; and Margaret Talbot, "Brain Gain: The Underground World of Neuroenhancing Drugs," *New Yorker*, April 27, 2009, http://www.newyorker.com/magazine/2009/04/27/brain-gain.
32. Olivier Thibault, "Viagra Turns 20: Chronicle of a Global Success," *Medical Xpress*, March 26, 2018, https://medicalxpress.com/news/2018-03-viagra-chronicle-global-success.html.
33. Carl Elliott, "American Bioscience Meets the American Dream," *The American Prospect*, May 31, 2003, http://prospect.org/article/american-bioscience-meets-american-dream.
34. Peter Conrad, *The Medicalization of Society: On the Transformation of Human Conditions Into Treatable Disorders* (Baltimore, MD: Johns Hopkins University Press, 2007), 56.
35. Klaus W. Lange et al., "The History of Attention Deficit Hyperactivity Disorder," *Attention Deficit Hyperactivity Disorder* 2, no. 4 (2010): 241–55, https://www.ncbi.nlm.nih.gov/pmc/articles/PMC3000907/.

8

Living in Infamy

Mass Shootings as Enduring Expressions of Masculinity

Learning Objectives

1. Identify limitations in the view that mass shootings are caused by mental illness.
2. Describe how mass shootings reflect the social construction of gender.
3. Define the Columbine effect and explain its significance.
4. Compare mass shootings to riots.
5. Explain how media coverage contributes to the likelihood that mass shooters will live in infamy.

Making Sense of the Senseless: Exploring What Drives Mass Shooters to Commit Acts of Destruction

8.1 Identify limitations in the view that mass shootings are caused by mental illness.

Think of individuals who've committed atrocities so large that they will never be forgotten. One of the following names may come to mind: Adolf Hitler, Osama bin Laden, Joseph Stalin, Mao Zedong, or Jeffrey Dahmer. Adam Lanza also belongs on this list. On December 14, 2012, the twenty-year-old shot and killed his mother at home and then drove to Sandy Hook Elementary School in Newtown, Connecticut, where he murdered twenty children and six staff members before turning the gun on himself.

A vigil for victims of the shooting at Sandy Hook Elementary School in Newtown, Connecticut.
WENN Rights Ltd / Alamy Stock Photo

As a result of committing this heinous act, Lanza gained instant notoriety. The same has happened to the perpetrators of other rampages in the U.S. over the past several years—a burgeoning list that includes Omar Mateen, who killed forty-nine people in Orlando in 2016; Stephen Paddock, who gunned down fifty-eight people in Las Vegas in 2017; and Nikolas Cruz, who murdered seventeen people in Parkland, Florida, in 2018. Often for many days following these massacres, there's media coverage of every conceivable story angle, including the minute-by-minute unfolding of horror, the body count, and details of the shooter's personal life. Because of this coverage, the shooters—who often end their rampage by turning the gun on themselves—live on in infamy.

iStockphoto.com/waewkid

FIRST IMPRESSIONS?

1. Why do you think mass shootings receive so much media attention?
2. What do you believe are the reasons a person would initiate a mass shooting?
3. Why is committing an atrocity a pathway toward forever being remembered?

On sixteen occasions during his presidency, Barack Obama consoled loved ones of mass shooting victims. Here, he is singing "Amazing Grace" in memory of South Carolina state senator Clementa Pinckney, who was killed during a 2015 church rampage in Charleston.
Joe Raedle/Getty Images

The FBI defines a **mass shooting** as the murder of four or more people in succession with a firearm.[1] One such shooting occurs in the United States about every two weeks. Whereas our nation has 5 percent of the world's population, 31 percent of all such shootings happen here.[2] This chapter focuses on **public mass shootings**, which are the shootings often in the news and publicized on social media because they occur in places where anyone may happen to be at a given moment, such as schools, malls, workplaces, restaurants, airports, theaters, and houses of worship.

Because of how often public mass shootings occur, references to particular events may seem dated. Although it's a challenge to keep up with the latest data pointing to the scope of this social problem, a glimpse at the period from 1982 to 2016 offers a snapshot. During this period, there were eighty-five public mass shootings in the United States, with well over half (forty-eight) occurring after 2006. Since the early 1980s, deaths from public mass shootings have risen significantly (see Figure 8.1).[3] You may be familiar with some of these events, given the significant media coverage they've received (see Figure 8.2 on page 154). Throughout this chapter, I interchangeably use the terms *mass shooting*, *rampage*, and *massacre* to refer to these public tragedies.

In all but three of the eighty-five mass shootings between 1982 and 2016, the perpetrators were boys or men. Media coverage typically portrays these assailants through an individual perspective—as mentally ill males addicted to violent video games who are easily able to obtain firearms with a rapid killing capability. You may know males who fit this description; millions of boys and men in the United States do. Because very few of them go on rampages, the individual perspective offers a narrow portrait of the males who commit these heinous crimes. A study of thirty-seven mass shootings over a twenty-six-year period found that the perpetrators did *not* in fact fit a single profile.[4] Fewer than 5 percent of the males who initiate gun violence suffer from mental illness.[5] In most cases, the assailants also had little exposure to violent media. And while the availability of assault-style firearms obviously magnifies the destruction, we still need to understand what propels only certain boys and men to use these weapons to inflict mass destruction. To gain this understanding, we need to look beyond the media coverage.

FIGURE 8.1 • An Epidemic of Carnage

Over the past few decades, there has been a dramatic rise in the annual number of deaths from mass shootings.

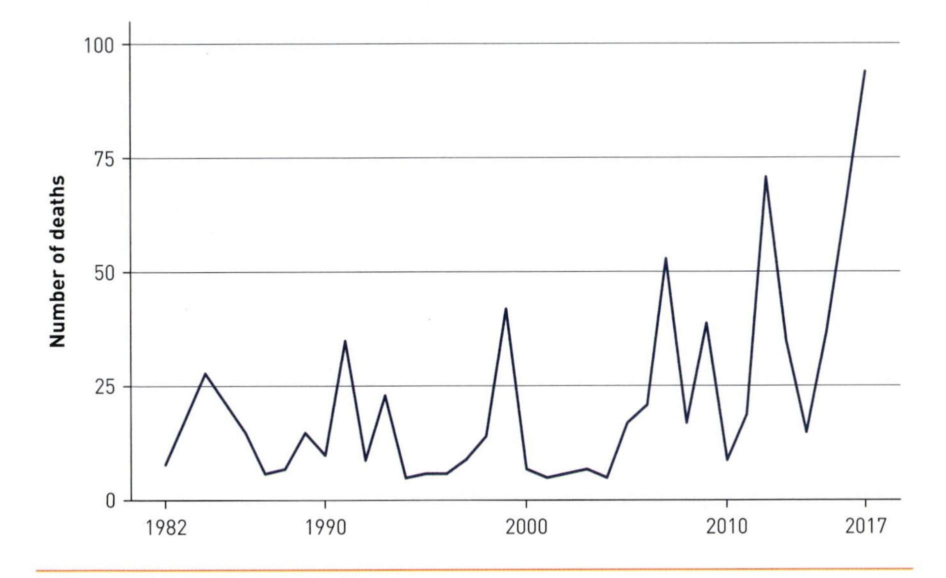

Source: Based on Mark Follman, Gavin Aronsen, and Deanna Pan, "U.S. Mass Shootings, 1982–2019: Data from *Mother Jones*' Investigation," *Mother Jones*, August 4, 2019, http://www.motherjones.com/politics/2012/12/mass-shootings-mother-jones-full-data.

There's certainly value in seeing mass shootings through an individual perspective. However, believing mental illness is the main cause stands in the way of fully recognizing what motivates the perpetrators. While their heinous acts lead reasonable people to view them as deranged and beyond the pale of humanity, what if they aren't necessarily madmen driven by the desire to kill as many people as possible? How might their indiscriminate violence actually *reflect*, rather than deviate from, social norms?

After the 2018 shooting at Marjory Stoneman Douglas High School in Parkland, Florida, several survivors spearheaded a national movement for gun control and thousands attended "March for Our Lives" demonstrations.
Citizen of the Planet/Alamy Stock Photo

Let's use the sociological perspective to investigate these questions, paying close attention to the most significant characteristic mass shooters share in common: their gender. This perspective explores how these rampages reflect traditional expectations our society places on males about how to be "real men." The fantasy gunmen have about living in infamy long after their

FIGURE 8.2 ● Rampage Nation

Here's a look at some of the deadliest mass shootings across the U.S. since 1999.

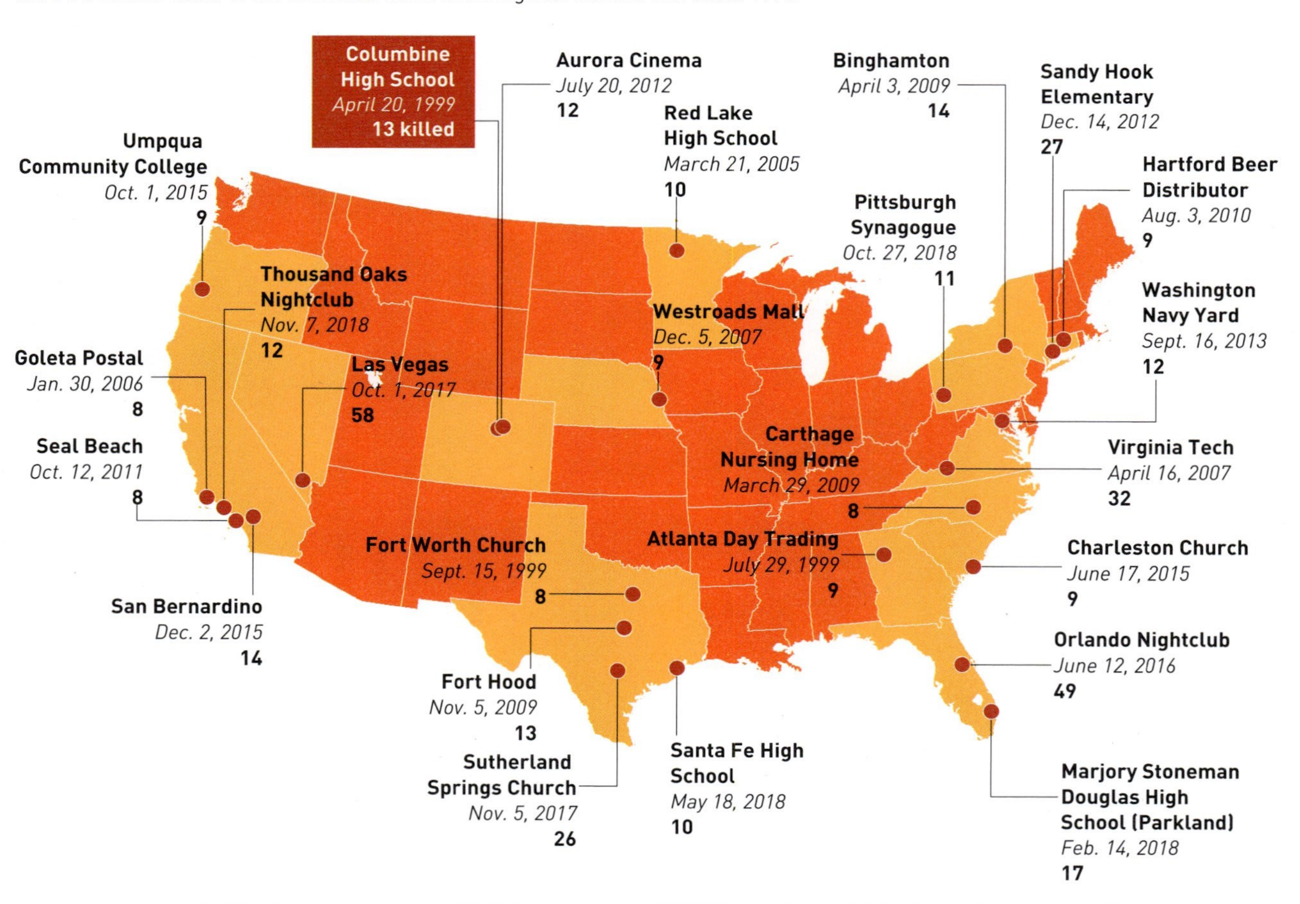

Source: AFP AFP/Newscom.

rampage is a crucial reason driving them to inflict widespread destruction. Doing so ensures that their final act in this world is so egregious that their power to carry it out will never be forgotten.

Seeking Revenge: Viewing Mass Shootings as Efforts by Disrespected Males to Get Even

8.2 Describe how mass shootings reflect the social construction of gender.

"People constantly make fun of my face, my hair, my shirts. I'm going to kill you all. You've been giving us shit for years."[6] Dylan Klebold uttered these chilling words in a video made the night before he and fellow Columbine High School senior Eric Harris massacred fifteen people at school in 1999, including themselves. Popular kids had often bullied the two boys. When Klebold undressed in the locker room for gym class, these kids teased him for having a deformity on his chest. In a book published a few years after the massacre, Klebold's friend, Brooks Brown, wrote,

A memorial at Columbine High School is dedicated to the fifteen people killed there in 1999.
iStockphoto.com/BanksPhotos

> At lunchtime, the jocks would kick our chairs, or push us down onto the table from behind. They would knock our food trays onto the floor, trip us, or throw food as we were walking by. When we sat down, they would pelt us with candy from another table. In the hallways, they would push kids into lockers and call them names while their friends stood by and laughed.[7]

A member of the football team acknowledged bullying Harris and Klebold: "But what do you expect with kids who come to school with weird hairdos and horns on their hats? It's not just the jocks; the whole school's disgusted with them. . . . If you want to get rid of someone, usually you tease 'em. So the whole school would call them homos."[8] Using homophobic slurs was a clear and intentional way of emasculating Harris and Klebold.

Fifteen years after the Columbine shooting, twenty-two-year-old Elliott Rodger killed six people in Isla Vista, near the campus of the University of California–Santa Barbara, and then fatally shot himself. In a video he uploaded to YouTube the day before, Rodger detailed his plans for the attack and explained why he was doing it:

> Well, this is my last video, it all has to come to this. Tomorrow is the day of retribution, the day in which I will have my revenge against humanity, against

> all of you. For the last eight years of my life, ever since I hit puberty, I've been forced to endure an existence of loneliness, rejection and unfulfilled desires all because girls have never been attracted to me. Girls gave their affection, and sex and love to other men but never to me.[9]

These words attest to Rodger's identification as an **incel**, an online subculture of heterosexual males who see themselves as victims because of their involuntary celibacy. They believe that since they deserve unlimited access to female bodies, violence toward girls and women is legitimate revenge for having been denied their entitlement to sexual pleasure.[10] For this reason, as part of his rampage Rodger targeted members of a sorority.

As with the Columbine and Isla Vista rampages, many other massacres have been in response to bullying or feelings of sexual rejection. Killing innocent people is, of course, never justified. Still, you may be able to relate to the level of pain and isolation these males experienced in the days and weeks before they acted out in rage. Because most people who feel marginalized don't go on rampages, it makes sense to believe those who do are psychologically troubled. Even though some indeed are, there's much more to their motivations than meets the eye.

Whereas *gunman* is a common word, *gunwoman* isn't.
iStockphoto.com/OSTILL

While girls and women also experience exclusion and/or suffer from mental illness, it's noteworthy that compared to men they rarely respond to their pain through violence (see Figure 8.3). Women are more inclined to talk about their emotions as a way to understand themselves and work toward feeling better. On the other hand, many boys learn to regard discussing feelings as a sign of weakness and view violence, alternatively, as how "real men" address pain. It's no coincidence that the people who've committed history's worst atrocities have all been men.[11]

To explain why boys and men perpetrate most violent crimes, let's turn the focus away from these crimes and think about the **social construction of gender**, which is the idea that a society views certain roles, but not others, as acceptable for individuals with particular types of bodies. The United States and many other parts of the world have a **gender binary** classification, whereby a person may legitimately identify as either male or as female and the valued characteristics of each identity lie in opposition to one another. These characteristics include the following:

- "Male": Strong, independent, assertive, dominant, aggressive
- "Female": Submissive, dependent, passive, sensitive, nurturing

Over your lifetime, the acceptance and inclusion of transgender people has increased in some pockets of American society. However, *transgender* is still not a legally recognized category. Moreover, from the very moment children are born, most parents socialize them as either "male" or "female." Besides

FIGURE 8.3 • Gender and Violence

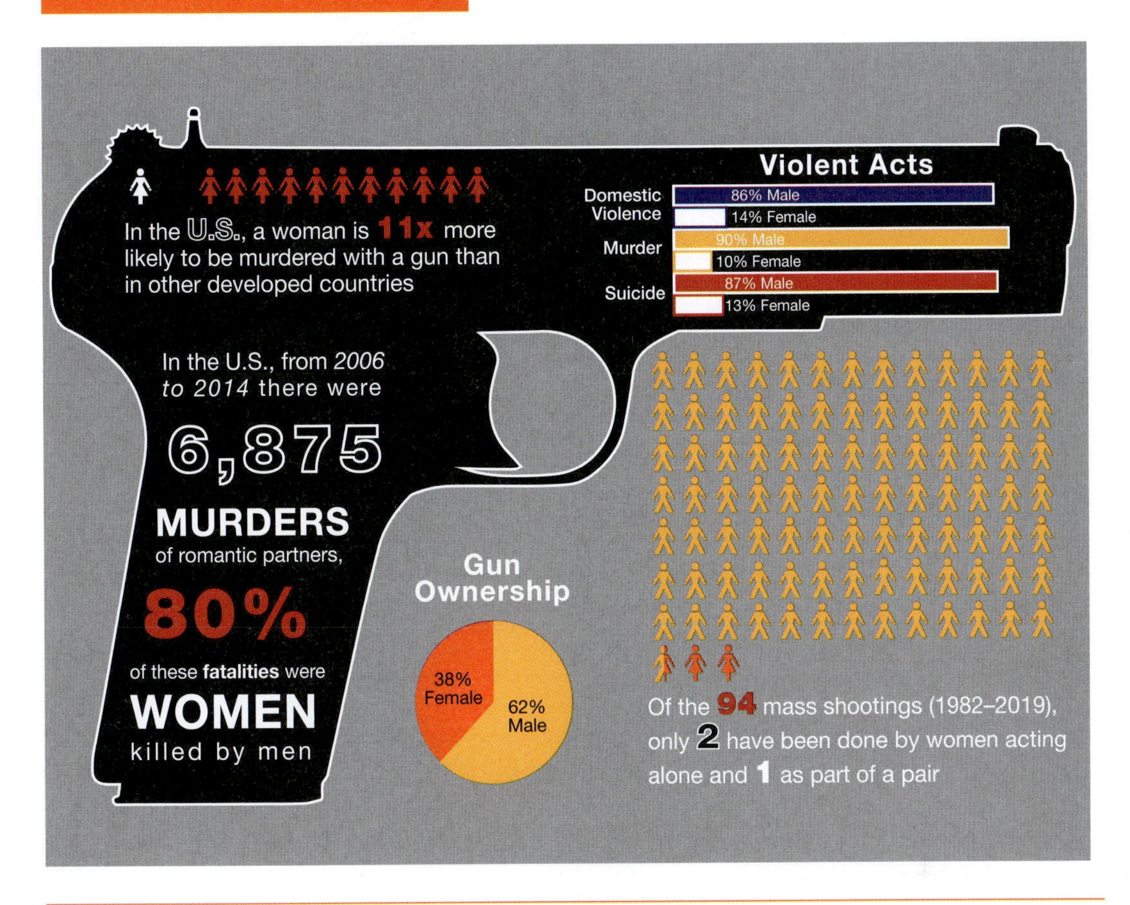

Source: Adapted from *Mother Jones* 2019, Gender and Gun Control, Pew Research Center, and Nonviolence NY.

within the family, can you think of other places in American society where people reinforce the gender binary?

The **Man Box** is a useful way to visualize the prevalent expectations in American society about what it means to be "manly" and, therefore, to see why males perpetrate most violent crimes (see Figure 8.4 on page 158). By thinking about the Man Box, you can begin to understand why some males derive social status by going on shooting rampages. These boys and men believe that their maleness hinges on acting abusively toward others and often toward themselves too—a view known as **toxic masculinity**.

Here's a jarring thought: Initiating a rampage is a way for males who feel emasculated to assert that they fit in and prove that they are "normal" men. For boys and men who closely identify with the Man Box, being presumed gay and bullied is a piercing source of disrespect that deserves revenge. Like the Columbine gunmen

FIGURE 8.4 ● The Man Box

Many males are consumed with trying to live up to rigid ideas about masculinity.

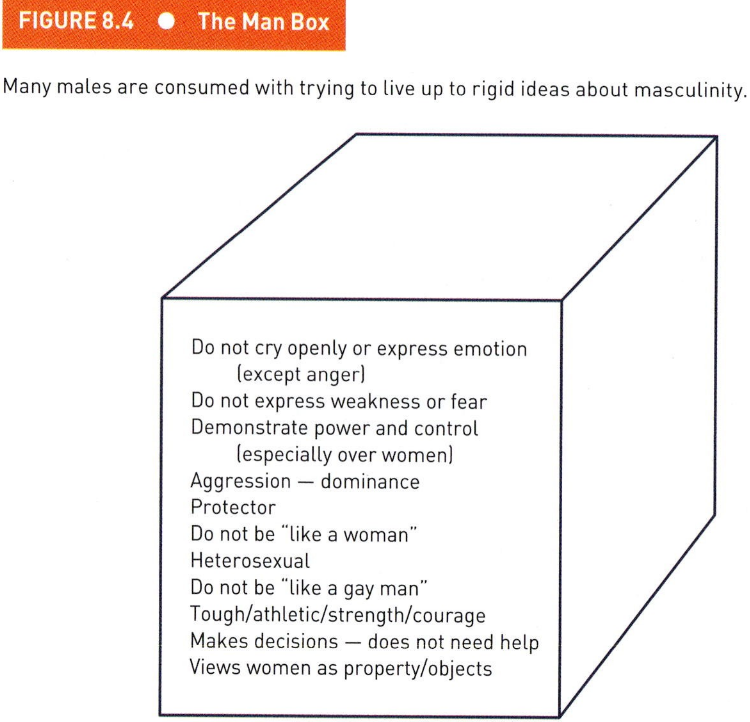

Source: Tony Porter, *Breaking out of the Man Box* (New York: Skyhorse Publishing, 2016).

Some countries, like Samoa, formally recognize a third gender. Pictured here are the Fa'Afafine.
Olivier CHOUCHANA/Gamma-Rapho via Getty Images

Dylan Klebold and Eric Harris, many other perpetrators of mass shootings were ridiculed with homophobic comments prior to initiating the massacre.[12] The same is true for some heterosexual males who experience rejection by females. A study of mass shootings over a thirty-year period found twenty-three cases resembling the Isla Vista rampage where the perpetrator specifically targeted girls and women as retribution for being denied sex. Therefore, it's crucial that we see mass shootings as efforts by certain boys and men to get even with those who have assaulted their masculinity. These massacres highlight the entitlement built into some heterosexual males' expectations about what it means to "be a man." This entitlement lies at the root of all gender violence (see Chapter 9).[13]

There are many places where boys and men get affirmation for toxic masculinity. When I was growing up, Sylvester Stallone played Rocky Balboa, a champion boxer.

In another popular role, he was Rambo, a former prisoner of war who turned violent as a civilian. Both films had multiple sequels crossing generations; the latest Rocky film was *Creed II* (2018), and the latest Rambo film was *Last Blood* (2019). Another icon of toxic masculinity is the video game Grand Theft Auto, which encourages virtual acts of violence against women. Players can fondle a stripper, kill a prostitute after sleeping with her, and rape other players.[14] If you're unfamiliar with this game, it may horrify you to learn such details. It's even more chilling to acknowledge that the option to engage in these sexist and violent behaviors is part of why the game is so popular. In 2018, many lists of the top video games placed Grand Theft Auto second only to Fortnite, which also glorifies male violence.[15] Boys who play these games are apt to grow up viewing aggression as a definitive part of who they are.

Can you think of other ways American culture glorifies toxic masculinity? This question underscores that males who go on rampages get reinforcement for their belief that they're entitled to address feelings of emasculation through violence. These males do not deviate from societal norms but rigidly embrace the characteristics most valued as masculine. In their comprehensive study of mass shootings, sociologists Rachel Kalish and Michael Kimmel wrote, "If young men are surrounded by messages telling them that real men are strong, tough, and violent, and that they do not back down to threats, then using lethal violence to prove one's masculinity is not only expected, it supports those very values."[16]

These massacres should sound an alarm in all of us, and not only because they cause so much destruction and tragedy. We must also recognize how this violence is symptomatic of an even larger, yet rarely acknowledged social problem: the violent expectations our society places on boys and men. While most of the males who encounter obstacles in their efforts to fit within

The Rambo films illustrate the validation our society gives to males who act violently.
TriStar/Getty Images

A childhood ritual for many American boys is to go with their fathers to buy firearms.
Arterra Picture Library/Alamy Stock Photo

Media images of violent males typically portray members of minority groups. By diverting attention from the reality that males of *all* races commit violent crimes, these images reinforce the sense of entitlement held by White boys and men.
iStockphoto.com/ephraimphotography

the Man Box suffer quietly, others are inclined toward desperate acts of retribution, including wreaking havoc on innocent people in order to make a powerful and enduring statement that they are "real men."

It's revealing that news coverage rarely explores the role of gender in these atrocities. Since males are expected to be aggressive ("Boys will be boys"), seemingly there's no reason to analyze why the perpetrators are typically male. Yet we're seeing that these rampages have *everything* to do with gender. Heterosexual males who feel disrespected because they are bullied as gay or rejected by girls are prone to seeing violence as a justifiable response. However, one manifestation of male superiority is their power to divert public attention from the gendered roots of violent crime. Because the alarming statistics in Figure 8.3 rarely receive media coverage, most people come to expect and often condone violent behavior in boys and men.[17]

The Columbine Effect: How Online Networks Enable Aggrieved Males to Plot the Next Rampage

8.3 Define the Columbine effect and explain its significance.

Of all the mass shootings that have occurred in the United States over the past several decades, the 1999 Columbine massacre stands out because it was the first to go viral. It received significant media coverage and became imprinted online in a way no prior rampage had. This was partially a matter of timing, as internet use had just started to grow significantly. The perpetrators, Eric Harris and Dylan Klebold, created a website where they posted videos of themselves testing their weapons and explained how the massacre would avenge the pain that popular kids at their high school had inflicted on them.

The dark web—sites that can't be found via traditional search engines and that require a secret key or password to access—are often where potential mass shooters learn how to outdo the carnage from past massacres.
Dmitry Molchanov/Alamy Stock Photo

In reporting about the online presence Harris and Klebold had cultivated, news coverage produced constructive public discussion about the boys' motives. But because the reporting also fueled other aggrieved males' interest in discovering what the two boys had posted online, the Columbine massacre created a blueprint for future rampages. Over the more than twenty years since Harris and Klebold carried out their rampage, there has been mounting evidence of what is known as the **Columbine effect**, or the tendency for gunmen to model their attack on the Columbine shooting. Many perpetrators have left behind evidence that they studied and imitated the tactics of Harris and Klebold (see Figure 8.5).[18]

There are now innumerable places online where people can share ideas about how to plan and carry out a mass shooting. On YouTube, for example, there's a

FIGURE 8.5 ● Copycat Killings

The 1999 Columbine massacre created a rampage blueprint that many aggrieved males over the years have followed.

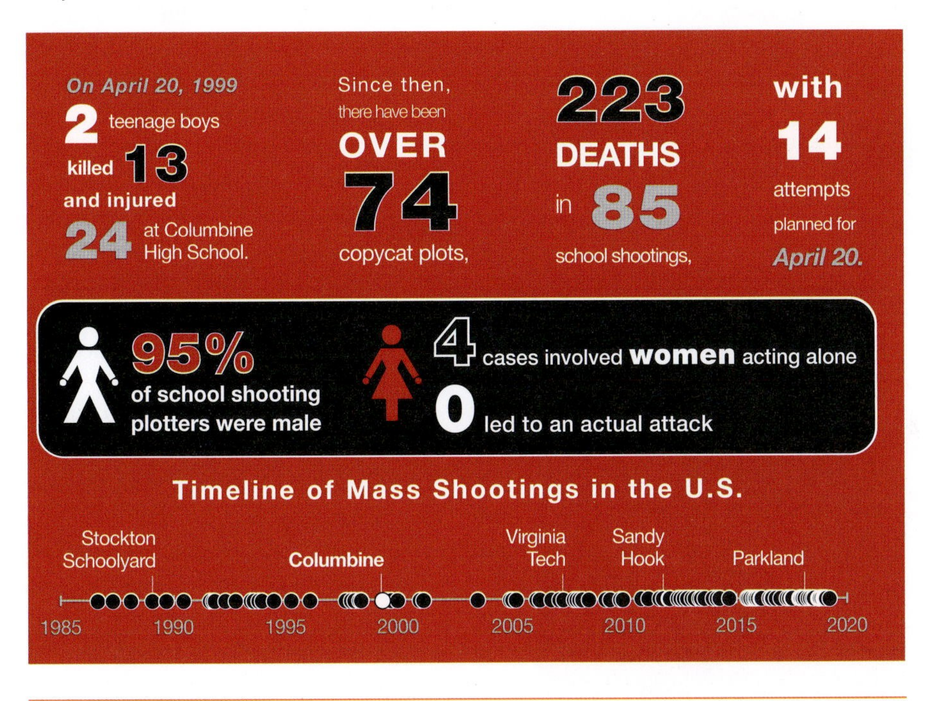

Source: Mark Follman and Becca Andrews, "How Columbine Spawned Dozens of Copycats," *Mother Jones*, October 5, 2015, https://www.motherjones.com/politics/2015/10/columbine-effect-mass-shootings-copycat-data.

network of people who produce and watch rampage videos.[19] Elsewhere, the sharing occurs via encrypted websites inaccessible by traditional search engines.

Online networks not only reinforce toxic masculinity but racism too. Attacks in recent years, particularly those on mosques, synagogues, and Black churches, indicate that a growing subset of mass shootings are motivated by **White supremacy**, which is the belief that Whites belong to the superior race and therefore are justified in subordinating people of other races.[20]

These online networks validate a script for future rampages that's available to anyone at any time. Therefore, it's not surprising that mass shootings often occur in clusters. In the weeks after the Columbine massacre, the National School Safety Center tracked at least three thousand similar threats. The copycat effect is most pronounced during roughly the thirteen days after a rampage.[21]

Elliott Rodger, the perpetrator of the 2014 rampage in Isla Vista, California, was part of the online mass shooter network. He left detailed descriptions of his motives

In 2019, a gunman killed twenty-two people at a Walmart in El Paso, Texas. Motivated by white supremacy, he wrote in a manifesto that his aim was to kill as many Hispanics as possible because they were infiltrating a country where they didn't belong.
Mario Tama/Getty Images

and tactics in a video he posted on YouTube and in a lengthy written manifesto. When I combed through media coverage of the shooting, I found no indication that Rodger explicitly modeled his tactics on the Columbine massacre. That's revealing since he carried out his attack fifteen years afterward and still the two shootings had many details in common. By the time Rodger went on his rampage, the specifics of Harris and Klebold's massacre had been so widely shared online and replicated in subsequent mass shootings that these details had become a familiar battlefield script. Rodger didn't have to reference Columbine directly in order to have been significantly influenced by it.

An aerial view of the huge media presence following the Columbine massacre. News coverage of mass shootings contributes to the awareness of future gunmen that carrying out a rampage will immortalize them.
AP Photo/Ed Andrieski

Here lies the power of viewing mass shootings through a sociological perspective. Over time, the impetus to initiate a rampage has been increasingly predicated on how an online community embracing toxic masculinity has validated this destructive behavior as an immortalizing way to prove one's mettle. This perspective enables us to recognize why we can't fully understand these tragedies just by focusing on assailants' individual characteristics, such as their troubled mental health, fascination with violent media, or desire to possess deadly weapons. We must also recognize how, in the years since the Columbine massacre, many males who do not necessarily share these characteristics have become connected to one another.[22]

For someone who's an active participant in the online mass shooter network, media coverage of a massacre may be *the* critical factor tipping them from thinking about following suit to actually plotting an attack. While the blueprint for doing so has existed online for many years, news reporting provides an impetus to formulate actual plans by spotlighting the latest shooter's expertise in carrying out a rampage. Whenever a person aspires to learn a skill, they're inclined to mimic others who've successfully honed it. There's something eerily "normal," therefore, about mass shooters. Just like you and me, they take cues from experts about how to accomplish their desired goals.[23]

Violence without Outrage: How Mass Shootings Resemble Riots

8.4 Compare mass shootings to riots.

Sometime before the end of the school year in 2014, seventeen-year-old John LaDue was planning a massacre at his Waseka, Minnesota, high school. Fortunately, a woman tipped off law enforcement about his suspicious behavior after she saw him trespassing through her backyard wearing an oversized backpack. During the ensuing three-hour interrogation, LaDue told police officers that he was making bombs at a local storage facility and had purchased several firearms. He kept a notebook under his bed where he detailed his plans. He would first shoot his parents and sister. Then he'd detonate a pressure cooker between periods at school. As people were fleeing, he would throw pipe bombs and Molotov cocktails. Finally, he'd kill himself.

LaDue defied the conventional wisdom about mass shooters. He wasn't mentally ill, didn't play video games or listen to violent music, and hadn't been bullied. The fact that he appeared to like his family made his intention to murder them the most chilling detail of all. His story reveals that a male no longer has to feel disrespected to plan a rampage. All he needs is a well-defined, socially validated script he's seen others put into action. We'd be overlooking this crucial point if we were to focus solely on the details of LaDue's plot, which was intended to closely resemble the destruction inflicted by many other mass shooters over the past few decades. The truth is that he differed in a crucial way from the other gunmen discussed in this chapter. Eric Harris, Dylan Klebold, and Elliott Rodger were all sufficiently aggrieved that their pain motivated them to exact revenge on the people who'd bullied or rejected them. LaDue had no such motive, and yet he was planning a significantly *more* destructive rampage than his predecessors had carried out.[24]

John LaDue wrote in his notebook that he was plotting a rampage so that he could live in infamy as a "real man."
iStockphoto.com/lolostock

To understand this seeming contradiction, we need to recognize how a person's decision to initiate a mass shooting resembles the impetus to participate in a riot. To explore this parallel, let's think about an amateur video that became national news a couple of months before I graduated from college. Shortly after midnight on March 3, 1991, a plumbing parts salesman awoke to police sirens outside his Los Angeles apartment and started recording the scene. His jarring footage showed four White police officers relentlessly beating a Black man named Rodney King after they'd stopped him for a traffic violation. A year later, after an all-White jury acquitted the officers on all criminal charges, the most destructive riot in the U.S.

A person didn't have to believe the 1992 L.A. riot was justified in order to have taken part in it.
Steve Grayson/WireImage

since the Civil War broke out on the streets of South-Central L.A. Fifty-eight people died and another 2,383 were injured. Damage to 1,100 buildings totaled $785 million.[25]

If in 1992 you'd been a young adult living in this L.A. neighborhood, would you have been one of the people to hurl a brick through a store window? Think for a moment before you answer. Consider that many of the thousands of individuals who took part in the mayhem were otherwise law-abiding citizens. It wasn't that their views suddenly changed and they now saw violence as acceptable. Rather, they participated because others' actions gave them the license to join in regardless of how they actually felt about what they were doing.

We can think of those who were the first to break windows, set fires, and loot property as *low-threshold* rioters. These people believed that the acquittal of the four police officers was so unjust that it demanded a violent response. Then there were those for whom it took this first group to act lawlessly for them to follow suit. If a riot were taking place all around you, can you imagine how you might participate even though doing so probably goes against your beliefs? The significance of this second wave of rioters is that their lawless behavior made it more thinkable for an even wider group of *high-threshold* bystanders to participate in the violence. This process continued as more and more people lost their inhibitions and started acting lawlessly. At these later stages in the evolution of the riot—if not sooner—I too would have joined the mayhem.[26]

Riots and mass shootings, of course, differ in significant ways. Millions of people believed that the acquittal of the four police officers who beat Rodney King was a social injustice, and therefore they could understand why the L.A. riot subsequently erupted. There is no similar justification for initiating a mass shooting. Moreover, whereas the L.A. riot reflected powerless people trying to right the abuses of those in authority (police), mass shooters are perpetrated by powerful people (heterosexual males) who believe they're entitled to act violently to uphold their privilege.

Despite these differences, the comparison between mass shootings and riots is useful. Like high-threshold rioters, John LaDue wasn't the sort of person who would have plotted a rampage unless many others had created a template for him to do so. Dozens of prior gunmen, including Eric Harris, Dylan Klebold, and Elliott Rodger, had paved the way and fueled his motivation to inflict even more destruction than they had. His aim was to follow their lead. He didn't need to feel emasculated or be vengeful in order to want to outdo them; all he needed was for them to have given him the license to plot the next massacre. LaDue's story reveals how the Columbine effect has changed over the years. "The problem [of mass shootings nowadays]," according to journalist Malcolm Gladwell, "is not that there is an endless supply of deeply disturbed young men who are willing to contemplate horrific acts. It's worse. It's that young men no longer need to be deeply disturbed to contemplate horrific acts."[27]

Males like John LaDue who don't feel aggrieved may still plan to carry out a violent act because they're part of a social network where others give them the license to do so.
Aelitta/Shutterstock.com

Had the police not foiled LaDue's plot to wreak havoc at his high school, the rampage would have occurred just a few weeks after Elliott Rodger's in California. Undoubtedly, hordes of journalists would have flocked to the crime scene and reported about copycat killings. But LaDue's plan to imitate prior gunmen was only the tip of the iceberg. Just as viruses often mutate as they spread, the same is true of the epidemic of mass shootings. It's clear from watching Rodger's YouTube video and reading his manifesto that he felt pained and vindictive. Like Harris and Klebold, he had a low threshold for violence and believed that initiating a rampage would be a way to inscribe his manly heroism in people's minds. LaDue, however, had a much higher threshold. He plotted a massacre not because he felt deeply disrespected but because the online mass shooter network had shown him that he could gain notoriety by following a tradition forged by so many prior gunmen.

Cementing Their Legacy as "Real Men": How Rampage Reporting Immortalizes Toxic Masculinity

8.5 Explain how media coverage contributes to the likelihood that mass shooters will live in infamy.

I bet some of the mass shooters mentioned in this chapter—Dylan Klebold and Eric Harris (Columbine), Elliott Rodger (Isla Vista), Adam Lanza (Newtown), Omar

These are some of the forty-nine people killed in the 2016 shooting at Pulse, an LGBTQ nightclub in Orlando, Florida.
iStockphoto.com/JannHuizenga

A year after the 2017 mass shooting in Las Vegas that killed fifty-eight people, a public exhibit was created that displayed portraits of the victims.
Ethan Miller/Getty Images

In dramatizing the devastation rampages create, news coverage masks the role it plays in enabling mass shooters to live in infamy.
Richard Levine/Alamy Stock Photo

Mateen (Orlando), Stephen Paddock (Las Vegas), and Nikolas Cruz (Parkland)—may have already been familiar to you. But how many recent Nobel Prize winners can you name? If you can't think of any, please don't feel badly. It's a commonality you and I share. This lack of knowledge isn't a personal shortcoming, but it indicates how news reporting reflects our society's value system. Journalists pay much greater attention to people who are destructive than to those who are productive. While getting an "A" in your Social Problems class is something to feel really good about, it's not going to get you on TV. Chillingly, opening fire in your classroom just might.[28]

To underscore that our society places higher value on disgrace than accomplishment, I've deliberately not included in this chapter photos of any mass shooters. Showing you their faces would further imprint into your memory the heinous acts they committed. Instead, I'd like you to think about the innocent people whose lives were cut short simply for being in the wrong place at the wrong time. In the wake of such tragedies, we often tell ourselves we'll never forget these victims and the agony they endured. That's easier said than done. The truth is we're likelier to remember the shooters since reporters dramatize their destruction.[29]

Not showing you photos of mass shooters highlights that this chapter has addressed a significant social problem that may have been invisible to you. I'm not just referring to mass shootings but also how our culture affirms the efforts of violent males to live in infamy. It wouldn't be surprising if you've never thought about the problem in this way before. The invisibility of this perspective is a by-product of how journalists typically report about rampages. By leading audiences to focus on the level of destruction, they divert attention from the role they play in immortalizing the perpetrators of these tragedies.

These gunmen are culturally astute. They know that by committing spectacular acts of violence, they'll forever remain in our minds. They're keenly aware that news sources will sensationalize their carnage, providing repeated play-by-play accounts of it. The shooters gain further affirmation within the online mass shooter network. Having the opportunity to get all this attention feeds their sense that the destruction they inflict, which often ends in suicide, will go down in history as an enduring sign of their masculinity. These shooters guarantee that their names will forever be attached to their heinous acts and that in their final moments on earth, they will have once and for all proven that they're "real men."[30]

The sociological perspective toward mass shootings sidesteps the thorny debate about Americans' constitutional right to bear arms. From this perspective, the root of these shootings is toxic masculinity, not access to guns.
iStockphoto.com/vicm

This sociological explanation builds upon the individual perspective, which views mass shootings as a reflection of untreated mental illness. Of course, mental illness is sometimes a factor in these shootings. However, as antiviolence activist and educator Jackson Katz points out, "Even if some of these violent men are or were 'mentally ill,' the specific ways in which mental illness manifests itself are often profoundly gendered."[31] In other words, mental illness manifests itself differently in males and females because of the social construction of gender. Further, Katz asks, "Why is depression in women much less likely to contribute to their committing murder than it is for men?"[32] The answer, we've seen, is that depressed males are likelier to have internalized toxic ideas about how to deal with their pain. Focusing on gender, therefore, is essential for getting to the root of mass shootings. Perpetrators are typically males because our society teaches them it's acceptable to address feelings of vulnerability through violence.

What Do You Know Now?

1. Based on your life experience, how accurately do you believe the Man Box concept captures societal expectations of what it means to be a "real man"? Discuss the sources of pressure that you believe are most influential in giving boys and men the feeling that they must try to fit within this box.
2. In what sense are mass shootings "normal" in that they reflect conformity to masculine expectations that each of us condones, if not affirms outright?
3. Why is the Columbine effect significant? How is John LaDue's foiled 2014 plot to blow up his high school an illustration of this effect?
4. What do mass shooters understand about our value system—and particularly how the news media uphold it—that enables them to inscribe their names in our minds forever while the people they victimize often become forgettable?

5. Given the news media's likelihood of enabling mass shooters to live in infamy, do you think reporters restricting how they cover rampages could diminish this likelihood? Why or why not?

Key Terms

Mass shooting 152
Public mass shooting 152
Incel 156
Social construction of gender 156
Gender binary 156
Man Box 157
Toxic masculinity 157
Columbine effect 160
White supremacy 161

Visit **edge.sagepub.com/silver** to help you accomplish your coursework goals in an easy-to-use learning environment.

Notes

1. The FBI defines *mass shooting* interchangeably with *mass murder* and *serial murder*; see *Serial Murder: Multi-disciplinary Perspectives for Investigators* (Washington, DC: Federal Bureau of Investigations, 2008).
2. Data about the frequency of mass shootings in the United States relative to other countries come from James N. Meindl and Jonathan W. Ivy, "Mass Shootings: The Role of the Media in Promoting Generalized Imitation," *American Journal of Public Health* 107, no. 3 (2017): 368–70. For data about the frequency of mass shootings, see Media Education Foundation, "What Percentage of Violence Is Committed by Men?" 2013, http://www.mediaed.org/handouts/WhatPercentageOfViolenceIsCommittedByMen.jpg.
3. Data about the frequency of public mass shootings are from an ongoing study by *Mother Jones* magazine that has tracked all such events since 1982; see Mark Follman, Gavin Aronsen, and Deanna Pan, "U.S. Mass Shootings: Data from *Mother Jones'* Investigation," *Mother Jones*, August 4, 2019, http://www.motherjones.com/politics/2012/12/mass-shootings-mother-jones-full-data.
4. The U.S. Secret Service and the U.S. Department of Education jointly conducted this study; see U.S. Secret Service and U.S. Department of Education, "The Final Report and Findings of the Safe School Initiative: Implications for the Prevention of School Attacks in the United States," 2004, https://www2.ed.gov/admins/lead/safety/preventingattacksreport.pdf. For an elaboration of how often mass shooters deviate from this individual profile, see Alex Mesoudi, "Mass Shooting and Mass Media: Does Media

Coverage of Mass Shootings Inspire Copycat Crimes?" *International Human Press*, 2013, http://dro.dur.ac.uk/12822/1/12822.pdf?DDD5+qhbg69+gpdg62+qhbg69+dul4eg.

5. Evidence about the rarity of mental illness among gunmen comes from Tage Rai, "The Myth That Mental Illness Causes Mass Shootings," *Behavioral Scientist*, October 13, 2017, http://behavioralscientist.org/myth-mental-illness-causes-mass-shootings/.
6. Quoted in Rachel Kalish and Michael Kimmel, "Suicide by Mass Murder: Masculinity, Aggrieved Entitlement, and Rampage School Shootings," *Health Sociology Review* 19, no. 4 (2010): 451–64. Quote appears on pages 452–3.
7. Quoted in Brooks Brown and Rob Merritt, *No Easy Answers: The Truth behind Death at Columbine* (Herdon, VA: Lantern Books, 2002), 50.
8. Quoted in Elliott Aronson, *Nobody Left to Hate: Teaching Compassion after Columbine* (New York: Henry Holt, 2001), 71–2.
9. The transcript of Elliott Rodger's video is from CNN, May 28, 2014, http://edition.cnn.com/2014/05/24/us/elliot-rodger-video-transcript/index.html. Rodger also wrote a 107,000-word manifesto that was widely circulated online. It can be found at https://www.documentcloud.org/documents/1173808-elliot-rodger-manifesto.html.
10. Jia Tolentino, "The Race of the Incels," *New Yorker*, May 15, 2018, https://www.newyorker.com/culture/cultural-comment/the-rage-of-the-incels.
11. For discussion of why many men regard conversation about feelings as not masculine, see Terrence Real, *I Don't Want to Talk about It: Overcoming the Secret Legacy of Male Depression* (New York: Scribner, 1997), 113–36.
12. For discussion of the link between antigay bullying and mass shootings, see Michael S. Kimmel and Matthew Mahler, "Adolescent Masculinity, Homophobia, and Violence: Random School Shootings, 1982–2001," *American Behavioral Scientist* 46, no. 10 (2003): 1439–58.
13. For discussion of mass shootings as a response to sexual rejection by females, see Jessie Klein, *The Bully Society: School Shootings and the Crisis of Bullying in America's Schools* (New York: NYU Press, 2012), 58–65.
14. Abby Johnston, "The Grand Theft Auto Rape Modification Is Disgusting," *Bustle*, August 22, 2014, https://www.bustle.com/articles/36896-the-grand-theft-auto-rape-modification-is-disgusting.
15. There is no single or definitive list of the top video games. Though different lists rank according to sales or fan-based popularity, in 2018 Grand Theft Auto was consistently ranked at #2 and Fortnite #1.
16. Kalish and Kimmel, "Suicide by Mass Murder," 458.
17. Jessie Klein, "Teaching Her a Lesson: Media Misses Boys' Rage Relating to Girls in School Shootings," *Crime, Media, Culture* 1, no. 1 (2005): 90–7.
18. Ralph W. Larkin, "The Columbine Legacy: Rampage Shootings as Political Acts," *American Behavioral Scientist* 52, no. 9 (2009): 1309–26; Nathalie E. Paton, "Media Participation of School Shooters and Their Fans: Navigating between Self-Distinction and Imitation to Achieve Individuation," in *School Shootings: Mediatized Violence in a Global Age*, eds. Glenn W. Muschert and Johanna Sumiala (Bingley, United Kingdom: Emerald Group Publishing Limited, 2012), 203–29.
19. Atte Oksanen, James Hawdon, and Pekka Räsänen, "Glamorizing Rampage Online: School Shooting Fan Communities on YouTube," *Technology in Society* 39 (2014): 55–67.
20. David Neiwert, "The New Age of Chain Terrorism: White Far-Right Killers Are Inspiring Each Other Sequentially," *Daily Kos*, April 29, 2019, https://www.dailykos.com/stories/2019/4/29/1853869/-The-new-age-of-chain-terrorism-White-far-right-killers-are-inspiring-each-other-sequentially?detail=emaildkre.
21. For discussion of the copycat effect after the Columbine massacre, see Jennifer L. Murray,

"Mass Media Reporting and Enabling of Mass Shootings," *Cultural Studies ↔ Critical Methodologies* 17, no. 2 (2017): 114–24, 120.

22. Selina E. M. Doran, "'You Made Me What I Am. You Added to the Rage': School Shooters in the United States and the Cultural Script of Vengeance," Unpublished manuscript, http://www.inter-disciplinary.net/wp-content/uploads/2011/06/SelinaDoranRevengepaper3.pdf, 4–5.
23. For an analysis of people's psychological tendency to mimic others who have the knowledge they seek, see Alex Mesoudi, "Mass Shooting and Mass Media: Does Media Coverage of Mass Shootings Inspire Copycat Crimes?" *International Human Press*, February 11, 2013, http://ithp.org/articles/mediacopycatshootings.html.
24. Malcolm Gladwell, "Thresholds of Violence: How School Shootings Catch On," *New Yorker*, October 19, 2015, https://www.newyorker.com/magazine/2015/10/19/thresholds-of-violence.
25. Richard I. Kirkland, "What Can We Do Now?" *Fortune*, June 1, 1992: 41–8.
26. Mark Granovetter, "Threshold Models of Collective Behavior," *American Journal of Sociology* 83, no. 6 (1978): 1420–43.
27. Gladwell, "Thresholds of Violence."
28. Jennifer Lynn Murray, "The Mass Killer's Search for Validation through Infamy, Media Attention and Transcendence," in *The Death and Resurrection of Deviance: Current Ideas and Research*, eds. Michael Dellwing, Joseph A. Kotarba, and Nathan W. Pino (New York: Palgrave, 2014), 235–52, 238.
29. Lionel Shriver, "Dying to Be Famous," *New York Times*, March 27, 2005, https://www.nytimes.com/2005/03/27/opinion/dying-to-be-famous.html.
30. Murray, "The Mass Killer's Search," 245.
31. Jackson Katz, "Memo to Media: Manhood, Not Guns or Mental Illness Should Be Central in Newtown Killing," *Huffington Post*, February 17, 2013, https://www.huffingtonpost.com/jackson-katz/men-gender-gun-violence_b_2308522.html.
32. Ibid.

iStockphoto.com/waewkid

9

#MeToo

Why Gender Violence Is Everyone's Problem

Learning Objectives

1. Describe how #MeToo has given publicity to forms of exploitation most people used to conceal.
2. Define different types of gender violence and illustrate how each reflects the perpetrator's sense of entitlement.
3. Identify why blaming survivors of gender violence is a way of further victimizing them.
4. Explain how gender violence is a by-product of conventional ideas about masculinity.
5. Recognize that while only certain people are to blame for gender violence, all of us are responsible for it.

Publicizing Trauma: How Social Media Has Brought Gender Violence out of the Shadows

9.1 Describe how #MeToo has given publicity to forms of exploitation most people used to conceal.

At 4:21p.m. on October 15, 2017, actress Alyssa Milano (@Alyssa_Milano) tweeted, "If all the people who have been sexually harassed or assaulted wrote 'Me too' as a status, we might give people a sense of the magnitude of the problem."[1] By the following morning, nearly forty thousand people had replied to her post.[2] Milano's call to action occurred ten days after the *New York Times* published an article detailing accounts by numerous women that Hollywood producer Harvey Weinstein had coerced them to perform malicious sexual acts. The platform #MeToo created has emboldened people to share similar accounts of exploitation, often inflicted by a friend, co-worker, or family member. Not surprisingly, the spotlight has been on people already in the public eye—a list that includes Louis C.K., Ryan Seacrest, Kevin Spacey, Bill Cosby, Matt Lauer, Al Franken, Charlie Rose, Morgan Freeman, Marshall Faulk, and Donald Trump. Although some people have come to these men's defense, for many of the accused the egregious behavior has ruined their careers and/or their reputations.[3]

#MeToo has exposed victimization often hidden in plain sight.
Ingimage Ltd/Alamy Stock Photo

Peter Cripps/Alamy Stock Photo

People of varying ages and backgrounds have posted #MeToo testimonials. Here's a sample from Twitter:

> I was 15 he was 24. I said stop! He kept going. The worst part it was my brothers friend and he believed him over me.
>
> Me too, he was 56 and I was 17.
>
> I was 9. . .
>
> Me too, he was my stepfather.
>
> #me too. More times than I can count.
>
> #ME TOO by several family members.
>
> Me too. Christmas 2010. We were colleagues. He's a doctor, I'm a nurse. Guess who had no choice other than to quit?

Each post gives visibility to mistreatment that people previously kept private and may have presumed was simply their own fault. During the year following Alyssa Milano's #MeToo tweet, this hashtag was used nineteen million times on Twitter—an average of 55,319 times per day.[4]

FIRST IMPRESSIONS?

1. Do you know anyone who has written a #MeToo post? If so, what feelings did the person have about making that post?
2. What sorts of replies did the post elicit, either on social media or in person?
3. Have you noticed ways that #MeToo posts have changed people's awareness of the types of humiliating behavior the posts expose? If so, how?

When boys and men experience gender violence, they may downplay their pain because of the pressure they feel not to appear "unmanly" by showing weakness.[6]
iStockphoto.com/skynesher

#MeToo posts characterize **gender violence**, harm inflicted by people in powerful positions that reinforces norms about appropriate male and female behavior. Heterosexual boys and men are typically the perpetrators of gender violence. Their actions illustrate **toxic masculinity**, which is the idea that being a "real man" hinges on acting abusively toward others, and often toward oneself too. "If you have a mother or a girlfriend or eyes," wrote journalist Moises Velasquez-Manoff, "it's hard not to be aware of the aggressive entitlement that many men feel toward women's bodies"—and toward the bodies of others with lesser power: girls, gay males, and transgender people.[5]

In addition to fueling gender violence, male entitlement also inhibits remorse about the pain such violence causes. This pain isn't necessarily physical. Consider the mental and emotional injuries women endure when heterosexual men expect them to act submissively or appear "sexy." These expectations can demean women by giving them the message that their worth hinges on how well they satisfy male desires and that personal qualities like leadership, intellect, and judgment don't matter.[7]

Tarana Burke used "me too" several years before anyone knew the seismic impact a hashtag could have.
dpa picture alliance/Alamy Stock Photo

Whereas inappropriate behavior by male celebrities has gotten media attention for many years, the #MeToo movement has uncovered the more typical examples of gender violence perpetrated by ordinary men. The limited statistical data that exist about victimization rates lend support to anecdotal evidence that minority groups experience gender violence more frequently than Whites (see Figure 9.1). Therefore, it's no wonder that it was a woman of color—Black civil rights activist Tarana Burke—who

FIGURE 9.1 ● Victimized Because of Their Gender

These are the percentages of females from different racial groups who have experienced sexual assault or rape.

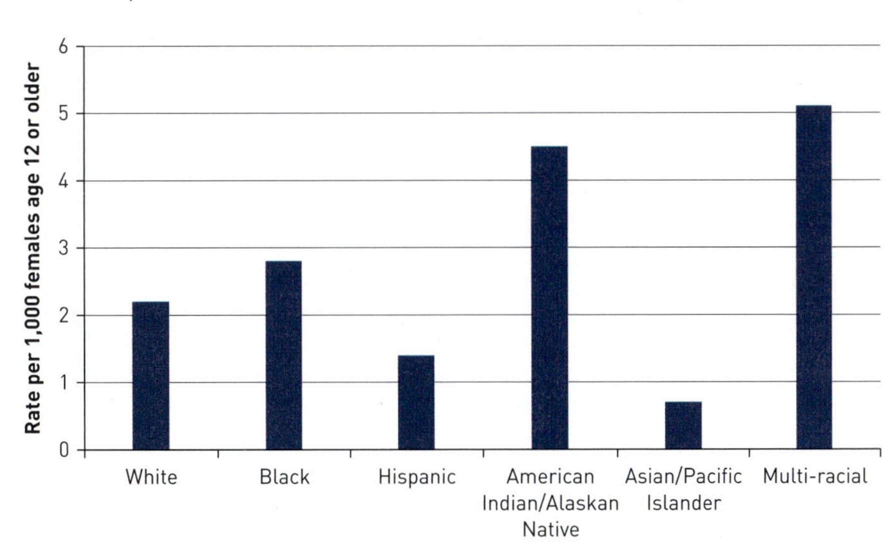

Source: Adapted from Michael Planty, Lynn Langton, Christopher Krebs, Marcus Berzofsky, and Hope Smiley-McDonald, "Female Victims of Sexual Violence, 1994–2010," Bureau of Justice Statistics, March 2013, https://www.bjs.gov/content/pub/pdf/fvsv9410.pdf.

first uttered "me too" publicly. White actress Alyssa Milano often gets the credit, however, because it was her doing so on social media that launched a social movement.[8]

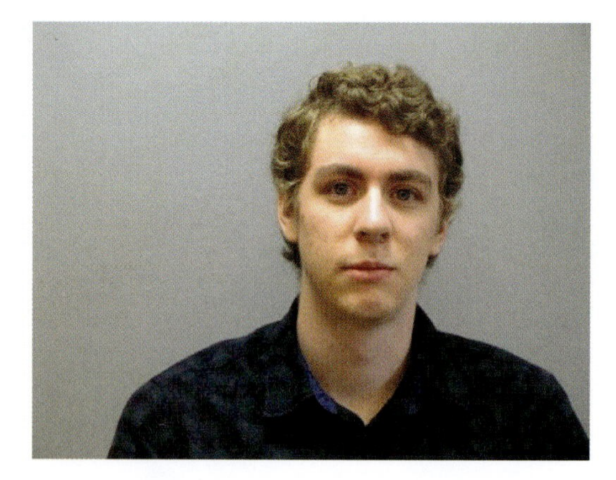

Viewing a man like Brock Turner as a monster is a convenient yet shortsighted way to make sense of sexual behavior that lies beyond the pale of human decency.
Greene County Sheriff's Office via AP, File

The individual perspective toward gender violence views the perpetrators as uniquely different from other males—as "damaged men." It's a compelling view for explaining the behavior of a guy like Brock Turner who behaves unconscionably. On the evening of January 18, 2015, two bikers discovered the nineteen-year-old Stanford University student forcing himself on a half-naked, unconscious woman outside his fraternity house. Since he acted alone, he's obviously the sole person to blame. It's convenient, moreover, to view gender violence as limited to a subset of deviant males like Turner. This individual perspective enables both women and the "good guys" who don't behave abhorrently toward females, gays, or transgender people to believe that they bear no responsibility for the problem.

This chapter takes a less convenient path. It explores various types of gender violence and exposes the physical, mental, emotional, and economic injuries they inflict. We'll see why assigning violent boys and men sole responsibility for this problem limits our understanding of why it occurs so frequently. By highlighting the social forces that lead some males to act in inappropriate—and sometimes criminal—ways toward individuals with less power, the sociological perspective reveals just how *many* people play a role in reinforcing the beliefs that lie at the root of gender violence. We'll see that such violence is not a deviation from social norms but a reflection of them.

Exhibiting Entitlement: Gender Violence as a Display of Male Power

9.2 Define different types of gender violence and illustrate how each reflects the perpetrator's sense of entitlement.

The ascendance of #MeToo came on the heels of #NeverthelessShePersisted, a movement that arose eight months earlier during the confirmation hearings for Jeff Sessions as attorney general. Senator Elizabeth Warren took issue with Sessions' record on civil rights and had the evidence in hand to support her case. When she started reading a letter written thirty years earlier by Martin Luther King Jr.'s widow, Coretta Scott King, Senate Majority Leader Mitch McConnell stopped her. He mentioned that he had warned her not to read the letter aloud, yet "nevertheless she persisted." A new hashtag was born. Supporters appropriated McConnell's words as a rallying cry for women victimized by men in powerful positions. #NeverthelessShePersisted has become a slogan of resilience for survivors of gender violence in all its various forms. *Survivor* is more fitting than *victim* as a way of conveying that, despite the many obstacles they face, a person can move beyond having been exploited.[9]

Many survivors of gender violence have embraced the slogan "Nevertheless She Persisted" as a way to publicize the wide spectrum of abuse they've experienced.
Shelly Rivoli/Alamy Stock Photo

Silencing

Women whose lives are far removed from politics can still relate to the dehumanization Elizabeth Warren endured. Across a range of workplaces, women are prone to **silencing**, men ignoring, censoring, or reprimanding them simply because they have spoken what's on their minds. An assertive man is seen as strong; a brazen woman as a bitch (see Figure 9.2). While men also experience workplace penalties if they come across as too abrasive, a woman's perceived worth to a company drops on average more than twice as much as a man's when others perceive her as exercising too much power. In addition to the risk of being

FIGURE 9.2 • Double Standard

Women who exhibit the very qualities at work that enable men to get ahead may face criticism and be denied promotion.

MEN ←		WOMEN →
Direct	=	Abrasive
Disruptor	=	Disruptive
Passionate	=	Emotional
Takes control	=	Bossy
Assertive	=	Pushy
Honest	=	Judgmental
Takes his time	=	Takes too long
Quick	=	Impulsive
Analytical	=	Needs to follow her gut
Follows his gut	=	Irrational
Expert	=	Show-off

Source: "Men v. Women Feedback Cheat Sheet," *The Cooper Review*, August 22, 2017, https://medium.com/conquering-corporate-america/men-vs-women-feedback-cheat-sheet-55655e659f53.

In school, girls typically participate in class less often and for shorter periods of time than boys.
iStockphoto.com/Sergio Lacueva

passed over for a raise or demoted, silencing can lead women to question their validity and self-worth.[10]

The forces that contribute to silencing take shape early in girls' lives and are most pronounced at school. In coed classrooms, they're likeliest to participate when the teacher's prompt calls for a brief response, whereas boys are more willing to make open-ended remarks. This inequality may contribute to girls' questioning the worth of their ideas and boys' growing up to become men who believe they're entitled to silence women like how Mitch McConnell cut off Elizabeth Warren mid-sentence while she was speaking on the floor of the Senate.[11]

Sexual Harassment

Any woman who has waited tables knows that even though being friendly is part of the job, making small talk can subject her to abusive treatment. If a customer does something offensive, should she speak up and risk losing her tip—or even worse, her job? An eye-opening *New York Times* investigation accompanied by a jarring video exposed the everyday mistreatment experienced by women who work for tips. It's all too common for customers to make crude comments, touch them, or proposition them for sex. Some men snap photos while waitresses are bending over to take their food orders and then post the pictures of exposed cleavage on social media. These are all examples of **sexual harassment**, which occurs when a person asserts their power over someone else by making sexual innuendos or unwanted advances.[12]

Since waitresses depend on tips, making small talk with customers subjects them to sexual harassment. Often they can't go to their boss for support because they risk getting fired for speaking out.
iStockphoto.com/SDI Productions

Sexual harassment most often occurs in the workplace. There are three contexts where the majority of women experience this mistreatment. The first, as we've seen, is restaurants, bars, and other settings where earnings mostly come from tips. The second is environments where women work as chambermaids, janitors, or in other low-wage service jobs. Women in these positions face a significant risk of sexual harassment because they work alone and have few opportunities to find other jobs. And finally, women in male-dominated occupations like construction or finance encounter colleagues who may harass them as a way of signaling that they shouldn't be doing "manly" work.[13]

Women in jobs where they publicly exhibit authority risk facing particularly malicious forms of sexual harassment. Consider Julia Gillard, who was president of

Australia from 2010 to 2013. In his campaign to unseat her from power, Tony Abbott stood in front of Parliament next to signs saying "Ditch the Witch" and "Man's Bitch."[14] Female journalists are another case in point. It's common for men to post comments on their social media pages that don't merely criticize their reporting but make crude references to their bodies or sexuality. Nearly half of female journalists routinely experience this sort of degradation. Women who write about gender violence, not surprisingly, receive the most threatening comments of all.[15]

Women who speak their minds on social media risk encountering sexual harassment, not for what they say but simply because they said it.
digitallife/Alamy Stock Photo

Given this persistent threat, women are much less likely than men to post comments on news sites, and they may resort to using pseudonyms when they do. Research indicates that commenters with female names comprise on average just a third—and in some cases, as little as 3 percent—of those who post most frequently. This means that when journalists turn to online forums for a snapshot of public opinion, women's voices and perspectives are significantly underrepresented.[16]

Street Harassment

A video that went viral of a woman walking around New York City on a typical workday shows many unfamiliar men making unprovoked comments about her body. During the time she traversed the city on foot, there were more than one hundred such comments. Many men also winked or whistled at her. The video depicts **street harassment**, which occurs when a person receives unwanted sexual attention in public from unfamiliar men. It's a form of sexual harassment that warrants its own discussion because of the many ways a person can feel violated in public space.

Because girls and women so often encounter street harassment, it would be easy to dismiss it as simply a fact of life. Doing so adds insult to injury by reinforcing the acceptability of this demeaning behavior.
iStockphoto.com/JackF

As Figure 9.3 indicates, most women have experienced one or more types of street harassment. Girls often encounter it for the first time at a young age. It may be overtly demeaning slurs about their bodies or phrases that seem harmless to the man saying them, such as "You look beautiful" or "Smile, baby." Yet such utterances are anything but innocuous. Since girls and women get bombarded with media messages telling them that their bodies exist for male pleasure, it's degrading for a stranger to say these types of things. Such comments can make girls and women feel that they do not have or deserve equal access to public space.[17]

Figure 9.3 (on page 180) reveals that males encounter street harassment too, though less frequently. Male victims often identify as LGBTQ. The most common

FIGURE 9.3 ● Publicly Humiliated

Women experience many types of street harassment, and with considerable frequency.

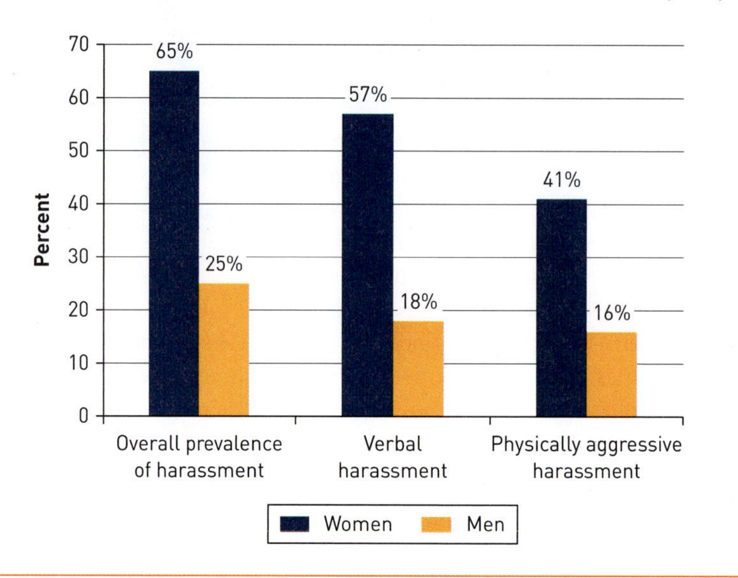

Source: Stop Street Harassment, "Statistics," 2019, http://www.stopstreetharassment.org/resources/statistics/.

type of harassment is homophobic or transphobic slurs, which reflect boys' and men's power to assert that it's unacceptable to deviate from **heteronormativity**, or the view that heterosexuality is the only legitimate expression of desire. Given how often girls, women, and LGBTQ people experience street harassment, it's no wonder they may feel unsafe in public.

Sexual Assault and Rape

A few years before the rise of #MeToo, Columbia University senior Emma Sulkowicz started carrying a fifty-pound mattress wherever they went—to class, the library, the dining hall, and out with friends. Emma explained in a video that this was a fitting final project for their visual arts degree. Emma's aim was to demonstrate the trauma of **sexual assault**, which occurs when a person asserts their power by physically making sexual advances on someone else without their consent, and **rape**, which occurs when the sexual coercion involves oral, anal, or vaginal penetration.

The mattress was the same size as the ones in Columbia's dorms, where, Emma claimed, a male acquaintance raped them. Carrying the mattress gave Emma public acknowledgment of their trauma, which few sexual assault survivors receive. Given that deep feelings of shame cause many survivors to hide the pain, Emma got the chance to share a personal story of exploitation and raise awareness about a problem that afflicts one in five college women in the U.S. and one in four gender-nonconforming students like Emma. These numbers are even higher than the research indicates since many survivors don't report their experiences. You may be familiar with these statistics as they're often publicized on college campuses.[18]

Emma had the mattress with them during the entire school year and walked across the stage with it at graduation to receive their diploma.
Andrew Burton/Getty Images

Because these numbers are also the subject of debate, we need to look more closely at the realities of sexual assault. Let's probe the validity of the one-in-five statistic concerning college women. First, think about the role of drinking. If you've ever had a few too many at a party, you know that alcohol can blur a person's judgment. This can mean some women waking up with the knowledge that what they supposedly agreed to do the night before while drunk was actually rape. And second, consider false accusations. A 2014 *Rolling Stone* article chronicled the account of a University of Virginia student who claimed that at a party several fraternity brothers had gang-raped her as part of an initiation ritual. There was uproar when a follow-up investigation discredited key details of her story.[19]

Why do some people believe sexual assault can't take place if drinking is involved?

iStockphoto.com/Django

Intoxication and fabricated victimization are topics that may have crossed your mind when you've heard about campus sexual assault and rape. These topics foster doubt in some people's minds about the severity of the problem. But they shouldn't. In truth, these issues are smokescreens; they divert attention from the underlying truth about consent: It's a black and white issue. If a person hasn't explicitly communicated in the moment that they're interested in sex, they're not—period.

The "Yes Means Yes" movement promotes that only by saying yes does a person give consent. This movement is involved in a campaign to pass laws holding perpetrators accountable for making sexual advances on people unable to give their consent because alcohol has impaired their ability to do so.
HelloWorld Images/Alamy Stock Photo

Let's first consider the role of alcohol. Research indicates that a significant reason sexual assault is so prevalent on college campuses is because the power imbalance between men and women escalates when one or both people have been drinking. As a guy consumes more alcohol, the likelihood that he will make sexual advances on a woman without her consent increases. And the more intoxicated he is, the more forceful his behavior is likely to be.[20]

Although misreporting by *Rolling Stone* of a gang rape occurring at the University of Virginia's Phi Kappa Psi fraternity received significant publicity, false allegations of sexual assault are actually rare.
Jay Paul/Getty Images

Likewise, don't let publicized cases of fabricated sexual assault accounts mislead you into thinking the problem is exaggerated.[21] It not only runs rampant but 95 percent of college women who experience sexual assault don't report it to their resident assistant or campus police because they feel ashamed about what happened. They may believe others will discredit their story if they were intoxicated or will blame them for underage drinking. They also have reason to think there's no point in coming forward because their rapist will face little accountability. Recall the story earlier of Brock Turner, who assaulted an unconscious woman outside his fraternity house at Stanford University. He served just three months in prison, and oftentimes perpetrators of sexual assault receive no punishment at all.[22]

Of the relatively few women who do go public with an allegation of sexual assault, only about 5 percent base their case on fabricated evidence. Because false accusations unfairly stain men's reputations, they're certainly a serious injustice. Yet this injustice shouldn't be our primary concern here. The truth is that the overwhelming majority of sexual assault cases involve women making *legitimate* claims.[23]

Intimate Partner Violence

Ray Rice was one of the best running backs in the National Football League during his six-year career, but that's not why his name is familiar to many people. The Baltimore Ravens terminated Rice's contract after a video went viral of him assaulting his fiancé, Janay Palmer, in the elevator of an Atlantic City casino. **Intimate partner violence** occurs when one person in a romantic relationship uses their power to weaken, shame, and humiliate their lover (see Figure 9.4). The violence isn't necessarily physical; for example, one partner may continually insult the other, withhold essential financial support, or make the other feel and believe they're crazy (a tactic known as **gaslighting**). *Intimate partner violence* is a more fitting term than *domestic violence* since, as Rice's actions indicate, the victimization may occur in public as well as at home. On any given day across the U.S., hotlines for survivors of intimate partner violence receive about twenty thousand calls.[24]

Whereas Ray Rice's violent behavior toward Janay Palmer got enormous media attention, most people who abuse their romantic partners aren't celebrities and there are no security cameras to document the violence.
Kenneth K. Lam/Baltimore Sun/MCT via Getty Images

FIGURE 9.4 • Displaying Entitlement

Whether the abuse occurs physically, mentally, emotionally, or verbally, intimate partner violence reflects the perpetrator's belief that they have the right to exercise control over their partner.

Source: National Center on Domestic and Sexual Violence, "Power and Control Wheel," http://www.ncdsv.org/images/PowerControlwheelNOSHADING.pdf.

One in three heterosexual women experiences intimate partner violence during her lifetime. This problem occurs in all types of relationships. Gays and lesbians who internalize homophobia may project that bias onto their partners by making disparaging comments or inflicting physical force. Transgender people are especially vulnerable to abuse from a lover. Nearly half of them experience it, typically when their partner is not also trans. The partner may withhold hormonal medication from the person who's transitioning, sabotage their ability to pay for gender confirmation surgery, criticize them for not being a "real" man or woman because their genitals don't match their gender identity, or threaten to reveal their partner's birth-assigned sex to coworkers, family members, or others who may be judgmental. All instances of intimate partner violence reflect power imbalances between the people involved in the romantic relationship.[25]

Adding Insult to Injury: Why Blaming Gender Violence Survivors for Their Victimization Contributes to the Problem

9.3 Identify why blaming survivors of gender violence is a way of further victimizing them.

At a time when accounts of gender violence are frequently in the news and posted on social media, it may be hard to imagine that for most of American history this social problem wasn't a topic of public concern. That changed because of several developments initiated by feminists beginning in the 1960s: the founding of the first shelter for survivors of intimate partner violence, the passage of laws that allowed courts to consider abuse as grounds for divorce, and the creation of rape crisis centers for survivors of sexual assault. #MeToo is just the latest development in a massive movement spanning many decades to publicize the pain and humiliation that survivors of gender violence experience.[26]

Over the past half century, activism publicizing the traumatic effects of gender violence has been part of a wider effort to highlight everyday indignities women experience.
Diana Walker/Liaison

Even though speaking openly about gender violence gives visibility to the victimization, something is still often missing from the ways people talk about this social problem. Consider a statement like "Mary is a battered woman." The focus is entirely on Mary; there's no mention of the person who hurt her. Notice how simple it is to erase the perpetrator from public discussion of intimate partner violence. When people initially speak about this issue, they might say something like "John beat Mary." But then, they're quick to move to the passive voice: "Mary was beaten by John." This diverts attention from John to Mary, making it easy for John to drop out of the sentence entirely: "Mary was beaten." Mary's identity is now detached from John's actions, prompting us to say, "Mary is a battered woman." How might the perpetrator similarly get erased from conversation about sexual assault?[27]

Removing perpetrators from discussions of gender violence absolves them of responsibility and may lead people to ask questions like "How was she dressed at the party?" This question implies that women are solely responsible for what happens to them in public and therefore undeserving of protection from danger or support in the wake of harm. It's no wonder survivors may not report incidents of gender violence; they often don't believe they'll receive validation. On the other hand, some people may cast blame on survivors who don't speak out because not doing so ensures that the criminal justice system cannot hold perpetrators accountable. It seems survivors are damned if they do and damned if they don't.[28]

Sexual assault survivors often fear that bringing a criminal case against their rapist will lead judges, juries, lawyers, or police to accuse the survivor of having contributed to their own victimization.
iStockphoto.com/RichLegg

Exposing the sociological reasons for survivors' silence reveals why it's shortsighted to blame these people for their own victimization. In a study of sexual assault survivors who didn't report the crime, respondents indicated that they felt doing so would have intensified the shame the assault produced in them and led to reliving the pain and humiliation, or **revictimization**. They believed going to the authorities would have made them seem blameworthy for not having been sufficiently mindful of how they dressed, spoke, or behaved.[29]

There's a similar story concerning intimate partner violence. Some people might wonder why a person in danger doesn't just leave. The story is never that simple. For starters, a survivor may not necessarily see their partner's violent behavior as out of the ordinary, given the widespread belief in our society that violence is a legitimate expression of masculinity. Therefore, survivors may be willing to put up with mistreatment. They may also be financially incapable of leaving if they have no immediate way to support themselves otherwise. The most significant barrier to escaping from an abusive partner is fear. Survivors may stay if they view doing so as the lesser of two evils. Fleeing may put them in even greater danger if their partner finds them, which is a real possibility given the vast array of

From a sociological perspective, survivors of intimate partner violence rarely flee from their abuser because the terror they're experiencing constrains their ability to make this choice.
Zoonar GmbH/Alamy Stock Photo

tracking apps that can help in locating a missing person. There's a risk that if the person is found, the abusive partner will become even more violent toward them.[30]

"Be a Man": Violence as a Socially Acceptable Expression of Masculinity

9.4 Explain how gender violence is a by-product of conventional ideas about masculinity.

The summer before #MeToo swept across American society, my family visited China. Among the many reasons I found it to be a fascinating trip, one that surprised me was the hordes of motorcycles whizzing by on city streets. This wasn't something I'd read about in a guidebook, nor would have expected to find noteworthy. Yet I was mesmerized and couldn't stop thinking about motorcycles long after returning home. It certainly wasn't because I wanted to saddle up on one and hit the open road. After all, I don't drive due to my visual impairment. In retrospect, it seems obvious that I fixated on the two-wheeled vehicles dotting the streets of Beijing and Guangzhou because this scene starkly differed from the familiar scene in the U.S. of men riding solo. I frequently saw Chinese women driving motorcycles. Sometimes couples rode with one or more children in tow on a scooter hardly bigger than a bicycle.

Riding a Harley-Davidson doesn't cause a guy to act violently. However, the popularity of Harleys reflects how American society encourages boys and men to embrace a set of behaviors that may include violence.
iStockphoto.com/ermess

After the trip, I started to connect the dots and came to see how these differences expose the roots of gender violence. In many cultures—certainly in the U.S., though perhaps not in China—motorcycles aren't simply a mode of transportation. They also reflect valued masculine traits like independence, risk-taking, and toughness. Moreover, just about everywhere American boys spend time—at home, in school, on the sports field, or hanging out amongst friends—they hear the message that violence is "manly." Receiving positive reinforcement for trying to win at all costs or for relentlessly exhibiting strength teaches them it's acceptable to assert power over others. It's no wonder that the boys and men who perpetrate gender violence have internalized toxic masculinity, believing their manliness hinges on objectifying girls and women. These males are particularly inclined to become violent when educational attainment, occupational success, and other legitimate expressions of status are unattainable to them.[31]

It would be convenient to downplay the many ways American culture teaches males that violence is "manly" and to attribute responsibility for rape or sexual harassment solely to the individuals who perpetrate these crimes. When you think about gender violence, it makes sense if the images that come to mind are of men who whistle at women walking down the street or who hit their girlfriends. After all, these guys are the public face of this social problem and fit the conventional wisdom that the only people responsible for it are the perpetrators.[32]

Violence isn't an inborn masculine trait. Boys grow to value it based on the activities adults encourage them to pursue.
iStockphoto.com/jacoblund

Sometimes I find myself embracing this individual perspective. When I hear about incidents of gender violence, I take pride in knowing I'm someone who would never use my power in these malicious ways. Yet sociology has taught me to widen my understanding of this social problem by recognizing the invisible ways my life connects to broader forces in our society. As a result, I now understand that *anyone* may reinforce the cultural messages that underlie gender violence. I also acknowledge just how much validation I receive merely for being a man who doesn't exploit others. The bar for being a "good guy" is indeed pretty low.

I gained this sociological wisdom during college when I started thinking back on the period of my life between roughly ages ten to seventeen. During those years, I spent several hours each week glued to the TV cheering on the New York Jets. They usually lost, but I enjoyed watching them do battle against the other team. Although I no longer closely follow professional football, I still tune in once a year for the Super Bowl. Doing so validates the spectacle of men violently tackling—and sometimes brutally injuring—one another. The truth is I support an activity that reinforces toxic masculinity.[33]

The role of cheerleaders is to foster team spirit. But they also give off the message that football's entertainment value hinges on the objectification of women's bodies.
Scott Halleran/Stringer/Getty Images

Beyond the head-butting that takes place on the field, there are many other ways that professional football legitimizes gender violence. What do boys learn when they see cheerleaders with lots of cleavage provocatively moving their bodies to inspire fans to root for teams comprised entirely of men? Or when the few women announcers and the many women in beer commercials dress in ways that cater to male heterosexual desires? What messages might fans take away in knowing that one of the few players the league has punished for intimate partner violence—Ray Rice—just so happens to be one of the only players whose abusive behavior got caught on video?[34]

Football is hardly the only setting in U.S. culture where people reinforce ideas that lie at the root of gender violence. There's lots of other evidence of how people who don't directly perpetrate this social problem still indirectly contribute to it.

For example, consider the millions of males who listen to Rush Limbaugh, the host of one of the most popular talk-radio shows in the country. Here are some of the stances he's taken on the air:

- When legal rights activists Sandra Fluke tried to impress upon Congress in 2012 that health insurance plans must cover birth control, Limbaugh called her a "slut" and a "prostitute."[35]
- He questioned why people made such a big deal over Ray Rice's assaulting Janay Palmer. According to Limbaugh, "Well, how bad could it have been if she said yes to the proposal? How bad could the guy be if she went ahead and married him?"[36]
- After Ohio State University instituted a policy that people must explicitly give one another consent before becoming sexually involved, he commented, "How many of you guys in your own experience with women have learned that no means yes, if you know how to spot it?"[37]

Despite the offensiveness of these statements, it's not simply the case that listening to Rush Limbaugh causes a person to demean women. I have a friend whose father is a kind, loving, church-going man and also an avid fan of the show. I doubt he's ever behaved abusively toward his wife, to whom he's been married for over fifty years. However, the takeaway is that the support my friend's father and countless other listeners give to Limbaugh's ideas validates toxic masculinity, which lies at the root of gender violence in American society.

Critics often highlight the crude lyrics in hip-hop because pointing out violent themes in songs written and/or performed by Black men reinforces public beliefs that these men are dangerous (see Chapter 3 for a larger discussion of these beliefs).
iStockphoto.com/innovatedcaptures

A more relatable example is listening to popular music. It's easy to find songs that contain lyrics depicting gender violence. Here's a sampling:

- "Put Molly all in her champagne, she ain't even know it. I took her home and I enjoyed that, she ain't even know it." (Rocko, "U.O.E.N.O.")
- "Tryin' to send the b--ch back to her maker. And if you got a daughter older then fifteen, I'mma rape her. Take her on the living room floor, right there in front of you. Then ask you seriously, whatchuwanna do?" (DMX, "X Is Coming")
- "Now I gotsa to give your mother----kin a-- a beatin. I punched her in the ribcage and kicked her in the stomach. Take off all my mother----kin jewelry, b--ch runnin. I stomped her and I kicked her and I punched her in the face." (Kool G, "Hey Mister Mister")

While these songs are all by rappers, hip-hop certainly isn't unique among musical genres in validating toxic masculinity.

If you start paying attention to lyrics of songs that have been popular during your lifetime, you'll see that many different musical genres reinforce toxic masculinity. A favorite song of mine before I started listening closely to the words was the 2014 hit "Animals" by Maroon 5: "Baby I'm preying on you tonight. Hunt you down eat you alive. Just like animals." Likewise, the title of Robin Thicke's dance party favorite "Blurred Lines" tells boys and men it's okay to view the murkiness of sexual consent while drinking as entitling them to use female bodies as they so desire: "I know you want it. You're a good girl. Can't let it get past me. You're far from plastic alright. Talk about getting blasted. I hate these blurred lines. I know you want it." Because there are so many songs like these, it's easy to sing along without even noticing that they validate toxic masculinity.[38]

Music you've grown up with isn't unique in its validation of toxic masculinity. Consider the message in these lyrics from the 1954 hit song "Baby Let's Play House," which Elvis Presley covered: "Now listen to me, baby. Try to understand. I'd rather see you dead, little girl, than to be with another man."
Gary Null/NBC/NBCU Photo Bank via Getty Images

If you listen to popular music, tune in to Rush Limbaugh, or watch football, you support forms of popular culture that condone males' mistreatment of women as a way of proving they're "real men." I'm not blaming you but asking you to become aware of, and take responsibility for, your actions. I too give license to toxic masculinity, even though my intention—both in this book and in my teaching—is to foster awareness about the roots of gender violence. Many people who repudiate such violence may still validate toxic masculinity in ways they haven't considered. Their validation underscores that gender violence stems from *shared* ideas about the meaning of masculinity. The responsibility for this social problem extends far beyond the guys who whistle at women walking down the street or drop drugs in women's drinks at bars and then rape them.

Seeing Gender Violence as Everyone's Problem

9.5 Recognize that while only certain people are to blame for gender violence, all of us are responsible for it.

In Joe Ehrmann's eyes, the most destructive words a boy hears while growing up are "Be a man." It may surprise you that Ehrmann, who played professional football for thirteen years, would hold this view. He appeared to epitomize the strength and toughness that American society prizes in men—towering above 6′0″, weighing over 250 pounds, and intimidating quarterbacks with his blindside hits. But looks can deceive. When his teenage brother developed terminal cancer, Ehrmann felt emotionally unequipped to handle the trauma. Over the next several years, his conception of

Although Joe Ehrmann's teammates told him not to cry after his brother's death, trying to "man up" only intensified his sadness.
George Gojkovich/Getty Images

what it meant to be a man radically changed. He's now a motivational speaker who highlights how the ideas boys learn about manliness lead some of them to become violent toward others as well as inflict damage on themselves.[39]

Ehrmann's message reminds us that getting to the root of gender violence hinges on widening the lens beyond perpetrators. The males who make crude comments to women or take advantage of them while drunk are hardly the only ones responsible for this malicious behavior. Perhaps you've never before considered how your *own* beliefs or actions contribute to silencing, sexual harassment, street harassment, sexual assault, rape, or intimate partner violence. If not, you now have reason to look inward. The impetus for doing so is high. Who among us doesn't know someone—a friend, family member, or coworker—who's experienced one or more of these types of exploitation? #MeToo arose as a way to publicize the epidemic of gender violence that plagues American society. The sociological perspective exposes the value in expanding the meaning of this hashtag to encompass people's collective role in the persistence of this social problem. So who's responsible for gender violence? Most people, including me too.

What Do You Know Now?

1. For each of the types of gender violence discussed in this chapter, how do the perpetrator's actions reflect entitlement?
2. Why is it shortsighted to blame gender violence survivors for contributing to their own victimization?
3. What evidence does the chapter offer to indicate that gender violence stems from widely shared ideas about what it means to be a man?
4. How might your own behaviors reinforce the masculine norms that lie at the root of gender violence?
5. In a TED Talk, activist Tony Porter commented, "My liberation as a man is tied to your liberation as a woman." Based on what you learned in this chapter, what do you think Porter means?

Key Terms

Gender violence 174
Toxic masculinity 174
Silencing 177
Sexual harassment 178
Street harassment 179
Heteronormativity 180
Sexual assault 180
Rape 180
Intimate partner violence 182
Gaslighting 182
Revictimization 185

Visit **edge.sagepub.com/silver** to help you accomplish your coursework goals in an easy-to-use learning environment.

Notes

1. Alyssa Milano's tweet and the replies to it can be found at https://mobile.twitter.com/Alyssa_Milano/status/919659438700670976.
2. This figure is reported in Heidi Stevens, "#MeToo Campaign Proves Scope of Sexual Harassment, Flaw in Mayim Bialik's Op-ed," *Chicago Tribune*, October 16, 2017, http://www.chicagotribune.com/lifestyles/stevens/ct-life-stevens-monday-me-too-mayim-bialik-1016-story.html.
3. *USA Today* compiled a list of 150 famous men who were the targets of sexual misconduct allegations during the four-month period between October 2017 and January 2018 at https://www.usatoday.com/pages/interactives/life/the-harvey-weinstein-effect/. The story that broke the Harvey Weinstein scandal is Jodi Kantor and Megan Twohey, "Harvey Weinstein Paid Off Sexual Harassment Accusers for Decades," *New York Times*, October 5, 2017, https://www.nytimes.com/2017/10/05/us/harvey-weinstein-harassment-allegations.html.
4. Monica Anderson and Skye Toor, "How Social Media Users Have Discussed Sexual Harassment since #MeToo Went Viral," Pew Research Center, October 11, 2018, https://www.pewresearch.org/fact-tank/2018/10/11/how-social-media-users-have-discussed-sexual-harassment-since-metoo-went-viral/.
5. Moises Velasquez-Manoff, "Real Men Get Rejected, Too," *New York Times*, February 24, 2018, https://www.nytimes.com/2018/02/24/opinion/sunday/real-men-masculinity-rejected.html.
6. Heather R. Hlavka, "Speaking of Stigma and the Silence of Shame: Young Men and Sexual Victimization," *Men and Masculinities* 20, no. 4 (2017): 482–505.
7. Dacher Keltner, "Sex, Power, and the Systems That Enable Men Like Harvey Weinstein," *Harvard Business Review*, October 13, 2017, https://hbr.org/2017/10/sex-power-and-the-systems-that-enable-men-like-harvey-weinstein.
8. For discussion of more frequent gender violence victimization among women of color than White women, see Tarana Burke, "#MeToo Was Started for Black and Brown Women and Girls. They're Still Being Ignored," *Washington Post*, November 9, 2017, https://www.washingtonpost.com/news/post-nation/

wp/2017/11/09/the-waitress-who-works-in-the-diner-needs-to-know-that-the-issue-of-sexual-harassment-is-about-her-too/?utm_term=.e3f7db614170.

9. Susan Chira, "Elizabeth Warren Was Told to Be Quiet. Women Can Relate," *New York Times*, February 8, 2017, https://www.nytimes.com/2017/02/08/opinion/elizabeth-warren-was-told-to-be-quiet-women-can-relate.html.
10. Marianne Cooper, "For Women Leaders, Likability and Success Hardly Go Hand-in-Hand," *Harvard Business Review*, April 30, 2013, https://hbr.org/2013/04/for-women-leaders-likability-a; Kathy Caprino, "Gender Bias Is Real: Women's Perceived Competency Drops Significantly When Judged as Being Forceful," *Forbes*, August 25, 2015, https://www.forbes.com/sites/kathycaprino/2015/08/25/gender-bias-is-real-womens-perceived-competency-drops-significantly-when-judged-as-being-forceful/#3278d0cc2d85.
11. Mary W. Bousted, "Who Talks? The Position of Girls in Mixed Sex Classrooms," *English in Education* 23, no. 3 (1989): 41–51; Judith Baxter, "Jokers in the Pack: Why Boys Are More Adept than Girls at Speaking in Public Settings," *Language and Education* 16, no. 2 (2002): 81–96.
12. Catrin Einhorn and Rachel Abrams, "The Tipping Equation," *New York Times*, March 12, 2018, https://www.nytimes.com/interactive/2018/03/11/business/tipping-sexual-harassment.html.
13. Chai R. Feldblum and Victoria A. Lipnic, "Select Task Force on the Study of Harassment in the Workplace," U.S. Equal Employment Opportunity Commission, June 2016, https://www.eeoc.gov/eeoc/task_force/harassment/report.cfm.
14. James Massola, "Julia Gillard on the Moment That Should Have Killed Tony Abbott's Career," *Sydney Morning Herald*, June 23, 2015, https://www.smh.com.au/politics/federal/julia-gillard-on-the-moment-that-should-have-killed-tony-abbotts-career-20150622-ghug63.html.
15. James Gardner, "Nearly Half of Women Journalists Face Abuse: Survey," *iheartradio*, November 24, 2017, http://www.iheartradio.ca/cfax-1070/news/nearly-half-of-women-journalists-face-abuse-survey-1.3461275; Rachel Schallom, "Women in Public-Facing Journalism Jobs Are Exhausted by Harassment," *Poynter*, June 28, 2018, https://www.poynter.org/news/cohort-women-public-facing-journalism-jobs-are-exhausted-harassment.
16. Fiona Martin, "Getting My Two Cents Worth In," International Symposium on Online Journalism, April 15, 2015, https://isojjournal.wordpress.com/2015/04/15/getting-my-two-cents-worth-in-access-interaction-participation-and-social-inclusion-in-online-news-commenting/.
17. Stop Street Harassment, "Unsafe and Harassed in Public Spaces: A National Street Harassment Report," April 2014, http://www.stopstreetharassment.org/wp-content/uploads/2012/08/National-Street-Harassment-Report-November-29-20151.pdf.
18. For research confirming the one-in-five statistic, see Charlene L. Muehlenhard et al., "Evaluating the One-in-Five Statistic: Women's Risk of Sexual Assault while in College," *Annual Review of Sex Research* (2017): 549–76. For research supporting the one-in-four statistic, see Jake New, "The 'Invisible' One in Four," *Inside Higher Ed*, September 25, 2015, https://www.insidehighered.com/news/2015/09/25/1-4-transgender-students-say-they-have-been-sexually-assaulted-survey-finds.
19. The false accusation at the University of Virginia is reported in Sabrina Rubin Erdely, "A Rape on Campus: A Brutal Assault and Struggle for Justice at UVA," *Rolling Stone*, November 19, 2014. After the discrediting of the student's story, *Rolling Stone* removed the article from its website. An online version can still be found at http://web.archive.org/web/20141119200349/http://www.rollingstone.com/culture/features/a-rape-on-campus-20141119.

20. Antonia Abbey et al., "The Relationships between the Quantity of Alcohol Consumed and the Severity of Sexual Assaults Committed by College Men," *Journal of Interpersonal Violence* 18, no. 7 (2003): 813–33.
21. For discussions of how rarely false accusations occur relative to credible accounts of sexual assault, see Joanne Belknap, "Rape: Too Hard to Report and Too Easy to Discredit Victims," *Violence against Women* 16, no. 12 (2010): 1335–44.
22. Emanuella Grinberg and Catherine E. Shoichet, "Brock Turner Released from Jail after Serving 3 Months for Sexual Assault," CNN, September 2, 2016, https://www.cnn.com/2016/09/02/us/brock-turner-release-jail/index.html.
23. Belknap, "Rape: Too Hard to Report."
24. National Coalition Against Domestic Violence, "Statistics," https://ncadv.org/statistics.
25. Ibid.; Colleen Styles-Shields and Richard A. Carroll, "Same-Sex Domestic Violence: Prevalence, Unique Aspects, and Clinical Implications," *Journal of Sex and Marital Therapy* 41, no. 6 (2015): 636–48; Adam F. Yerke and Jennifer DeFeo, "Redefining Intimate Partner Violence: Beyond the Binary to Include Transgender People," *Journal of Family Violence* 31 (2016): 975–9.
26. For discussion of how the feminist movement gave visibility to gender violence, see Susan Schechter, *Women and Male Violence: The Visions and Struggles of the Battered Women's Movement* (Cambridge, MA: South End Press, 1982), 53–80 and Chad Sniffen, "History of the Rape Crisis Movement," California Coalition against Sexual Assault, November 1, 2009, https://www.sfwar.org/pdf/History/HIST_CALCASA_11_09.pdf.
27. This discussion is based on a TED Talk by antiviolence educator Jackson Katz. He draws on the work of linguist Julia Penelope to show how everyday talk about gender violence erases the perpetrator from our minds.
28. Belknap, "Rape: Too Hard to Report," 1335.
29. Chelsea Spencer et al., "Why Sexual Assault Survivors Do Not Report to Universities: A Feminist Analysis," *Family Relations* 66, no. 1 (2017): 166–79; Belknap, "Rape: Too Hard to Report," 1336–8.
30. Karen G. Weiss, "'Boys Will Be Boys' and Other Gendered Accounts: An Exploration of Victims' Excuses and Justifications for Unwanted Sexual Contact and Coercion," *Violence against Women* 15, no. 7 (2009): 810–34; Jinseok Kim and Karen A. Gray, "Leave or Stay? Battered Women's Decision after Intimate Partner Violence," *Journal of Interpersonal Violence* 23, no. 10 (2008): 1465–82; J. Michael Cruz, "'Why Doesn't He Just Leave?' Gay Male Domestic Violence and the Reasons Victims Stay," *Journal of Men's Studies* 11, no. 3 (2003): 309–23.
31. Joshua Robert Adam Maynard, "Between Man and Machine: A Socio-historical Analysis of Masculinity in North American Motorcycling Culture," Master's Thesis, Queen's University, 2008, http://qspace.library.queensu.ca/bitstream/handle/1974/1198/Maynard_Joshua_RA_200805_MA.pdf?sequence=1&isAllowed=y; Esperanza Miyake, "No, It's Not My Boyfriend's Bike," *New York Times*, August 4, 2018, https://www.nytimes.com/2018/08/04/opinion/sunday/women-on-motorcycles.html; Mark Totten, "Girlfriend Abuse as a Form of Masculinity Construction among Violent, Marginal Male Youth," *Men and Masculinities* 6, no. 1 (2003): 70–92.
32. Jackson Katz, *The Macho Paradox: Why Some Men Hurt Women and How All Men Can Help* (Naperville, IL: Sourcebooks, 2006), 20–1, 28.
33. Jesse A. Steinfeldt et al., "Bullying among Adolescent Football Players: Role of Masculinity and Moral Atmosphere," *Psychology of Men and Masculinity* 13, no. 4 (2012): 340–53.
34. Teresa C. Younger, "The NFL Must Stop Promoting Sexism—and Start Promoting More Women," *Huffington Post*, September 18, 2014, https://www.huffingtonpost.com/

teresa-c-younger/the-nfl-must-stop-promoting-sexism_b_5836738.html.

35. Sarah O'Leary, "Gender Slurs: Why What Rush Limbaugh Said Is So Damning to Women," *Huffington Post*, March 13, 2012, https://www.huffingtonpost.com/sarah-oleary/why-what-rush-l_b_1337183.html.
36. Quoted in "Rush Limbaugh on Ray Rice Domestic Abuse: 'How Bad Could It Have Been?'" *Sports Illustrated*, July 29, 2014, https://www.si.com/nfl/2014/07/29/rush-limbaugh-ray-rice-domestic-violence.
37. Quoted in Rex H. Huppke, "Sorry, Rush Limbaugh, but No ALWAYS Means No," *Chicago Tribune*, September 17, 2014, http://www.chicagotribune.com/news/opinion/huppke/chi-limbaugh-consent-nfl-20140917-story.html.
38. For a discussion of the racism in singling out gender violence themes in rap, see Amy Binder, "Constructing Racial Rhetoric: Media Depictions of Harm in Heavy Metal and Rap Music," *American Sociological Review* 58 (1993): 753–67.
39. NPR: All Things Considered, "The Three Scariest Words a Boy Can Hear," July 14, 2014, https://www.npr.org/2014/07/14/330183987/the-3-scariest-words-a-boy-can-hear.

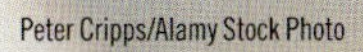

Peter Cripps/Alamy Stock Photo

10

Bearing Witness to Inhumanity

Making Sense of Cruelty to Animals

Learning Objectives

1. Identify the different types of animal cruelty and discuss why they affect so many people on such a deep level.
2. Describe how animal mistreatment reflects social norms.
3. Explain why the cruelties of factory farming are invisible to most meat eaters.
4. Compare successful efforts to combat animal cruelty.
5. Recognize the importance of using the sociological perspective to understand animal cruelty.

Feeling Their Pain: Exploring Why Animal Cruelty So Deeply Affects Us

10.1 Identify the different types of animal cruelty and why they affect so many people on such a deep level.

In 2017, the owner of the Westport Tenant Farm in Massachusetts and its twenty-six tenants were indicted on multiple charges of animal cruelty after investigators discovered that over 1,400 animals had been kept in overcrowded and filthy conditions without adequate food or water. Many of the cows, horses, goats, and dogs were barely alive. "This situation is unparalleled to anything I've seen in my 37 years as an animal law enforcement officer," commented a spokesperson from the Animal Rescue League of Boston.[1] Three years earlier and seventy-five miles to the north, the police had rescued eighty-one birds, seventy-seven cats, twenty-seven dogs, and an array of other animals that had been living with Leonard and Gaye Miville in their modest single-family house. *Barely living* is a more accurate description. Crates and cages were stacked floor to ceiling and feces were strewn everywhere. The fetid air was difficult to breathe.[2]

The buildup of manure at the Westport Tenant Farm was so deep that it had rotted pigs' hooves.
iStockphoto.com/arisara1978

These are examples of **animal cruelty**, which occurs when humans inflict pain, suffering, physical injury, or death on members of other species. Here are the common forms this behavior takes:

- **Neglect:** Providing inadequate care for domestic animals. The Westport Tenant Farm is an example.
- **Hoarding:** Neglect that stems from a clinical condition where people stockpile possessions of all kinds. Leonard and Gaye Miville are an example.

Sarah McLachlan's hit song "Angel" is the soundtrack to a jarring ad that has raised over $30 million for the Society for the Prevention of Cruelty to Animals.[4]
Rockstar Photography/Alamy Stock Photo

There are about two thousand reported cases of animal hoarding per year in the U.S. involving approximately 200,000 animals.[3]

- **Abuse:** Hitting, kicking, or choking a domestic animal in order to assert dominance over it.
- **Fighting:** A bloodsport where two animals—often dogs or roosters—are put in an enclosed area to battle until one is maimed or killed.
- **Dirty play:** When someone beats, throws, burns, poisons, or drowns a wild animal for the fun of it.
- **Food production:** Raising animals in filthy, overcrowded conditions in order to quickly produce vast quantities of inexpensive meat.
- **Entertainment:** Keeping large mammals in unhealthy conditions so they can thrill audiences, such as at circuses, marine theme parks, zoos, and aquariums.

You may have heard stories about animal cruelty on social media. Many celebrities have spoken out against it, including Pink, Ellen DeGeneres, Alec Baldwin, and Leonardo DiCaprio. This social problem tugs at our heartstrings in ways that may feel even sharper to you than other issues described in this book. After all, so many of us live with animals and regard them as family members.

Sitting next to me as I write these words is Cinnamon, a twenty-eight-pound mixture of a Cavalier King Charles spaniel and cocker spaniel. She's the first pet I've ever had. I have no idea why I went so many years without one. She'd only been with my family two weeks when I told my wife, Nancy, that I never wanted to live without a dog again. This was a complete reversal from just a few months earlier, when I still clung to the foolish thought that having a dog would be too much work and too expensive. Our daughter, Arielle, had been asking for years, but I kept coming up with new ways to say no. When she turned twelve, I figured maybe it was time to open my mind. When we're sitting on the couch with Cinnamon, I sometimes turn to Arielle and ask her jokingly why she made me wait so long to get a dog!

Arielle helped me discover one of the finest things in life!
Ira Silver

FIRST IMPRESSIONS?

1. In thinking about examples of animal cruelty that you've heard about and the stories that opened this chapter, why do you think people mistreat animals?
2. Why is animal cruelty an issue many celebrities publicize?
3. Why does this issue tug so strongly at many people's heartstrings?

Maybe the bond I feel with Cinnamon is instinctual. Since we're hardwired to feel a connection to certain types of animals, we have the capacity to see our relationships with them as mirrors for better understanding ourselves.[5] As Mahatma Gandhi, the leader of the Indian independence movement, famously observed, "The greatness of a nation and its moral progress can be judged by the way its animals are treated."[6] Therefore, animal cruelty nakedly exposes our capacity for inhumanity.

We're innately drawn to animals with large heads, big eyes, and soft bodies—the sorts of features that resemble our own babies.
iStockphoto.com/HadelProductions

If you shuddered while reading the stories that opened this chapter, you may think of people who act cruelly to animals as deranged or disturbed. Research supports this individual perspective, describing animal cruelty as impulsive behavior that serves no purpose beyond venting anger or asserting power. For example, studies of children who engage in dirty play characterize them as having antisocial personality disorder. Seeing this social problem from an individual perspective enables people to make sense of a world where a dog like Cinnamon receives warmth and affection (and lots of treats!) in my home but might be abused if she lived elsewhere.[7]

Most people are more likely to see McDonald's as a place to make a quick stop for a tasty meal than as a participant in animal cruelty.
iStockphoto.com/RiverNorthPhotography

If we look at animal cruelty instead from a sociological perspective, we can recognize that it often has broader significance. Food production is the most obvious example. As we'll see, most livestock in the U.S. live in deplorable conditions before they're sent to the slaughterhouse. And yet you may never have thought much about this cruelty. That's because many people—and perhaps you—have an interest in keeping it hidden since meat eating is such a conventional part of American culture.

In this chapter, we'll use the sociological perspective to look broadly at animal cruelty.

When you tell your kids someday about the orca shows at SeaWorld, they may wonder how people could have ever thought it was acceptable to keep large mammals in captivity in order to entertain people.
Vlad Ghiea/Alamy Stock Photo

Our aim is to uncover forces that underlie the different manifestations of this social problem. Whether we're investigating people who eat hamburgers or those who organize dog fights, the source of animal cruelty stems from factors beyond the actions of any particular individual. We will highlight that there are similar reasons behind the choice to eat meat and the decision to stay up late at night to watch two pit bulls spar until one of them is maimed or killed. Both behaviors reflect social norms.

Lots of recent examples highlight how looking at animal cruelty sociologically can lead people to change their ways. For instance, consider that your parents grew up at a time when it was common to regard elephant tricks and whale flips as wholesome family entertainment. However, in the wake of massive publicity given to the mistreatment of performance animals in circuses and marine theme parks, interest in such theatrics has declined. These examples reveal that taking a hard look at what leads us to inflict harm on animals can be a pathway toward treating them better.

Gory Thrills: Making Sense of the Pleasure Some People Get from Abusing Animals

10.2 Describe how animal mistreatment reflects social norms.

Football fans remember Michael Vick as an exceptional quarterback who played at Virginia Tech and then went on to have a strong professional career spanning fifteen years. Many others know about him not because he was one of the most storied players of his generation but because of his antics off the field. Vick is perhaps the most well-known person ever to have been involved in, and served time in prison for, dogfighting. He founded Bad Newz Kennels in 2001, the same year the Atlanta Falcons selected him first overall in the National Football League draft. The Virginia dogfighting enterprise bred and trained pit bulls for matches that occurred late at night and often lasted several hours. The dogs routinely suffered severe injuries.

Lucas is one of twenty-two pit bulls rescued from Michael Vick's dogfighting ring. He received refuge at a shelter in Kanab, Utah.
Carol Guzy/The Washington Post via Getty Images

The pit bulls that Vick and his associates believed weren't fighting viciously enough were hanged, shot, or electrocuted. In 2007, police found fifty abused dogs on Vick's property and arrested him. He pled guilty and served twenty-one months in prison. Despite

FIGURE 10.1 • Dogs Doing Battle

Although dogfighting has been outlawed in most of the world, there are places that it remains legal. Even in countries where it's a crime, dogfighting still occurs, in some contexts openly and in others discreetly.[9]

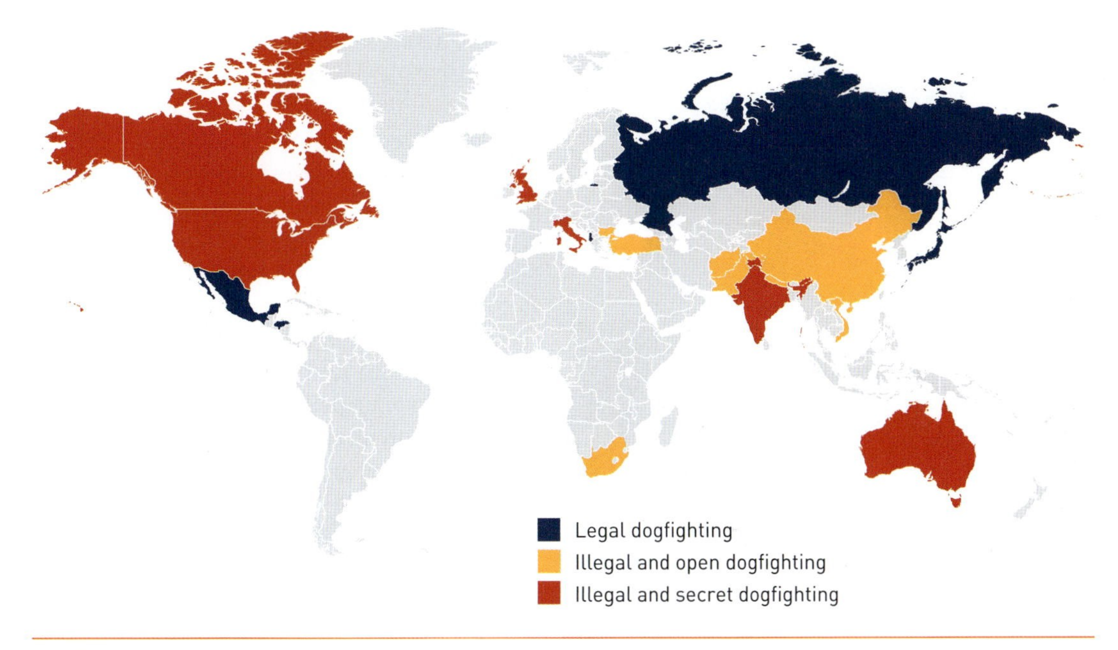

Source: Tom Symonds," Inside the Illegal World of Organised Dogfighting." BBC, February 13, 2019, https://www.bbc.com/news/uk-46991589. "Inside the Underground World of Dog Fighting," National Public Radio, July 19, 2007, https://www.npr.org/templates/story/story.php?storyId=12098479. Orhan Yilmaz, Fusun Coskun, and Mehmet Ertugrul, "Dog Fighting in the World." Retrieved from Academic.edu.

the negative publicity Vick has received ever since, the widespread publicity of his story actually made dogfighting *more* popular in the United States. It also occurs in several other parts of the world (see Figure 10.1).[8]

There were protests outside the courthouse during Michael Vick's sentencing.
Ben Gray/MCT/Newscom

Let's use the sociological perspective to explore why some people train dogs to brutalize one another and other people find it exciting to watch this carnage. Seeing dogfighting in this way shifts attention away from the particular individuals involved and instead toward the social forces that contribute to the appeal of dogfighting for them.

This sociological explanation will begin to reveal itself when you consider that it's primarily men who support dogfighting. As many as ten thousand American men may be involved in dogfighting.[10] Whereas from an individual perspective these men appear to be dark symbols of humans' capacity for brutality, from a sociological

Spectators often place bets on which dog will prevail.
AP Images/Wang zirui - Imaginechina

Cockfighting, perhaps the world's oldest spectator sport, is like dogfighting in that watching roosters do bloody battle to the bitter end affirms men's sense of masculinity.
Muellek Josef/Shutterstock.com

Some kids feed squirrels and then torment them by grabbing their tails or poking their eyes. The sociological perspective enables you to recognize how this cruel behavior plays a key role in these kids' transition to adulthood.
Gokhan Balci/Anadolu Agency/Getty Images

perspective the dogs are the key symbols. Their ferociousness in fighting till the bitter end reflects two qualities the men involved with dogfighting have learned throughout their lives to define as masculine: being ruthlessly competitive and acting violently. Even though they may otherwise be law-abiding citizens, the men view their connection with this illegal form of entertainment as legitimate because it reinforces these qualities. Therefore, the sociological perspective enables you to see that dogfighting isn't senseless cruelty but rather a reflection of societal norms.[11] (See Chapters 8 and 9 for fuller discussions of these norms.)

The same is true for other forms of animal abuse. Sociologist Arnold Arluke interviewed college students who used to—and in some cases still did—engage in dirty play. They'd throw rocks at skunks, play Frisbee with frogs, or burn insects. Seemingly, the only way to understand these acts is through an individual perspective and therefore with condemnation: These young people are impulsive, destructive, and antisocial. However, Arluke learned during his research that teens' cruelty toward animals was neither haphazard nor purposeless but served important functions.

How can this be? Well, let's step back for a moment from the details of the teens' malicious behavior and recognize the social forces that give rise to it. Consider what is perhaps the central developmental goal of adolescence: to prepare for adulthood. The sociological perspective highlights the process of accomplishing this goal instead of judging how particular teens undergo this preparation. From this perspective, dirty play enables certain teens to assert independence at a time of life when they crave it yet have limited autonomy. By planning which animals to torment and teaching themselves how to do so, they are asserting independence and figuring out things for themselves.

Even though according to societal norms about moral conduct these teens' cruelty toward animals is inappropriate, their behavior is conventional in a different sense: They are preparing for adulthood, as you did if you got a paid job at sixteen or filled out the FAFSA without your parents' help. Just as men who participate in dogfighting embrace widely shared masculine expectations, dirty play with animals in the backyard or woods enables teens, in Arnold Arluke's words, to "gain things from their acts that the larger society supports and defines as essential."[12] These behaviors are indicative of a *normal* process—how kids come of age by taking on adult roles and responsibilities.

Tastes So Cruel: Exposing the Hidden Workings of the Factory Farming System

10.3 Explain why the cruelties of factory farming are invisible to most meat eaters.

Visitors to the United States who are excited to taste American food often order a hamburger or a steak. When they go to the ballpark to get a feel for our national pastime, they discover that the experience of watching baseball isn't complete without sinking their teeth into a sizzling hot dog. Meat eating, indeed, is a central part of American culture. But besides vegetarians—who comprise about 3.4 percent of the U.S. population—few people give much thought to, or care to know about, the largest source of cruelty that humans inflict on animals.[13] Films like *Food, Inc.* and *Fast Food Nation* chronicle the devastating effects of the industrial production of beef, pork, poultry, and seafood, both in terms of how animals live and die. Full disclosure: In 2016, I gave up eating land animals, though I still eat eggs, dairy, and fish. My aim here is *not* to guilt you into changing your eating practices but to make you more aware of the animal cruelty that lies behind meat consumption.

While estimates vary, at least sixty billion animals are raised each year for Americans' food consumption—some fifty billion fish and ten billion land animals. Per capita worldwide fish consumption has been steadily on the rise since the 1960s (see Figure 10.2 on page 204), and there's been a similar trend for meat eating. Per capita consumption is projected to be significantly higher in 2025 than it is today (see Figure 10.3 on page 205).[14]

Thinking about what's involved in producing hot dogs can dampen the enjoyment of sinking your teeth into one while watching a ball game.
Kerin Forstmanis/Stockimo/Alamy Stock Photo

I teach a course called Animals & Us. It's popular because, well, many people love animals! Part of students' experience in the course is facing the realization that maybe they don't love *all* animals. In the section where we explore meat eating, I ask them to watch an undercover video clip of their choosing about **factory farming**, which is the method of raising massive quantities of livestock under tightly controlled conditions until slaughter. Students discover that for all of their brief years, chickens and pigs are confined indoors within tight spaces where they have

FIGURE 10.2 ● Seafood Galore

Around the world, fish consumption has grown considerably since the 1960s.

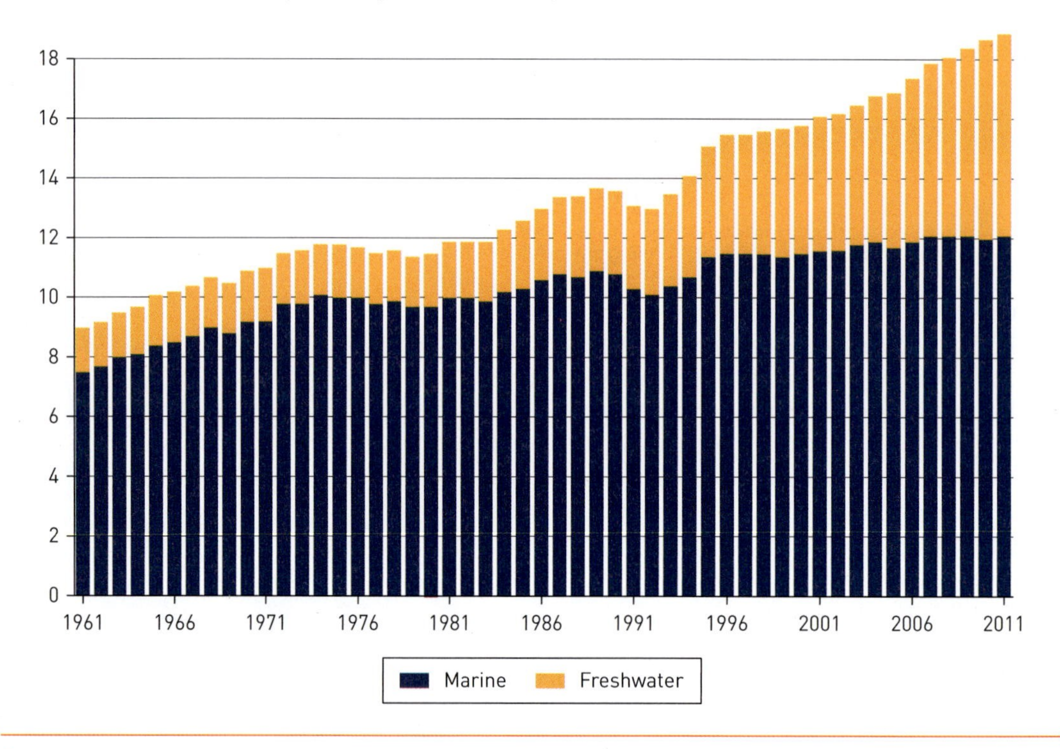

Source: Helgi Library, "How Much Fish Do We Eat per Person?" March 28, 2016, https://www.helgilibrary.com/charts/how-much-fish-do-we-eat-per-person.

minimal room to move and are prone to contracting diseases from their own excrement. Students also learn that although cattle graze outside for the first six months of their lives, they are prematurely weaned from their mothers and, like chickens and pigs, are then kept for the next fourteen months in filthy, overcrowded feedlots before being sent to the slaughterhouse.

Undercover videos expose the brutal conditions in which billions of farm animals live. They are encased in their own excrement with little room to move freely.
iStockphoto.com/PatrickPoendl

Watching these undercover videos is a tough assignment for my students to stomach because it makes visible cruelty that most of them have never seen before.[15] This assignment isn't meant to induce guilt or produce defensiveness. My aim is instead to encourage people to confront the reality that we learn not to see this animal cruelty. Some mention in class that they

FIGURE 10.3 ● Meat Eating on the Rise

Forecasts predict that global per capita consumption of land animals in 2025 will be even higher than it is today.

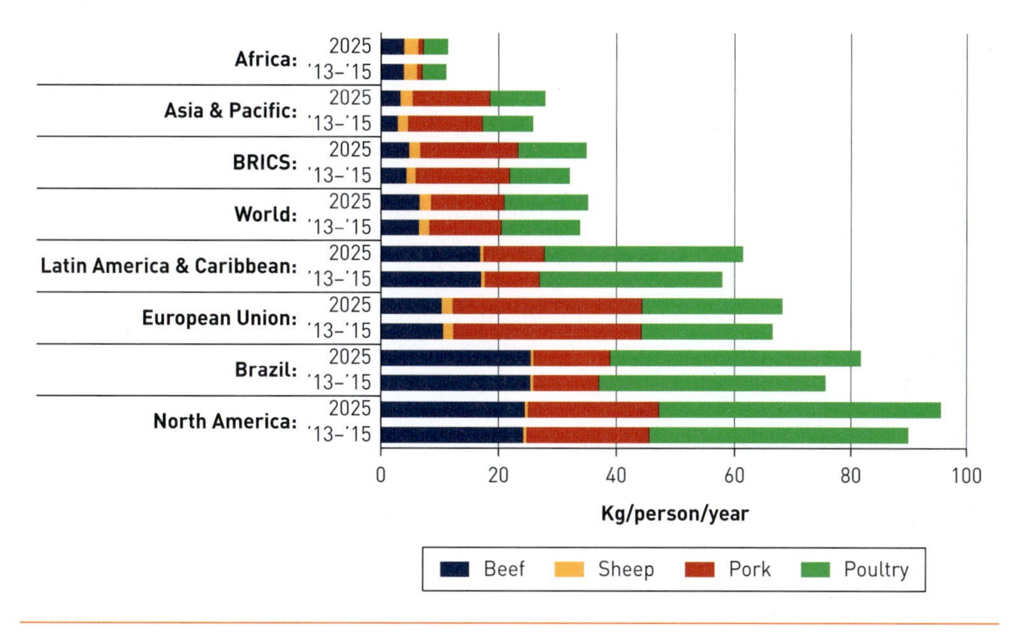

Source: Organisation for Economic Cooperation and Development, "OECD-FAO Agricultural Outlook 2016–25," OECD iLibrary, July 4, 2016, https://www.oecd-ilibrary.org/agriculture-and-food/oecd-fao-agricultural-outlook-2016_agr_outlook-2016-en. Reprinted with permission of WATT Global Media.

simply couldn't handle watching these videos. I acknowledge their discomfort and point out that not doing so underscores the very reason I created the assignment: to highlight that the brutal conditions in which most farm animals live are invisible, in part, because people choose not to bear witness to these conditions.

The assignment urges students to come to grips with an inconvenient truth: that by eating hamburgers or chicken wings, they contribute to these brutal conditions. Many of them experience **cognitive dissonance**, which is the uncomfortable feeling a person gets when their beliefs are out of sync with their behaviors. Most of us value the lives of other species and believe

Many people prefer not knowing about the animal cruelty involved in creating a beautiful table like this one.
iStockphoto.com/LauriPatterson

Meat companies regard the animals they raise not as sentient beings but as commodities.
Richard Levine/Alamy Stock Photo

Because they toil long hours in unsanitary conditions where pigs, chickens, and cows are regarded as units of production rather than as living beings, workers may become numb to how they treat these animals.
Gabe Palmer/Alamy Stock Photo

hurting them is wrong, yet the production of many of the foods we eat is by far the largest source of animal cruelty.

There are three ways people can respond to the cognitive dissonance of being a meat eater and watching the animal cruelty involved in producing hot dogs, fish sticks, and baby back ribs. The first is giving up meat. Very few choose this option because it involves making a major lifestyle change. More likely, people keep eating meat and tell themselves that it's okay because they're far removed from the places where the cruelty takes place. The third option is the easiest of all: denying the reality of this cruelty. An immediate way to do that would be if you stopped reading right now. You would have to acknowledge the hidden workings of the factory farming system and your role in it.

Let's now focus on the different groups that comprise the industrial system of meat production. For starters, there's the consumer—that's you and me. Of the meat products commonly eaten in the United States, such as hamburgers, hot dogs, steaks, chicken tenders, sausage, bacon, and fish sticks, most are produced by large agribusinesses that aim to sell their products at the lowest possible price while maximizing their profits. Their goal is to raise animals to be as large as possible as quickly as possible. Some of the animals become sick or die. Even though providing better living conditions might ensure that more animals produce high-quality meat, treating animals humanely is expensive and cuts into profits.[16]

If these details are making you nauseous, the obvious place to direct your anger is at the people who work in factory farms and meatpacking plants. They're the ones, after all, whom we see in the undercover videos inflicting pain on animals while seemingly having no regard for their suffering. Yet from a sociological perspective, the workers' behavior reflects the conditions of their labor rather than a personal disregard for animal welfare. Many meatpacking workers are undocumented immigrants and therefore have few options for earning a living. The companies that employ them pay low wages while exposing them to brutal conditions. Like the animals, the workers are victims of the factory farming system.[17]

The meat industry not only exercises power over its workers but also significantly influences the level of government oversight it receives. Companies spend millions of dollars each year endorsing political candidates who oppose placing regulations on agribusiness. These companies also pay lobbyists to advocate for laws that support meat production and to fight against laws that regulate the companies' profit-making abilities. Journalists also rarely expose factory farm animal cruelty because the news organizations that employ them are similarly beholden to the meat industry via advertising revenue.[18]

Therefore, as consumers enjoy the availability of cheap meat, eggs, and dairy products, they are often ill-informed that low prices are a by-product of animal mistreatment. This cruelty is the **elephant in the room**, the uncomfortable reality they'd prefer not to acknowledge while eating cheeseburgers, hot dogs, or milkshakes. The truth is consumers of animal products are complicit in a system that makes this cruelty invisible. They often don't see the brutality, let alone the role they play in it. This isn't something to feel guilty about but to recognize with eyes wide open.

Meat companies' influence over politicians explains why federal regulators weakly monitor these companies' treatment of livestock.
iStockphoto.com/David Tran

Confronting Our Inhumanity: A Look at Efforts to Cease the Practice of Animal Entertainment

10.4 Compare successful efforts to combat animal cruelty.

Here's another option I didn't mention earlier that you could take to respond to the cognitive dissonance of bearing witness to animal cruelty while continuing to eat meat: You could decide only to eat meat from animals raised in decent living conditions. In Massachusetts, the majority of citizens supported this option in 2016 by voting yes on Ballot Question 3, which "prohibits the sale of eggs, veal, or pork of a farm animal confined in spaces that prevent the animal from lying down, standing up, extending its limbs, or turning around."[19] California approved a similar measure in 2008.

It's tempting to write off these referendums as merely the sentiment of people living in two of the bluest states. Doing so, however, would miss the wider significance of these campaigns. Public concern about the inhumane treatment of factory farm animals has been rising nationwide. This concern has led many of the largest food retailers, including McDonald's and General Mills,

Opposition to the mistreatment of farm animals has become mainstream. Even Walmart has restricted its sale of certain products manufactured through cruel practices.
iStockphoto.com/shaunl

The killing of Cecil the lion fueled protests around the world against the trophy hunting of large mammals.
Michel Porro/Getty Images

to stop selling eggs from chickens confined to battery cages, where birds are so tightly packed together that they're unable to spread their wings. These changes reflect that resistance to animal cruelty no longer comes solely from radical groups like People for the Ethical Treatment of Animals (PETA).[20]

The animal welfare issue around which people have mobilized most strongly in recent years is the mistreatment of large mammals that do theatrical stunts to thrill humans. To appreciate how significantly public sentiment toward animal entertainment has changed, consider the way that Americans used to view one of the stars of traveling circuses: lions. Up until roughly the turn of the twentieth century, lions were primarily status symbols. Going on expeditions to hunt and collect lions was a way for men to demonstrate dominance over one of the fiercest creatures on earth. However, the rise of the conservation movement in response to the growing endangerment of many large mammals around the world has changed public sentiment. There's now strong interest in protecting lions from human exploitation.[21]

The uproar over the death of Cecil the lion in 2015 underscored the magnitude of this shift. Cecil lived in a Zimbabwean national park, beloved both by the tourists who came to gaze at him and the scientists who studied his behavior. But that summer, an American big-game hunter named Walter Palmer downed Cecil with an arrow and then killed him with his rifle. Palmer later expressed regret over what he'd done, claiming he didn't know that Cecil—whom he had lured out of the park with the scent of a carcass—had been protected. But it was too late for apologies. Cecil was dead, and the most powerful shot had come not from Palmer's gun but from the camera that had captured him smirking with his kill. The rapid dissemination of this glib image of cruelty toward an animal whose real-life popularity rivaled that of Disney's fictionalized Lion King stirred massive public outcry and calls for Palmer's arrest.[22]

The viral photo of Palmer grinning behind his kill has hardly been the only image of cruelty toward lions to produce outrage. Jan Creamer and Tim Phillips have also captured the inhumane treatment of these spectacular big cats. The photos and videos taken by this British couple may not be as familiar as images of Cecil, yet they've similarly fueled public opposition. Cofounders of the organization Animal Defenders International, Creamer and Phillips have spent years documenting the physical harms circus animals experience. Because of their photos and videos depicting trainers kicking and punching lions in order to get them to do tricks, thirty-three lions were rescued from inhumane treatment in Peruvian and Colombian circuses and now live in a South African wildlife sanctuary.[23]

An undercover investigation of Ringling Brothers and Barnum & Bailey Circus exposed similar brutality toward elephants. It found that trainers often used long, sharp metal hooks to get the elephants to do the sorts of stunts that dazzle audiences: playing musical instruments, riding tricycles, shooting hoops, and balancing on wooden barrels. Because this coercion involved inserting one end of the bell hook into an elephant's mouth or into the top of its ears and pulling the other end, the animals frequently developed puncture wounds. Train rides between cities were also brutal, averaging twenty-six hours. It's no wonder many of the adult elephants had misshapen or infected feet and suffered from musculoskeletal disorders. They also had arrested development. Whereas elephants in the wild sometimes nurse until age four and continue learning social skills from their mothers well into their teenage years, circus elephants were weaned after just a year.[24]

During travel days, elephants in the Ringling Brothers and Barnum & Bailey Circus were chained within cramped spaces for many hours at a time, preventing them from exercising or breathing fresh air.
Ricky Carioti/The Washington Post via Getty Images

Beginning around 2000, behind-the-scenes exposés of brutality toward circus animals started to attract attention. Americans gradually became more interested in the treatment and welfare of these animals than in their ability to entertain crowds. Negative publicity of the Ringling Brothers and Barnum & Bailey Circus exacerbated what had already been a long-running decline in attendance (see Figure 10.4 on page 210). Diminishing revenue prompted its parent company, Feld Entertainment, to announce in 2016 that elephants would cease performing. Now without its top attraction, the "Greatest Show on Earth" no longer had much luster, and ticket sales continued to plummet. In 2017, the company decided that after traveling for 146 consecutive years, it was time for the Ringling Brothers and Barnum & Bailey Circus to shut down.[25]

The elephants now live on a two-hundred-acre parcel of company land between Tampa and Orlando called the Center for Elephant Conservation. This preserve, where they'll spend the rest of their lives, is just a short drive from Florida's top attraction, which of course is Mickey Mouse and his friends at Disney World. However, these elephants have recently become a close second, a distinction that used to belong to a different large animal—the orcas in SeaWorld's Shamu show. Shamu was the name of a particular orca that lived at the San Diego theme park during the 1960s. After her death, she became the namesake of SeaWorld's orcas show there and at its two other locations, Orlando and San Antonio. These majestic black and white marine mammals wowed audiences while providing cool and refreshing spray that offered relief from the blazing sun.

My family and I enjoyed seeing the Shamu show during our December 2010 trip to Orlando. Vacations usually seem better with time, as fond memories crystalize and take on a twinge of nostalgia, but not this one. The day we were at SeaWorld, I didn't give much thought to the tragic news from ten months earlier when a trainer had died on the job. There were no shrines in Dawn Brancheau's memory or any other reference to this woman who had worked at the park for fifteen years. Her invisible presence allowed me to take pleasure in watching orcas spin in the air and to feel good

FIGURE 10.4 ● The Show No Longer Goes On

Ringling Brothers and Barnum & Bailey Circus discontinued its traveling performance after many cities recorded consecutive years of plummeting ticket sales.

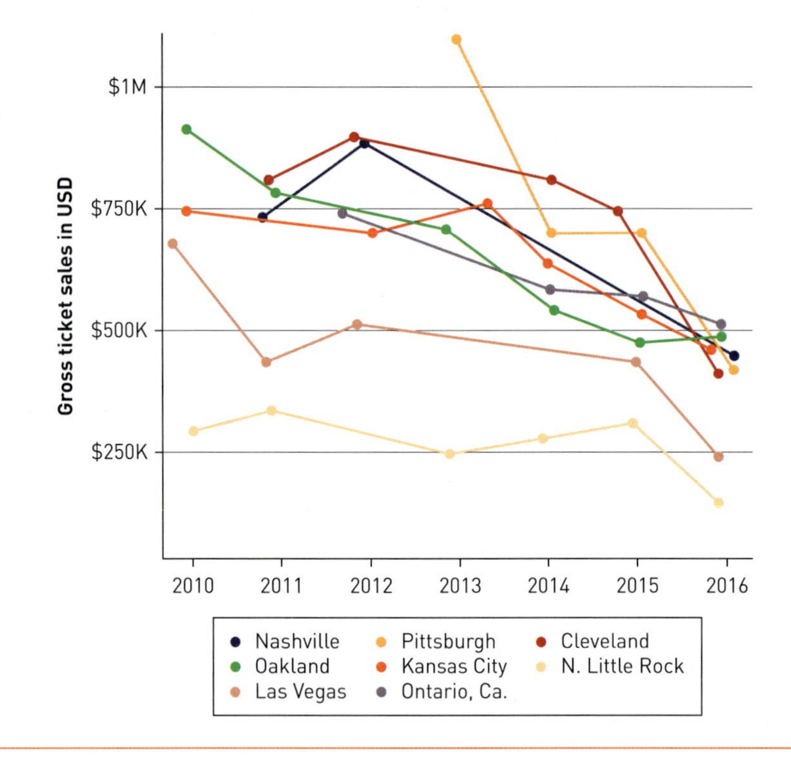

Source: Dave Brooks, "Anemic Sales Killed the Greatest Show on Earth," *Amplify*, January 16, 2017, http://www.ampthemag.com/the-real/anemic-sales-killed-greatest-show-earth.

knowing my ten-year-old son and eight-year-old daughter were having a good time. In retrospect, it was frighteningly easy to relegate Brancheau's death to the back of my mind and cling to the belief that it was an accident. She must have slipped and fallen into the water, I presumed, and what happened after that just proved why orcas are called "killer whales."

Little did I know at the time that Tilikum, the orca who killed Dawn, had claimed two prior victims, one in 1991 and the other in 1999. This information was public knowledge, but it wasn't part of public consciousness—at least not yet. That changed in 2013 with the release of the film *Blackfish*. This exposé of SeaWorld's cruelty toward orcas stirred outrage over the fact that one of the largest animals on earth lived in confined spaces in order to entertain tourists. It's one of the most riveting films I've ever seen. The harrowing footage prompts viewers to consider a wrenching question: How is it that these twelve-thousand-pound mammals are living in marine theme parks? This question had never before crossed my mind, and yet I now regarded the answer as the single most important detail about SeaWorld. Viewers see Tilikum at age two swimming off the coast of Iceland and then ominously watch as he is netted and taken away from his family members, who appear traumatized by his abduction.

For many years Tilikum and his trainer, Dawn Brancheau, delighted visitors at SeaWorld in Orlando.
WENN Rights Ltd/Alamy Stock Photo

This scene is one bookend of Tilikum's story; the other is the tragedy that unfolded on February 24, 2010. Dawn Brancheau was doing her routine training work with Tilikum after he'd performed in the "Dine with Shamu" show. She fed him fish, doused him with cool water, and gently stroked his head. Then Tilikum suddenly lunged for her ponytail. As Brancheau tried to break free, the orca pulled her into the pool. She was a strong swimmer and made several attempts to reach safety, but each time Tilikum continued to ensnare her in his grasp. Emergency personnel were eventually able to use a mechanical lift to raise Tilikum to the surface. When they pried open his jaws, they discovered part of Brancheau's arm detached in his mouth. She had already bled to death.[26]

This was *not* an accident, as many—myself included—had come to believe. Although SeaWorld had aggressively tried to present it that way, in truth Brancheau hadn't slipped into the water. It's practically unheard of for orcas to attack humans in the wild, and yet this one had now killed three people. Knowing Tilikum's life story explains why.

Tilikum's aggressive behavior began soon after a team of hunters trapped and took him away from his family.
iStockphoto.com/Ivkovich

After being captured while swimming in the ocean, he spent almost a year at an aquarium in Iceland and was then transferred to a marine theme park in Canada. He lived the next eight years there, joining a new hierarchy where the two dominant females often kept Tilikum in his place. The fact that the three orcas were limited to very tight quarters for over fourteen hours a day intensified their conflicts. The rest of the time they performed relentlessly, eight shows a day, seven days a week. The combination of long periods of confinement and rigid routine took its toll. Orcas' emotional health hinges on having lots of space to move and diverse kinds of stimulation. For Tilikum, being subordinate in the pecking order was the third strike against him. The situation was no better after he moved to SeaWorld in Orlando. He was once again subordinate to other orcas and suffered abuse that led to his mental instability.[27]

Blackfish caused a public relations nightmare for SeaWorld, which animal welfare organizations seized upon in their criticism of the theme park (see Figure 10.5). As a result, attendance significantly declined. People started regarding the Shamu shows as reflecting the cruelty of keeping large, intelligent, and emotionally complex mammals in captivity. Following the death of Tilikum in 2017 at age thirty-five, the company announced it was immediately discontinuing the orca shows at its San Diego theme park and would do the same at its Orlando and San Antonio locations by 2019. A critical segment of the American

FIGURE 10.5 ● Inhumane Treatment of Orcas

Even people who didn't see *Blackfish* caught wind of its stark messages about the cruelty of keeping some of the earth's largest and most intelligent mammals in captivity in order to entertain humans.

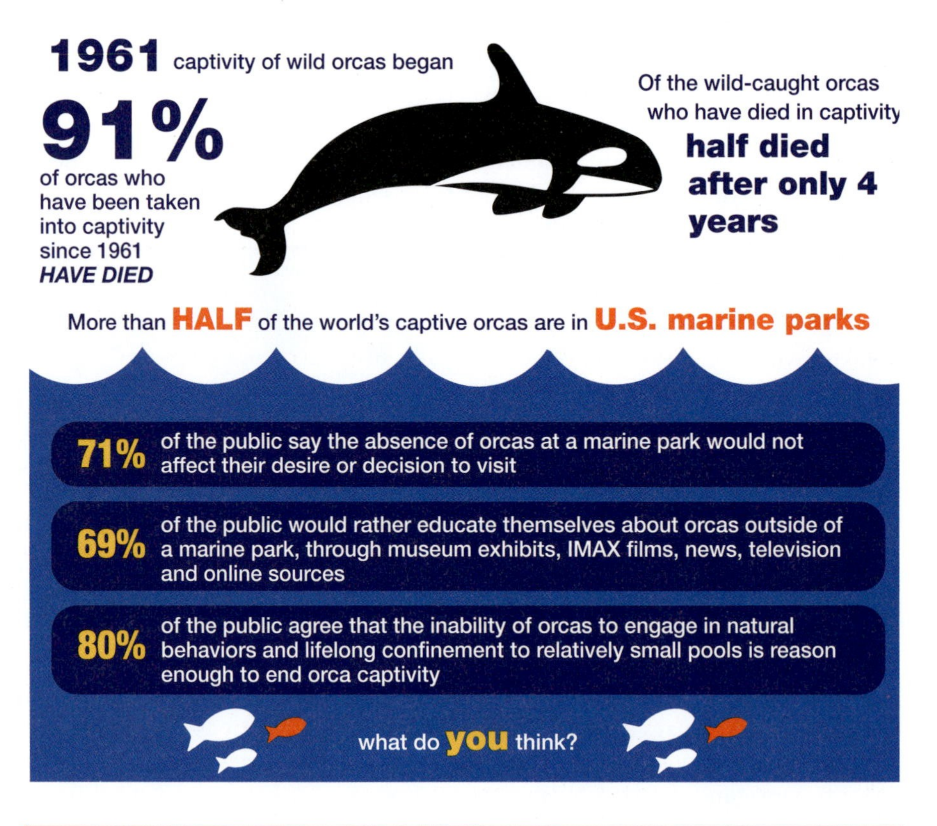

Source: Animal Welfare Institute, "Infographic for Animals in Captivity," https://awionline.org.

public had become aware of both the inhumane treatment of performance animals and of their own acquiescence to this cruelty via their desire to see animals dazzle them.[28]

Therein lies the sociological perspective toward the cruelty of animal entertainment. Several social forces produced this cruelty, including companies like Ringling Brothers and Barnum & Bailey Circus and SeaWorld, which for years have profited from subjecting large mammals to inhumane conditions; advertisers, who have depicted these places as wholesome family fun; and parents like me, who used to buy into this image. As a result, cruelty in animal entertainment has become a diminished part of the experience of going to the circus or visiting marine theme parks.

Someday we may look upon Harambe as having fueled similar sociological understanding. In the hours following the surprising result of the 2016 presidential election, an online rumor surfaced that Harambe had received ten thousand or more votes and that had those people chosen Hillary Clinton instead, she might have won. It was an unfathomable news story, and not only because the rumor was subsequently disproven.[29] Just a few months earlier, Harambe had been an unfamiliar name to most voters. He hadn't spent any time on the campaign trail. Besides, he was ineligible to become president; he wasn't at least thirty-five years old (the minimum age) and . . . well, he wasn't a human being! Harambe was a gorilla who lived in the Cincinnati Zoo until May 28, 2016, when a worker killed him after he grabbed a three-year-old boy who had entered his enclosure. Global outcry ensued over the decision to shoot Harambe.

The fact that some people wrote in Harambe's name on their presidential ballot was an indicator of the extent to which he had become a meme. Yet beneath the humor lay the same serious issue that animal mistreatment in circuses and marine parks had raised: the ethics of confining large mammals for their entire lives in order to entertain humans. It's possible that in the near future there will be

Cruelty toward elephants and other large mammals occurred for many years at the Ringling Brothers and Barnum & Bailey Circus, which began its 146-year run in 1871 and introduced the celebrated Jumbo the elephant in 1882.
Glasshouse Images/Alamy Stock Photo

SeaWorld's mistreatment of orcas began with the creation of its Shamu show in the 1960s. Serious criticism of the show didn't start until after the release of *Blackfish*.
Ian Dagnall/Alamy Stock Photo

Regardless of how many votes Harambe actually received, his popularity swelled when he became a symbol of public opposition toward zoos keeping large mammals in captivity.
John Leyba/The Denver Post via Getty Images

similar opposition toward zoos and aquariums as that mounted against Ringling Brothers and Barnum & Bailey Circus and SeaWorld.[30]

The Payoff of the Sociological Perspective: Seeing Who We Are through Our Relationships with Animals

10.5 Recognize the importance of using the sociological perspective to understand animal cruelty.

Most people who say they care about animal welfare are most likely to demonstrate that concern in how they treat their pets. This may describe you. But how might elements of your lifestyle also rely upon animal cruelty? This could be because of the foods you eat, the leather on your shoes, or prescription drugs you take that are produced through animal testing. My aim in asking this question isn't to make you feel uncomfortable or guilty but to get you to see that *all* of us are hypocrites when it comes to how we feel about versus how we act toward animals—me too.

The sociological perspective helps you to understand this hypocrisy by shifting your focus away from seeing animal cruelty in strictly individual terms and inviting you to focus on the hidden stories behind it. These stories include the men participating in dogfighting in order to embrace masculine norms, kids engaging in dirty play with wild animals as a way to come of age as emerging adults, factory farms treating livestock as commodities in order to produce cheap meat, and entertainment venues confining large mammals so that people may experience thrills during their leisure time. These hidden stories are the social forces that give rise to animal cruelty.[31]

The challenge in front of you is not to let your own blinders impede you from seeing these social forces. When you hear about animal mistreatment, the inclination is to view it simply as either a deviant act (which is how many people see dogfighting) or as essential for survival (which is how people often regard meat eating). The next time you read on social media about an incident of animal cruelty, think about possible sociological explanations that may account for this behavior—explanations that resemble the ones discussed in this chapter. As an illustration, let's return to the two vignettes that opened this chapter, the cases of the hoarding couple in Lynnfield, Massachusetts, and the farm tenants seventy-five miles to the south in Westport. What social forces might account for why these people mistreated animals?

While it's heart-wrenching to learn about the barbaric ways humans treat the members of other species, it can be uplifting to discover that seeing animal cruelty from a sociological perspective uncovers ways people can act more humanely. This perspective exposes how thinking broadly about animal mistreatment can enable us to better understand who we are and what we aspire to become. Just as animal cruelty reveals our inhumanity, understanding the roots of that cruelty can elicit more humane treatment.

What Do You Know Now?

1. How does the sociological perspective toward dogfighting shift attention away from the specific men who participate and focus instead on norms about what it means to be "masculine"?
2. If you feel rage in thinking about kids' dirty play with animals, it's important to recognize that this behavior has a broader meaning. In what sense is inflicting such cruelty on animals a way for kids to engage in the normal process of gradually taking on adult roles and responsibilities?
3. How do meat eaters take part in a hidden system of animal cruelty? Who are the major participants in this system? How does it feel to answer this question?
4. Why do you think people are more likely to be critical of animal cruelty at zoos, circuses, marine theme parks, and other places where animals entertain us than to be critical of inhumane factory farm conditions?
5. Is the decline in popularity of animal entertainment venues the beginning of a broad-based movement to increase animal welfare or merely a fad? Why?

Key Terms

Visit **edge.sagepub.com/silver** to help you accomplish your coursework goals in an easy-to-use learning environment.

Notes

1. Quoted in Michael Levson, "27 Indicted in Largest Animal Cruelty Case in New England," *Boston Globe*, March 31, 2017, http://www.bostonglobe.com/metro/2017/03/31/indicted-westport-farm-animal-cruelty-case/Hl4uT0LJkJH9XwhaqNg2fM/story.html.
2. Billy Baker, "199 Animals Removed from Lynnfield Home," *Boston Globe*, March 14, 2014, http://www.bostonglobe.com/metro/2014/03/10/animals-found-hoarded-lynnfield-house-where-woman-had-died/Uz4lkUVvgioCcu6Dyg2DjN/story.html.
3. Hal Herzog, *Some We Love, Some We Hate, Some We Eat: Why It's So Hard to Think Straight About Animals* (New York: HarperCollins, 2010), 138.

4. Stephanie Strom, "Ad Featuring Singer Proves Bonanza for the A.S.P.C.A," *New York Times*, December 25, 2008, http://www.nytimes.com/2008/12/26/us/26charity.html?_r=1.
5. Herzog, *Some We Love*, 39.
6. Quoted in People for the Ethical Treatment of Animals, "Can the Greatness of a Nation. . .?" https://www.peta.org/features/gandhi/.
7. Arnold Arluke, "Animal Abuse as Dirty Play," *Symbolic Interaction* 25, no. 4 (2002): 405–30; Roman Gleyzer, Alan R. Felthous, and Charles E. Holzer III, "Animal Cruelty and Psychiatric Disorders," *Journal of the American Academy of Psychiatry Law* 30 (2002): 257–65.
8. Christina M. Russo, "Michael Vick May Have Actually Made Dogfighting MORE Popular," *The Dodo*, January 21, 2016, https://www.thedodo.com/michael-vick-dogfighting-more-popular-1565236921.html.
9. NPR, "Inside the Underground World of Dog Fighting," July 19, 2007, https://www.npr.org/templates/story/story.php?storyId=12098479; Humane Society of the United States, "Mounting an Attack on Dogfighting in Mexico," July 20, 2016, https://blog.humanesociety.org/2016/07/hsi-mexico-dogfighting-campaign.html; League Against Cruel Sports, "Global Dog Fighting Network Exposed," https://www.league.org.uk/globaldogfighting; Orhan Yilmaz, Fusun Coskun, and Mehmet Ertugrul, "Dog Fighting in the World," 2015, https://www.academia.edu/16529039/Dog_Fighting_In_The_World; Hanna Gibson, "Detailed Discussion of Dog Fighting," Animal Legal & Historical Center, 2005, https://www.animallaw.info/article/detailed-discussion-dog-fighting#id-6.
10. Because dogfighting is illegal, the exact number of people involved is uncertain. This figure is an estimate based on reports in dogfighting publications as well as the number of dogs rescued with serious wounds. See The American Society for the Prevention of Cruelty to Animals, "A Closer Look at Dogfighting," https://www.aspca.org/animal-cruelty/dogfighting/closer-look-dogfighting.
11. Rhonda Evans, DeAnn K. Gauthier, and Craig J. Forsyth, "Dogfighting: Symbolic Expression and Validation of Masculinity," *Sex Roles* 39, no. 11/12 (1998): 825–38.
12. Arluke, "Animal Abuse as Dirty Play," 406.
13. Data about vegetarians are from 2015 and are based on an annual survey conducted by the Harris Poll on behalf of the Vegetarian Resource Group; see Charles Stahler, "How Often Do Americans Eat Vegetarian Meals? And How Many Adults in the U.S. Are Vegetarian?" *Vegetarian Resource Group Blog*, May 29, 2015, http://www.vrg.org/blog/2015/05/29/how-often-do-americans-eat-vegetarian-meals-and-how-many-adults-in-the-u-s-are-vegetarian-2/.
14. For data about meat consumption, see the Humane Society of the United States, https://www.humanesociety.org/sites/default/files/docs/table-us-per-capita-meat.pdf and United Poultry Concerns, "Average and Total Numbers of Animals Who Died to Feed Americans in 2008," October 22, 2009, https://www.upc-online.org/slaughter/2008americans.html.
15. Melanie Joy, *Why We Love Dogs, Eat Pigs, and Wear Cows* (San Francisco: Conari Press, 2010), 39–51.
16. Ibid., 39.
17. Ibid., 82–5.
18. Ibid., 89–90, 103–5.
19. "Massachusetts Minimum Size Requirements for Farm Animal Containment, Question 3 (2016)," *BallotPedia: The Encyclopedia of American Politics*, 2016, https://ballotpedia.org/Massachusetts_Minimum_Size_Requirements_for_Farm_Animal_Containment,_Question_3_(2016).
20. Jade Scipioni, "Even before Harambe, CEOs Forced to Acknowledge Animal Rights," Fox Business News, June 7, 2016, http://www.foxbusiness.com/features/2016/06/07/even-before-harambe-ceos-forced-to-acknowledge-animal-rights.html.
21. Stephen Fox, *The American Conservation Movement: John Muir and His Legacy* (Madison: University of Wisconsin Press, 1981), 148–82.
22. Mary Bowerman, "Minnesota Dentist 'Deeply' Regrets 'Taking' Cecil the Lion," *USA Today*, July 30, 2015, https://www.usatoday.com/story/news/nation-now/2015/07/28/

minnesota-dentist-walter-james-palmer-cecil-lion-africa/30785881/.

23. Bill Whitaker, "Saving the Lions," *60 Minutes*, March 4, 2017, http://www.cbsnews.com/news/disturbing-footage-of-circus-animal-abuse-leads-to-widespread-reform/.
24. Deborah Nelson, "The Cruelest Show on Earth," *Mother Jones*, November/December 2011, http://m.motherjones.com/environment/2011/10/ringling-bros-elephant-abuse.
25. Faith Karimi, "Ringling Bros. Elephants Perform Last Show." CNN, May 2, 2016, http://www.cnn.com/2016/05/01/us/ringling-bros-elephants-last-show/; Steph Solis, "Ringling Bros. Circus Closing after 146 Years," *USA Today*, January 14, 2017, https://www.usatoday.com/story/news/nation/2017/01/14/ringling-bros-circus-close-after-146-years/96606820/.
26. Tim Zimmerman, "The Killer in the Pool," *Outside*, July 30, 2010, https://www.outsideonline.com/1924946/killer-pool.
27. Ibid.
28. "SeaWorld San Diego Ending Killer Whale Show," *U.S. News & World Report*, January 7, 2017, https://www.usnews.com/news/business/articles/2017-01-07/seaworld-san-diego-ending-killer-whale-show.
29. Doug Criss, "No, Harambe Didn't Get 11,000 Votes for President," CNN, November 10, 2016, http://www.cnn.com/2016/11/10/us/harambe-votes-trnd/.
30. Andrew C. Revkin, "After Harambe's Death, Rethinking Zoos," *New York Times*, June 2, 2016, https://mobile.nytimes.com/blogs/dotearth/2016/06/02/after-harambes-death-rethinking-zoos/.
31. Herzog, *Some We Love*, 237–42.

11

"Better Safe Than Sorry"

Protecting Children from Strangers and Other Dangers

Learning Objectives

1. Describe how cases of sextortion typically unfold.
2. Explain why some kids are at a higher risk of online exploitation by a stranger than others, and discuss how this reflects inequalities in childhood experiences.
3. Identify the types of risks that can be healthy for kids to encounter as they grow up.
4. Recognize the ways in which interacting with strangers can be beneficial.
5. Compare public fears regarding the dangers kids face with the realities of these dangers.

Careful Who You Friend: Exploring the Predatory Behavior of Strangers Lurking on the Internet

11.1 Describe how cases of sextortion typically unfold.

"Want to make $3,000?" For an impressionable teen, this is a tough offer to refuse. The man had listed his name in his Facebook profile as "Tavus Cole." When the girl responded, he told her he worked for *Victoria's Secret* magazine and wanted to hire her as a model. After several days of back-and-forth messaging, the fifteen-year-old e-mailed him naked selfies. She was among the hundreds of girls whom this man, whose real name is Jordan James Kirby, victimized by **catfishing**, which involves creating a fake online identity in order to deceive others. Kirby used his phony persona to lure his victims into sharing nude photos—a practice known as **sextortion**. In 2015, the twenty-three-year-old California man received a sentence of twenty-nine years in federal prison for enticement of a minor and attempted production of child pornography.[1]

Jordan James Kirby and Jamy Church are the sorts of creeps lurking online whom you may have grown up fearing.
iStockphoto.com/sam thomas

Jamy Church's tactics were similar, but he instead exploited boys. Church created Facebook pages that appeared to represent high school girls, using names like "Hope Smith" and "Sara Chapman." The thirty-four-year-old Louisiana man, who at the time was coaching Little League baseball, friend requested boys ages thirteen to sixteen and, once accepted, sent them sexually explicit photos that were purportedly of these girls. In return, "she" asked each boy to send "her" photos of himself

iStockphoto.com/FatCamera

masturbating. In 2012, Church got a twenty-one-year prison sentence for enticing at least thirty-three minors in southwest Louisiana and Texas to engage in sexual activities and produce child pornography.[2]

You've probably heard repeatedly about the importance of being cautious around people you don't know. Parents and teachers have been instilling stranger-danger fears in kids since before I was born. Perhaps the most enduring cross-generational warning has been not to approach unfamiliar men offering candy. In recent years, adults who care for children have come to see the internet as a particularly scary place where strangers pose unique threats because of their anonymity. The many online dangers include pornography, sexting, identity theft, cyberbullying, and sextortion. As the stories of Jordan James Kirby and Jamy Church attest, a person can easily fabricate an identify and use it to exploit kids' vulnerabilities.[3]

FIRST IMPRESSIONS?

1. What thoughts came to mind as you read the stories of Jordan James Kirby and Jamy Church?
2. While you were growing up, in what situations and from whom did you hear stranger-danger warnings?
3. Have you ever thought about strangers in a way that equates them with something besides danger? If so, when? If not, why not?

Sextortion is an understudied problem. Since the government does not publish data about it, the little that's known comes from academic research:

- Almost a third (32 percent) of American teens have received a message via text, social media, or a gaming site from someone unfamiliar to them.[4]
- About 75 percent of the kids messaged are over thirteen, with fifteen-year-olds being the most frequently targeted group.[5]
- One in five teens has received a sexual solicitation from a stranger.[6]
- About 5 percent of youths ages twelve to seventeen have been a victim of sextortion.[7]

The most comprehensive study done about sextortion investigated seventy-eight prosecuted cases across twenty-nine states and U.S. territories. Every single perpetrator was male. In 65.4 percent of the cases, the perpetrator victimized ten or more people; in 33.3 percent of the cases, twenty or more people; and in 16.6 percent of the cases, one hundred or more. Whereas sometimes the victims were adults, in the majority of cases (71 percent), they were minors (see Figure 11.1). These data chillingly confirm why adults repetitively warn kids not to message or share personal information with people they don't know.

FIGURE 11.1 ● Sextortion

Here is what's known about cases where online predators use a false identity to lure a person to share nude photos with them.

Who Are The Victims?		Who Are The Perpetrators?		
71%	of cases involve only victims under the age of 18	**Every single prosecuted perpetrator is male**. They include college students, a State Department employee, even fathers and step-fathers of their victims.		
14%	of cases involve a mix of minor and adult victims	Sextortionists tend to be prolific repeat offenders. Among the cases we studied:		
12%	of cases involve only adult victims	25 cases involved at least	13 cases involved at least	13 cases involved more than
Nearly all adult victims are female, but both minor girls and boys are victimized.		**10** victims	**20** victims	**100** victims

Source: Benjamin Wittes, Cody Poplin, Quinta Jurecic, and Clara Spera, "Sextortion: Cybersecurity, Teenagers, and Remote Sexual Assault," Center for Technology Innovation at Brookings, May 2016, https://www.brookings.edu/wp-content/uploads/2016/05/sextortion1-1.pdf.

And yet, sometimes teens do. Friending a stranger allows the predator to engage in what is often an extended period of **grooming**, during which they offer gifts or flattery as a way of forging an emotional connection with the person they intend to victimize. The interaction typically starts off innocuously, but over time the victim becomes prone to sharing personal information and in some cases is willing to meet up in person. The predatory behaviors of Jordan James Kirby and Jamy Church are cautionary tales of what can happen if someone responds to an online message from a stranger and continues to interact with that person.

Sextortion is not a new social problem but an indicator of how stranger danger manifests itself within the limitless terrain of cyberspace.
Kiyoshi Takahase Segundo/ Alamy Stock Photo

The conventional wisdom is that teens who don't exercise vigilance online are susceptible to danger. While of course that's true, it's shortsighted to see sextortion simply as a problem of personal irresponsibility in how certain teens use the internet. By exploring this problem sociologically, we can discover that childhood inequalities influence how likely kids are to follow messages they've heard about ways to avoid stranger danger. We'll see that because during their childhoods some teens have experienced offline physical and psychological injuries, they are prone to not follow warnings about friending people unfamiliar to them or sharing personal information with them.

Although we all want to avoid danger, the sociological perspective exposes why it can be shortsighted to adhere strictly to the motto "Better safe than sorry."
iStockphoto.com/bluekite

This chapter also invites you to consider two other fascinating stories concerning the experience of growing up in a world fraught with danger. First we'll highlight the risks of overprotecting kids. Here the discussion moves beyond the dangers of spending time online and into topics that were as relevant when your parents and grandparents were growing up as they are today, such as how kids get to school and where they play when they come home. Then we'll take a close look at what it means to live in a world amidst people who are unfamiliar to us. While of course strangers often pose dangers, we'll also consider the positive roles they can have in the lives of children and adults alike.

Unequal Childhoods: Exposing Why Some Kids Are More Prone Than Others to Online Exploitation by Strangers

11.2 Explain why some kids are at a higher risk of online exploitation by a stranger than others, and discuss how this reflects inequalities in childhood experiences.

At age twenty, Ashley Reynolds went public with her story of the victimization that occurred when she was fourteen. She was among the hundreds of teenage girls Lucas Michael Chansler deceived by creating fake social media profiles and enticing these girls to send him nude selfies. In 2015, Chansler received a prison sentence of 105 years for exploiting as many as 350 minors across twenty-six states, three Canadian provinces, and the United Kingdom. He told authorities he targeted girls thirteen to eighteen because older females couldn't be as easily manipulated.[8]

After detaining Lucas Michael Chansler, law enforcement agents discovered that he had amassed roughly eighty thousand pornographic photos and videos.
Bob Daemmrich/Alamy Stock Photo

Using a profile picture of a teenage boy and the screen name "CaptainObvious," Chansler messaged Reynolds on MySpace, a popular social media site in the mid-2000s. She agreed to send him photos of herself posing in her bra. Chansler threatened to circulate them if she didn't send more—a form of sextortion similar to how Jordan James Kirby preyed on his fifteen-year-old victim. Reynolds obliged, thinking this would lead Chansler to stop harassing her. It didn't. His demands escalated to the point where he was insisting on sixty new photos a day. It was only when Reynolds' parents caught wind of what was happening that they provided the FBI with information from her computer that led to Chansler's arrest.[9]

Ashley Reynolds decided to speak out about the dangers of online predators by recording a video recounting her mistakes that enabled Chansler to victimize her. She also did an interview for a feature article in *Glamour* that highlighted how imprudent online behavior can lead strangers to exploit teens. The significant publicity Reynolds received reinforced the individual perspective that teen irresponsibility invites online exploitation. The publicity also masked how exceptional her story of victimization was.

Indeed, unlike most teens who encounter sextortion or other forms of online harassment by a stranger, Reynolds hadn't previously experienced the sorts of offline injuries that can lead a person to become curious about interacting with strangers. Research by a team of experts at the Crimes Against Children Research Center indicates that most often victims of internet predators already suffer from poor psychological health stemming from childhood trauma. The study, which investigated the lives of teens preyed upon online during the previous year, found that 96 percent had also been exploited in face-to-face settings over this same time period. They had witnessed family violence, been sexually victimized, been mistreated by a caregiver, or been physically assaulted (see Figure 11.2 on page 224). Online exploitation, therefore, is the **canary in the coalmine**—an alarm signaling even greater traumas most of these kids have experienced offline.[10]

Trauma often has enduring effects on a person. They may be emotionally withdrawn, have difficulty forming close attachments, or continually feel stressed. As a result, trauma victims are vulnerable to experiencing other kinds of exploitation too. Think, for example, about why trauma might compromise a person's judgment about whom they're interested in getting to know or befriend. For a child who has grown up with neglect or abuse, interacting with an unfamiliar adult may seem like an opportunity to gain attention they desperately crave. The fact that trauma survivors, relative to kids who've had emotionally secure childhoods, are especially vulnerable to exploitation via the internet highlights the roots of sextortion and other forms of online victimization. From a sociological perspective, the key issue here is *not* how kids respond if friended by a stranger; it's inequalities in the experience of childhood in American society. Because some kids encounter violence as they grow up, they become vulnerable to behaving in ways that make them prone to further victimization, especially online.[11]

Ashley Reynolds is an exception to this typical scenario in which strangers prey on kids online. She's a reasonably well-adjusted woman who, in hindsight, had the strength to admit publicly that she made a mistake when she was younger. For the majority of teens who interact with internet predators, however, their decision to do so doesn't simply reflect a momentary lapse of judgment but rather the reverberating effects of childhood trauma. In reinforcing the individual perspective that kids encounter online stranger danger because they're careless, Reynolds' story obscures the underlying reasons certain teens are particularly prone to exploitation by online predators. Kids who've experienced trauma and then been exploited online are unlikely to have the self-confidence as adults that Reynolds had to share their stories publicly.

FIGURE 11.2 ● Double Jeopardy

Here are the frequencies of offline exploitation experienced over a one-year period by teens who were also victimized online during this same time span.

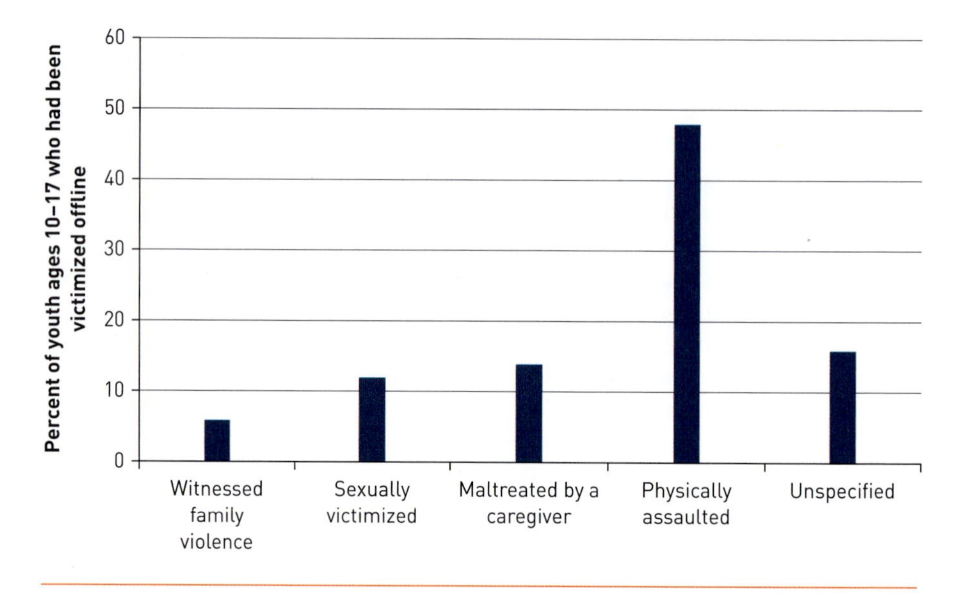

Source: Adapted from Kimberly J. Mitchell, David Finkelhor, Janis Wolak, Michele L. Ybarra, and Heather Turner, "Youth Internet Victimization in a Broader Victimization Context," *Journal of Adolescent Health* 48, no. 2 (2011): 128–34, https://www.mensenhandelweb.nl/en/system/files/documents/13%20jul%202015/Youth%20internet%20victimization%20in%20a%20broader%20victimization%20context.pdf.

One time when I taught about this issue, a student pointed out a key distinction between interacting with online strangers and being exploited by them. She posted the following comment to the class's discussion forum:

> I personally have talked to strangers on the internet because they cared more about my troubles than my parents did. When I felt that my parents were ignoring my problems, there were other people willing to listen and help me. I knew not to share information like my address, phone number, birthday, or anything like that and I knew not to try to actually meet these people; but I did develop short relationships with strangers online that benefitted me and my health. It is important to look past the fact that a teenager is simply talking to a stranger online and to dig deeper as to why they are talking to them. These interactions are often a cry for attention that goes unnoticed.[12]

Like all teens, she was experiencing physical and psychological changes she didn't quite understand. She wanted to talk about them but felt awkward doing so with her parents.

While friends were an option, she also wanted to get the perspective of an adult or someone anonymous. Friending online strangers, therefore, struck her as a reasonable option.

It may be hard for parents to fathom how strangers can be anything other than a danger. But to a teenager, the unfamiliarity of a stranger may be part of the appeal of interacting with them since it means the teen won't be judged. And if they know when and how to put on the breaks, teens can avoid being exploited. However, while my student could recognize how her online communication might turn dangerous if she continued it, teens who have experienced trauma may not be able to discern a distinction between curiosity-seeking messaging and exploitation.[13]

Like Ashley Reynolds' story, my student's reveals that online victimization of teens isn't fundamentally a by-product of the dangers that lurk on the internet. Seeing the victimization in this way runs the risk of making the internet into a **scapegoat**—something or someone that is unfairly given the blame for a social problem simply because it's an easy target. Instead of exclusively directing our attention to how teens use technology, let's also shine the spotlight on the array of offline injuries that leave certain teens susceptible to online exploitation.

Given that teens spend so much time on social media, their go-to sources for questions related to adolescent development may also be online.
iStockphoto.com/ViewApart

While parents are legitimately concerned about their child messaging with an unfamiliar adult, they may not recognize that the restrictive ways they parent may be driving their child toward this risky behavior.
iStockphoto.com/monkeybusinessimages

Overprotecting Children: The Unintended Consequences of Safeguarding Kids from Risk

11.3 Identify the types of risks that can be healthy for kids to encounter as they grow up.

At some point in your life, you may have heard your parents or grandparents say something like, "When I was growing up, kids would play outside after school and on weekends, but they don't do that anymore." Adults often have nostalgic memories of the way things were. But in this case, they're not exaggerating how childhood used to be. On average, kids spend 50 percent less time in unstructured outdoor play than they did when I was growing up in the 1970s and 1980s. Compared to prior generations, teens nowadays have much less freedom to move about as they please. These restrictions are one reason they spend a lot of time on social media. It's a space

It's uncommon to see kids playing outside without adult supervision.
iStockphoto.com/monkeybusinessimages

where they can be with their friends without their parents around them.[14]

Think about how you got to school when you were younger. In 1969, 89 percent of kids in grades K–8 who lived under a mile away either biked or walked. This group comprised 48 percent of all kids; another 12 percent were driven by a parent, 38 percent took the school bus, and the rest arrived by public transportation or some other means. By forty years later, these numbers had starkly changed. In 2009, just 35 percent of kids who lived under a mile away from school biked or walked. This group comprised 13 percent of how all kids got to school. Forty-five percent were driven by a parent, 39 percent took the school bus, and the rest either used public transportation or another means (see Figure 11.3). So if you didn't walk, you fit the norm for your generation. I *did* walk, which was typical for mine. Starting in kindergarten, my parents allowed me to be unsupervised during the seven-minute commute, which went partially through the woods. If I'd been born a few decades later, they might have gotten curious stares for allowing me to be alone in public at such a young age. While

FIGURE 11.3 ● Generational Differences in Growing Up

Notice the changes over a forty-year period in the ways kids got to school.

	Family vehicle	Walk or Bicycle	School bus	Transit	Other
1969	12%	48%	38%	1%	1%
2009	45%	13%	39%	2%	1%

Source: National Center for Safe Routes to School, "How Children Get to School: School Travel Patterns from 1969 to 2009," http://www.safekidsgf.com/Documents/Research%20Reports/NHTS%20School%20Travel%20Report%202011.pdf.

many factors may influence how a child gets to school, the data in Figure 11.3 highlight that changes over this forty-year period reflect growing parental concerns about kids' safety.[15]

Many parents nowadays won't let their kids be on their own in public even for brief periods of time; in some cases when they have, the police have deemed this to be child neglect. In 2016, a woman was arrested in South Carolina for allowing her nine-year-old nephew and three-year-old son to walk by themselves less than a quarter mile to a local McDonald's because the boys had to cross a busy street and pass several businesses.[16] In 2015, a Florida couple was charged with neglect after a neighbor saw their eleven-year-old son shooting baskets alone in the backyard for an hour and a half. He'd gotten home before his parents and was locked out.[17] Danielle and Alexander Meitiv received national attention when they were arrested in 2014 for allowing their two children, ages six and ten, to walk around their suburban Maryland neighborhood without an adult. The following year, the couple was arrested a second time for letting the children play at the local park unsupervised.[18] These are just a few of the many examples in recent years underscoring the belief that hypersupervision promotes safety and that not supervising kids is an indicator of irresponsible parenting.

In fifth grade, my friends and I were allowed to leave our elementary school during lunch period to walk by ourselves to a local pizzeria located on a busy street.
iStockphoto.com/Travel Wild

While these kids seem to be enjoying a beautiful day, have their parents acted irresponsibly by allowing them to be on the water unsupervised?
iStockphoto.com/Yobro10

Playgrounds are where we can see clear evidence of the adverse consequences of overprotecting kids. The one at my elementary school was typical for its day. Kids whizzed down slides, occasionally scraping their arms or legs if they didn't properly brace themselves toward the bottom. There was a castle I loved to climb—that is, until the time I slipped and tumbled onto the concrete. The playgrounds where you spent time as a child more closely resemble the ones that are common nowadays. Slides are graded to prevent kids from going dangerously fast. Rubber surfaces cushion anyone who might fall.

The change I noticed most when I used to take my two kids to playgrounds involved seesaws, which I learned had been renamed teeter-totters. As a kid, part of the thrill was in trying to get the person on the other side to fall suddenly. The name change reflected a redesign done to enhance safety. *Seesaw* referred to how the child higher up could tumble to the ground while the other child rapidly ascended upward.

When I was a young child in the 1970s, playground surfaces were either hard grass or concrete. Falling off a swing or tumbling from the monkey bars hurt a lot, especially if the surface was concrete.
iStockphoto.com/eyjafjallajokull

Playground surfaces nowadays have foam padding that provides protection for kids if they fall.
iStockphoto.com/SerrNovik

Given concerns nowadays about playground safety, it's amazing that at the Land Adventure Playground children are encouraged to play with fire.
Getty/jurgita.photography

American parents sometimes go to extreme measures to ensure that their children play safely.
Getty/D. Anschutz

Teeter-totter conveys how kids on both sides can gently float up and down. This long-time playground favorite now has a mechanism that prevents free falls.

Consider an unusual playground in Wales. The Land Adventure Playground is designed to offer a level of risk far greater than I got to experience growing up. Kids can swing across a creek on a frayed rope. There are hammers, saws, and other tools for building structures. Although paid adult "playworkers" watch from afar, they seldom meddle in kids' creative play or exploration. Children feel the excitement that comes from sensing there's danger in their midst and figuring out how to cope with it. This and similar adventure playgrounds across Europe have gotten lots of U.S. media coverage in recent years because the experience kids have sharply contrasts with the prevalent parenting philosophy that it's better to be safe than sorry.[19]

However, growing up in a society that constantly tries to minimize risks to kids may undermine their long-term interests. A growing body of research highlights that aggressive efforts to protect kids from danger can have a **boomerang effect**, hurting them in unforeseen ways. While it seems like a no-brainer that adults ought to do everything they possibly can to shield children from harm, overprotection carries the cost of kids growing up ill-equipped to handle life's many social and emotional challenges. Giving them chances, whether at the playground or on the internet, to feel like they're taking risks can be crucial for teaching them how to handle dangerous situations and lead healthy adult lives.[20]

Strangers with Benefits: Discovering the Advantages of Interacting with Unfamiliar People in Public

11.4 Recognize the ways in which interacting with strangers can be beneficial.

Imagine you've just boarded a bus, train, or plane. You find your seat, take out your phone, and open Snapchat. A few minutes later, a person sits down next to you and asks about your day. How does it feel to talk to them? When I present this scenario in class, many students shudder. At best, having a conversation with a stranger feels awkward. At worst, if it's a man, he may come across as the sort of person you've grown up fearing, particularly if you're a woman.[21]

Sometimes, however, there's nothing to fear but fear itself. Research indicates that interacting with strangers can produce the satisfaction of having connected with another person in an otherwise unemotional and anonymous setting. Consider an experiment where researchers offered a group of train commuters each a $5 Starbucks gift card if they had a conversation with the person next to them. These study participants reported having a more enjoyable ride than a second group of people who weren't offered the gift card and spent their commute as they pleased, such as on their phone or napping.[22]

If the results of this study surprise you, that's because they call into question messages you've heard your entire life about why you should harbor suspicions toward strangers. While it's unlikely for people raised on such messages to allow themselves to have positive interactions with strangers, these sorts of interactions are common in many parts of the world. In such places, the thought of talking to a stranger doesn't produce anxiety—a sign that kids grow up with different beliefs than what's common in the U.S. As a result, these kids learn to see strangers as potentially trustworthy.[23]

In Cameroon, mothers teach children from infancy to feel calm around, and accepting of, people unfamiliar to them.
iStockphoto.com/jclanciault

Although conversations among dog walkers are often brief and involve the same topics (breed, age, friendliness), these interactions are significant because they create a fleeting intimacy that's rare in public.
Alexander/Getty Images

Talking to a stranger for even a few minutes while taking public transportation can break down societal walls of mistrust.
iStockphoto.com/xavierarnau

But even for Americans raised on stranger-danger fears, interacting with unfamiliar people in public can be positive. We can begin to understand why by recognizing that people tend to put on their best face around those they don't know. They're not apt to complain or act temperamental, as can happen in the presence of friends or relatives. If the next time you're on a bus or train you strike up a conversation with the person sitting next to you, you may find yourself comfortable sharing with them details about your life. That's because they don't have a reason to judge you and you're unlikely ever to see them again.[24] I'm often reminded of the advantages of interacting with strangers when I'm walking in the neighborhood or at the park with Cinnamon, who's a twenty-eight-pound mixture of a cocker spaniel and Cavalier King Charles spaniel. Her presence fuels conversations that likely wouldn't take place otherwise. She makes people smile and let down their guard. Being in public with a baby can have the same effect.[25]

If you're wondering how you can reap the benefits of interacting with strangers without a dog or a baby, Kio Stark has some ideas for you. She's an author and speaker who offers useful tips for people who dread talking to strangers. Her advice is all the more unconventional when we consider that she has a young daughter she must teach to exercise discretion around unfamiliar adults. In a TED Talk, Stark described how being open to unfamiliar people is a crucial way to make our society more tolerant since stranger-danger warnings often stem from racial or ethnic biases. Even though these warnings may not explicitly mention such biases, kids may grow up holding particular suspicions toward people who look different from themselves. Indeed, one reason stranger-danger fears have endured over the years is because they seem generalizable rather than racially or ethnically targeted. Stark's message is timely given the strong anti-immigrant sentiment in the United States. She demonstrates an alternative to dividing the world into people like "us" whom we can trust and "them" whom we ought to fear. Her message is that we can instead evaluate strangers as we do the people we already know, based on their worth as *individuals*.[26]

The internet is creating new opportunities for positive interactions with strangers. Consider Airbnb, where people rent out travel accommodations in their homes. In 2015, my wife and I reaped the benefits of this opportunity when we visited Nova Scotia. Our host, Ann, showed us around her farm and invited us into her living room for a beer. We learned that she'd graduated from Framingham State University, where I teach. We talked for the next two hours like old friends. Two summers later, a different Airbnb host named CK became our instant confidant. After my son and I mistakenly got into an overpriced taxi at the Hong Kong airport, he instructed us via WhatsApp about how to avoid being scammed and waited for an hour outside his apartment for us to arrive. Because of this information he offered us, we felt he was someone we could trust.

Airbnb is more than just a site for booking travel accommodations. The user profiles and reviews on the site enable hosts and guests to vet one another and share limited amounts of personal information that can be the basis for forging mutual trust.
Andrew Harrer/Bloomberg via Getty Images

Other websites where people sell services similarly open up opportunities for positive interactions between strangers; examples include Rover (dog-sitting), Zilok (renting power tools), and Feastly (getting together to eat home-cooked meals). Then, of course, there are ridesharing sites like Uber and Lyft. A study of a different company popular in Europe called BlaBlaCar found that 88 percent of users highly trust drivers with complete profiles. This was a *greater* level of trust than people have in neighbors or colleagues and almost as high as they have in family members. These e-commerce sites that comprise the so-called sharing economy are among the few settings online where people place trust in strangers.[27]

However, the mobile devices we use to access these sites may inhibit people from interacting with strangers in public. When you're riding the bus or sitting in the park, isn't it more convenient to message with friends than engage in conversation with people nearby? Talking to strangers in public used to be simply a matter of being polite—unless, of course, the other person behaved inappropriately. While these types of interactions still occur, the person who initiates a face-to-face conversation may come across as impolite for interrupting others on their mobile devices. A tablet or smartphone seems to signal that although its user is physically present in a public place, others should treat them as if they're in a private space.[28]

Distancing ourselves from unfamiliar people by clinging to our phones can have dire consequences. If you're waiting on the corner to cross the street and the person next to you slips and falls, of course the right thing to do is offer help. However, in such situations there's an impediment to being a Good Samaritan. The **bystander effect** refers to public situations where a person needs help, yet someone who witnesses their distress chooses not to offer assistance because they believe that others will instead. Using phones and tablets in public may strengthen the bystander effect by making a person feel that they're not actually in public and therefore are not accountable to the strangers in their midst who need help. Having these devices may even lead some

The *New York Post* on December 4, 2012, showed a man who had been pushed onto the subway track and was about to be hit by an oncoming train. A bystander took the photo; in the predigital world they instead might have tried to save the man's life.
RICHARD B. LEVINE/ Newscom

people to respond voyeuristically to others' distress. In other words, they may see the distress as an impetus to take pictures and share them on social media rather than to do good.[29]

We're seeing that strangers come in many different forms. Sometimes they confirm our worst fears, such as when online predators prey on kids. At other times, they are in distress and need our help. Most of the time, strangers are neither *a* danger nor *in* danger. They're merely unfamiliar people with whom we share public space. Interacting with them can open up opportunities for fleeting moments of intimacy. Through such interactions, strangers can enable us to develop the good feeling that comes from having made a brief connection within an otherwise anonymous setting. Strangers offer lots of unforeseen benefits—that is, if you're willing to open yourself up to such a possibility.

Thinking Straight about Threats to Children

11.5 Compare public fears regarding the dangers kids face with the realities of these dangers.

Still, when it comes to strangers, most people think danger. Nothing is more terrifying to parents than an unfamiliar adult taking their child. It's among their top worries, eclipsing concerns about kids getting shot, attacked, or becoming addicted to drugs (see Figure 11.4). However, when a child goes missing, the culprit is rarely a strange man who has lured them with candy or climbed a ladder to abduct them while they're asleep. Table 11.1 indicates that much more often it's one of the parents who has taken the child amidst a custody battle. If a mother wants to keep her children safe, according to journalist Hanna Rosin, "her ironclad rule should not be: Don't talk to strangers. It should be: Don't talk to your father."[30]

These data align with what we saw earlier in this chapter when we exposed the childhood inequalities that lie at the root of sextortion and other forms of internet exploitation. Examining the lives of the teens most susceptible to online victimization exposes offline threats that come most often from family members, not strangers. A study of sextortion cases found, moreover, that about 60 percent did not originate with catfishing by an unfamiliar person. Instead, the exploitation began when a romantic relationship ended and one partner subsequently threatened to share sexually explicit photos taken of the other before the breakup.[31]

Indeed, it may surprise you that the internet can actually *protect* kids from stranger danger. Contrary to the prevailing narrative we often hear about the threats to life online, sex crimes against children have decreased since the late 1990s when kids first started spending time on the internet. Just as the web gives predators new

FIGURE 11.4 ● Anxieties about Kids' Safety

Here are the percentages of parents who say they worry that the following harms might happen to their children.

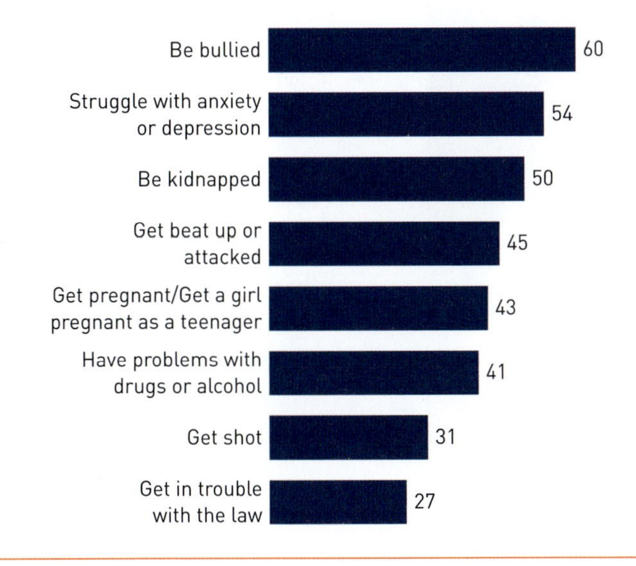

Source: Drew DeSilver, "Dangers That Teens and Kids Face: A Look at the Data," Pew Research Center, January 14, 2016, http://www.pewresearch.org/fact-tank/2016/01/14/dangers-that-young-people-face-a-look-at-the-data.

TABLE 11.1 ● Myth Busting

Kidnapping by a stranger is the least likely reason children go missing.

Reasons Kids Go Missing	Number (2015)
Kidnapping	332
Taken by a noncustodial parent	2,249
They have run away	302,000

Source: Adapted from Drew DeSilver, "Dangers That Teens and Kids Face: A Look at the Data." Pew Research Center, January 14, 2016, http://www.pewresearch.org/fact-tank/2016/01/14/dangers-that-young-people-face-a-look-at-the-data.

pathways for manipulation, it also enables law enforcement agents to catch these predators before they can inflict harm. And, like the rest of us, online predators leave digital footprints that can include self-incriminating evidence. Moreover, as teens migrated from chat rooms to social media sites during the 2000s, they

Because some of the time kids spend online displaces time that they'd otherwise physically be with their friends, the internet may keep them out of the sort of trouble that can arise when kids are hanging out in public.
iStockphoto.com/StefaNikolic

School assemblies about cybersafety are a setting where kids are reminded that the internet is a minefield of danger.
Enigma/Alamy Stock Photo

Because mobile devices constantly alert us about bad news, it's easy to believe that kids are growing up in a more menacing world than in the past.
iStockphoto.com/artiemedvedev

became more likely to limit their interactions only to the people they knew.[32]

In many ways, kids are safer nowadays than when your parents were growing up. Consider these trends:

- Reports of missing children have dropped by 40 percent since 1997.
- From 1993 to 2013, there was over a two-thirds decrease in children killed by automobiles while walking.
- Whereas in 1935 about 450 of every 100,000 children between ages one and four died, the mortality rate among this age group has fallen more than tenfold to 30 deaths for every 100,000 kids.[33]

Yet these facts may seem beside the point. After all, we hear so much more about harms to kids than we did before the digital age. Social media and innumerable websites spread information about these harms like wildfire. Even rare threats evoke alarming images that stir anxiety in kids and adults alike. It took just a couple of clicks for me to learn the terrifying stories of sextortion that opened this chapter. Although they're exceptional, they linger in our minds as reason to fear for kids' safety.

The very way I told those stories illustrates how publicity around relatively rare acts of stranger danger play on our emotions. Horror stories like these are more psychologically gripping than statistics indicating that kids actually face fewer risks to their safety nowadays than in past generations. It's no wonder why many parents restrict their kids' autonomy. As a father, I get where this impulse comes from. When our two kids were younger, Nancy and I attended school workshops about internet safety and spent many hours discussing what age was appropriate for them to have phones. We've had our share of worries about their whereabouts and have compared notes with friends about how to limit their web surfing and social media use. These concerns remain legitimate even if dangers to kids are no greater nowadays than in prior generations. Yet the cautionary tale in adhering too closely to "better safe than sorry" is that this behavior can, ironically, pose unintended dangers to kids as they grow up.[34]

What Do You Know Now?

1. What are the offline physical and psychological injuries certain kids have experienced throughout their lives that may lead them to become interested in messaging with strangers? What is it about having experienced these injuries that makes victims curious about interacting online with people they don't know?
2. How does the significant publicity given to a story like Ashley Reynolds' divert attention from seeing how these offline injuries contribute to kids' online exploitation?
3. In what sense can parental efforts to shield kids from danger unintentionally hurt them? What do we risk doing to kids when we raise them in environments devoid of risk?
4. In what ways can interacting with strangers sometimes be positive?
5. How persuasive do you find the argument that the internet can offer kids a safe haven from danger? Why do you think there is little publicity of this perspective but we so often hear that the internet can be a scary place for kids?

Key Terms

Catfishing 219
Sextortion 219
Grooming 221
Canary in the coalmine 223
Scapegoat 225
Boomerang effect 229
Bystander effect 231

Visit **edge.sagepub.com/silver** to help you accomplish your coursework goals in an easy-to-use learning environment.

Notes

1. Denny Walsh, "Paradise Man Sentenced for Online Solicitation," *The Sacramento Bee*, May 22, 2015, http://www.sacbee.com/news/local/crime/article21712116.html.
2. U.S. Immigration and Customs Enforcement, "Internet Predator Sentenced in Federal Court for Producing Child Pornography," April 26, 2012, https://www.ice.gov/news/releases/internet-predator-sentenced-federal-court-producing-child-pornography.
3. Mary Anne Stokes, "Stranger Danger: Child Protection and Parental Fears in the Risk Society," *Amsterdam Social Science* 1, no. 3 (2009): 6–24; Yvonne Jewkes, "Much Ado about Nothing? Representations and Realities of Online Soliciting of Children," *Journal of Sexual Aggression* 16, no. 1 (2010): 5–18.
4. Pew Research Center, "Teens, Privacy, and Online Social Networks," April 18, 2007, https://www.pewinternet.org/2007/04/18/friendship-strangers-and-safety-in-online-social-networks/.
5. Brian Hansen, "Cyber-Predators: Can Internet Child Sexual Exploitation Be Controlled?" *CQ Researcher* 12, no. 8 (2002), https://library.cqpress.com/cqresearcher/document.php?id=cqresrre2002030100; Justin W. Patchin and Sameer Hinduja, "Sextortion among Adolescents: Results from a National Survey of U.S. Youth," *Sexual Abuse*, 2018, https://doi.org/10.1177/1079063218800469.
6. Kimberly Mitchell, David Finkelhor, and Janis Wolak, "Risk Factors for and Impact of Online Sexual Solicitation of Youth," *Journal of the American Medical Association*, June 20, 2001, http://jamanetwork.com/journals/jama/fullarticle/193923.
7. Patchin and Hinduja, "Sextortion among Adolescents."
8. CBS News, "FBI Seeks 240 Victims of Man Imprisoned for Sextortion," July 8, 2015, http://www.cbsnews.com/news/fbi-seeks-240-victims-of-man-imprisoned-for-sextortion/.
9. Liz Brody, "Meet Ashley Reynolds, the Woman Fighting Sextortion," *Glamour*, July 7, 2015, http://www.glamour.com/story/ashley-reynolds-the-woman-fighting-sextortion.
10. Janis Wolak et al., "Online 'Predators' and Their Victims: Myths, Realities, and Implications for Prevention and Treatment," *American Psychologist* 63, no. 2 (2008): 111–28.
11. National Child Traumatic Stress Network, "Effects," https://www.nctsn.org/what-is-child-trauma/trauma-types/complex-trauma/effects.
12. I have quoted my student with her permission and with my promise to keep her identity anonymous.
13. Amy Adele Hasinoff, "Sexting as Media Production: Rethinking Social Media and Sexuality," *New Media & Society* 15, no. 4 (2012): 449–65.
14. Alliance for Childhood, "The Loss of Children's Play: A Public Health Issue," 2010, http://www.habitot.org/museum/pdf/play_research/Health_brief.pdf; Danah Boyd, *It's Complicated: The Social Lives of Networked Teens* (New Haven, CT: Yale University Press, 2014), 84–93.
15. Gill Valentine, *Public Space and the Culture of Childhood* (Burlington, VT: Ashgate, 2004), 15–30.
16. Lenore Skenazy, "Mom Arrested for Letting Kids Walk to McDonald's around the Corner," *Hit & Run Blog*, April 7, 2016, http://reason.com/blog/2016/04/07/mom-arrested-for-letting-kids-walk-to-mc.
17. Fox News, "Parents Charged with 'Neglect' after 11-Year-Old Plays in Yard for 90 Minutes," June 14, 2015, http://insider.foxnews.com/2015/06/14/florida-parents-charged-felony-neglect-after-11-year-old-son-plays-backyard-90-minutes.
18. Kelly Wallace, "Maryland Family under Investigation Again for Letting Kids Play in Park Alone," CNN, April 24, 2015, http://www.cnn.com/2015/04/13/living/feat-maryland-free-range-parenting-family-under-investigation-again/.
19. Hanna Rosin, "The Overprotected Kid," *The Atlantic*, March 19, 2014, http://www.theat

lantic.com/magazine/archive/2014/04/hey-parents-leave-those-kids-alone/358631/.

20. One illustration within this body of research is Ellen Beate Hansen Sandseter and Leif Edward Ottesen Kennair, "Children's Risky Play from an Evolutionary Perspective: The Anti-phobic Effects of Thrilling Experiences," *Evolutionary Psychology* 9, no. 2 (2011): 257–84.
21. J. R. Thorpe, "Why Don't We Like Talking to Strangers," *Bustle*, September 29, 2016, https://www.bustle.com/articles/186760-why-dont-we-like-talking-to-strangers.
22. Nicholas Epley and Juliana Schroeder, "Mistakenly Seeking Solitude," *Journal of Experimental Psychology: General* 143, no. 5 (2014): 1980–99.
23. For evidence that children outside the U.S. do not grow up learning to fear strangers, see Hiltrud Otto and Heidi Keller, "A Good Child Is a Calm Child: Mothers' Social Status, Maternal Conceptions of Proper Demeanor, and Stranger Anxiety in One-Year-Old Cameroonian Nso Children." *Psychological Topics* 1 (2015): 1–25, http://www.academia.edu/20976510/Cultural_Differences_in_Stranger-Child_Interactions_A_Comparison_Between_German_Middle-Class_and_Cameroonian_Nso_Stranger-Infant_Dyads.
24. Gillian M. Sandstrom, "Social Interactions and Well-Being: The Surprising Power of Weak Ties," *Personality and Social Psychology Bulletin*, April 25, 2014, https://pdfs.semanticscholar.org/822e/cdd2e3e02a3e56b507fb93262bab58089d44.pdf.
25. One study highlighting the role dogs play in fostering interactions among strangers is Lisa J. Wood et al., "More Than a Furry Companion: The Ripple Effect of Companion Animals on Neighborhood Interactions and Sense of Community," *Society & Animals* 15 (2007): 43–56.
26. Kio Stark, *When Strangers Meet: How People You Don't Know Can Transform You* (New York: TED Books, 2016), 38.
27. Frederic Mazzella et al., "How Digital Trust Powers the Sharing Economy," *Third Quarter* 30 (2016): 24–31.
28. Christine Rosen, "Our Cell Phones, Ourselves," *The New Atlantis: A Journal of Technology and Society* (Summer 2004): 26–45.
29. Christine Rosen, "Are Smartphones Turning Us into Bad Samaritans?" *Wall Street Journal*, October 25, 2013, https://www.wsj.com/articles/are-smartphones-turning-us-into-bad-samaritans-1382748843.
30. Rosin, "The Overprotected Kid."
31. Janis Wollak and David Finkelhor, "Sextortion: Findings from a Survey of 1,631 Victims," Crimes Against Children Research Center, University of New Hampshire, June 2016, https://rems.ed.gov/Docs/SextortionFindingsSurvey.pdf.
32. Lisa M. Jones, Kimberly J. Mitchell, and David Finkelhor, "Trends in Youth Internet Victimization: Findings from Three Youth Internet Safety Surveys 2000–2010," *Journal of Adolescent Health* 50, no. 2 (2012): 179–86.
33. A listing of these trends indicating greater child safety comes from Christopher Ingraham, "There's Never Been a Safer Time to Be a Kid in America," *Washington Post*, April 14, 2015, https://www.washingtonpost.com/news/wonk/wp/2015/04/14/theres-never-been-a-safer-time-to-be-a-kid-in-america/?utm_term=.68a298b69bc1.
34. For a discussion of how individual horror stories play on our emotions, see Paul Slovic, "'If I Look at the Mass I Will Never Act': Psychic Numbing and Genocide," *Judgment and Decision Making* 2, no. 2 (2007): 79–95, http://journal.sjdm.org/7303a/jdm7303a.htm.

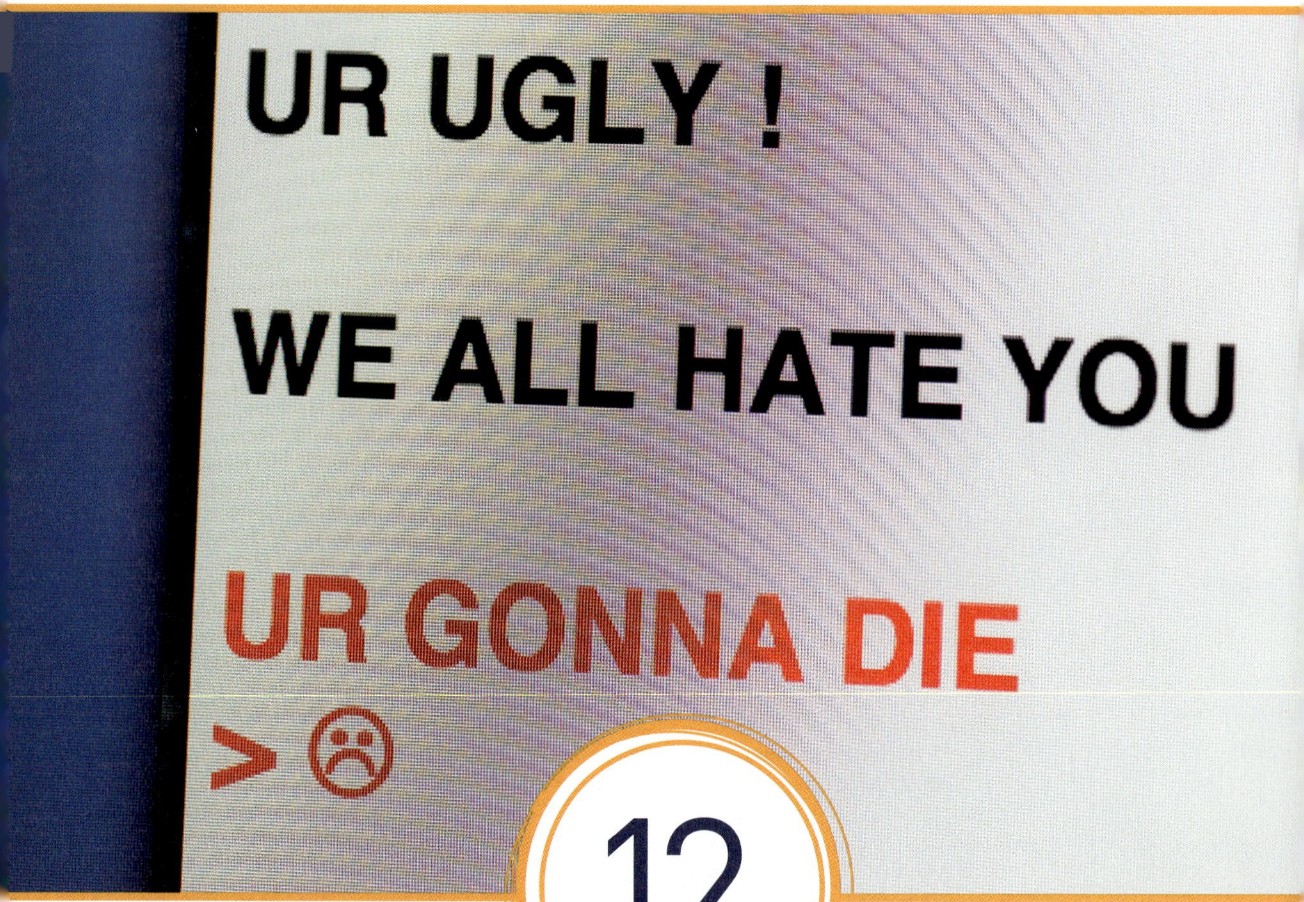

12

Have Kids Gotten Meaner?

An Up-Close Look at Cyberbullying and Suicide

Learning Objectives

1. Recognize the ways text-based communication has changed teens' capacity for meanness relative to prior generations.
2. Describe how cyberbullying is rooted in social norms among teens.
3. Explain how social forces, such as social class and social networks, influence a teen's susceptibility to developing suicidal thoughts.
4. Identify how focusing on the link between cyberbullying and suicide diverts attention from the root causes of both of these social problems.

Any Time and Place: How Teens Use Technology to Act Cruelly toward One Another

12.1 Recognize the ways text-based communication has changed teens' capacity for meanness relative to prior generations.

"I saw him making out with a dude," Dharun Ravi tweeted the evening of September 19, 2010. A first-year student at Rutgers University, Ravi had set up a webcam in his dorm room earlier that day. It was a prank so that he and his friends could spy on his roommate, Tyler Clementi. From down the hall, they watched Clementi intimately embracing a man he'd recently met. The group of friends peeked in a second time the next evening. Little could they have imagined that three days later Clementi would jump off the George Washington Bridge and plunge to his death. Prosecutors combed through Ravi's social media records in building their case that the webcam spying was an egregious form of bullying that outed Clementi as gay. Ravi was convicted on several counts in 2012, but a New Jersey court overturned the conviction. In 2016, he pleaded guilty to one count of attempted invasion of privacy.[1]

Almost exactly three years later, in Lakeland, Florida, twelve-year-old Rebecca Sedwick experienced the same fate after having been bullied online. Upon hearing about Sedwick's new boyfriend, two girls sent her Facebook messages calling her ugly and telling her to drink bleach and die. The morning of September 9, 2013, Sedwick changed her name on Kik Messenger to "That Dead Girl" and then leaped off a platform at an

Whereas webcams have many benign purposes, such as enabling grandparents to connect with grandchildren who live far away, the story that unfolded at Rutgers University highlights how webcams can also be used to invade others' privacy.
gary corbett/Alamy Stock Photo

Doug Steley C/Alamy Stock Photo

When Mark Zuckerberg came up with the idea for Facebook as a platform for people to connect with anyone they've ever known, he didn't envision how kids might use it to exploit one another.
Ian Dagnall/Alamy Stock Photo

abandoned cement plant.[2] Afterward, one of the girls wrote on Facebook, "Yes, I bullied REBECCA and she killed herself but IDGAF."[3]

Cyberbullying is when one or more people repeatedly ridicule another person via text or e-mail, post mean content about them on social media, or use technology in some other way to hurt them. Cyberbullying among teens increased 80 percent from 2007 to 2016. These were critical years since 2007 was when the first iPhone hit the market and when Facebook users surpassed the one million mark.[4] Just over a third of middle and high school students report having been cyberbullied at some point in their lives, and 16.9 percent say this occurred during the past month (see Figure 12.1).[5]

FIGURE 12.1 ● Cyberbullied

Here's a snapshot of the many ways teens experience cyberbullying.

	Percent
I have been cyberbullied (lifetime)	33.8
previous 30 days:	
I have been cyberbullied	16.9
Mean or hurtful comments online	22.5
Rumors online	20.1
Posted mean names or comments online about me with a sexual meaning	12.7
Threatened to hurt me online	12.2
Threatened to hurt me through a cell phone text	11.9
Posted a mean or hurtful picture online of me	11.1
Pretended to be me online	10.3
Posted mean names or comments online about my race or color	10.1
Posted mean or hurtful video online of me	7.4
Created a mean or hurtful web page about me	7.1
One or more of above, two or more times	25.7

Source: Sameer Hinduja and Justin W. Patchin, "Cyberbullying Victimization," Cyberbullying Research Center, 2016, https://cyberbullying.org/2016-cyberbullying-data.

These data suggest that there's reason for concern about the spread of bullying from the physical places where it occurred when I was a kid, such as school hallways and playgrounds, to the limitless terrain of cyberspace. With mobile devices, teens may exhibit cruelty 24/7. They may feel emboldened to send or post malicious messages because by hiding behind a screen, they don't have to confront—let alone feel remorse about—the emotionally destructive consequences of their actions. Given how easily they can use their mobile devices to cut one another down, it's reasonable to wonder whether kids are meaner nowadays than before the invention of these devices.

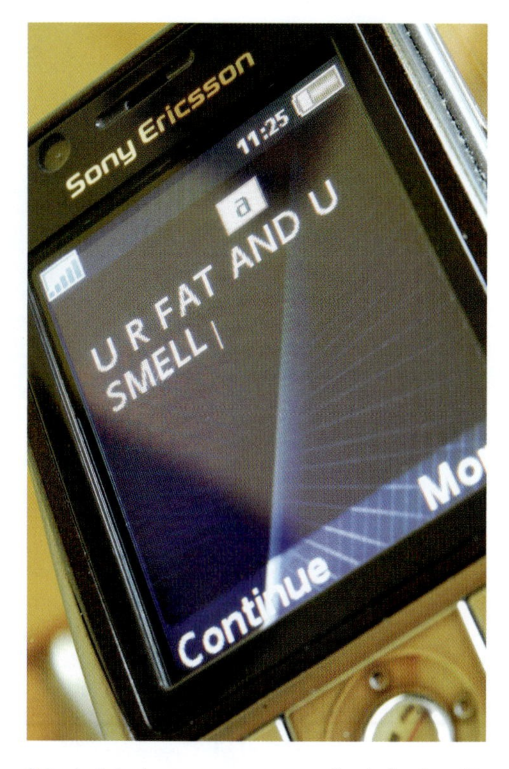

Cyberbullying became a common practice during the mid-2000s when phones started having texting capability.
Rawdon Wyatt/Alamy Stock Photo

FIRST IMPRESSIONS?

1. Why might a person be more inclined to behave meanly via a text or social media post than in face-to-face interaction?
2. How does the pain of cyberbullying defy the expression "Sticks and stones may break my bones but names will never hurt me"?
3. In what sense is it reasonable to believe that Tyler Clementi's and Rebecca Sedwick's deaths were the result of cyberbullying?

Despite the impression media reports give, victims of cyberbullying often do not experience suicidal thoughts, let alone act on them, and the majority of teens who die from suicide were not bullied. Indeed, Tyler Clementi and Rebecca Sedwick would be unfamiliar to most people if they hadn't subsequently taken their own lives after being cyberbullied. We also wouldn't know about them if they'd died from suicide but *hadn't* been cyberbullied beforehand. Nonetheless, stories about suicide in the wake of cyberbullying get lots of media attention. It's easy to draw a causal link between the two social problems, leading a reasonable person to believe that teens take their own lives because they've been cyberbullied.[6]

We can see this assertion of causality in the initial media reports about Tyler Clementi's death. Sources characterized Dharun Ravi—Clementi's roommate and ringleader of the webcam spying—as having forced Clementi to come out of the closet, which then propelled

Smartphones have made it easy to bully someone at any time and any place.
iStockphoto.com/Tero Vesalainen

Students walk past the Rutgers University dorm where Tyler Clementi shared a room with Dharun Ravi for about a month before Clementi died from suicide.
EMMANUEL DUNAND/AFP/Getty Images

him to jump to his death. Presumably, being publicly exposed as gay was so humiliating that Clementi couldn't bear living any longer. From this individual perspective, cyberbullying is a black-and-white issue; there is an identifiable perpetrator to blame and an innocent victim deserving of compassion. When cases of cyberbullying get publicity, it's easy to see the problem through this perspective, viewing the bully as a monster.[7]

This chapter uses the sociological perspective to delve beneath this characterization of cyberbullying in news reports. This perspective exposes the roots of this social problem in teen culture. Looking at the problem from this point of view reveals that the distinction between perpetrators and victims may be murky. Additionally, we'll unpack the link between cyberbullying and suicide, examining some of the underlying reasons why some young people take their own lives.

A Peek Inside Teen Culture: Uncovering Social Forces That Give Rise to Cyberbullying

12.2 Describe how cyberbullying is rooted in social norms among teens.

Most media accounts of Dharun Ravi's behavior during his first few weeks at Rutgers University portrayed him through an individual perspective as self-absorbed, arrogant, and homophobic. Examining the social forces that shaped how Ravi treated his roommate enables us to contextualize the events preceding Tyler Clementi's death. There's much more to the story than simply that Ravi exploited Clementi's vulnerabilities as a closeted gay male. We have ample evidence to consider since the two boys left a massive digital footprint that began the preceding summer when they first learned they would be roommates.

Whereas news reports portrayed Dharun Ravi as a self-absorbed homophobe, the sociological perspective exposes the social forces that lead some heterosexual teenage boys to act meanly toward their gay peers.
ASSOCIATED PRESS/Pool The Star-Ledger

I've often asked my students to evaluate Ravi's behavior relative to what they've noticed online. At one extreme, some might regard it as more malicious than any other behavior they've ever seen; at the other extreme, some might suggest it was just typical of how teens use technology. My students consistently rate his behavior as somewhere in the middle. In other words, even though the webcam prank was stupid and immature, it wasn't atypical relative to the behavior of other teens. Seeing Ravi in this way doesn't excuse his actions, but it does highlight why viewing cyberbullying simply as an expression of meanness is a limited way to understand this social problem. A richer approach is to focus on the forces within teen culture that give rise to cyberbullying.

Seeking Social Status

When he found out Tyler Clementi was his roommate, Dharun Ravi came to see him—in the words of journalist Ian Parker—as "material for a 'gay roommate' news scoop."[8] Parker's investigation of the events preceding Clementi's suicide revealed an underlying reason for Ravi's malicious behavior. As a heterosexual teenage guy who cared a lot about his image, Ravi wanted to do anything he could to show his bros that he fit in among them. It's common for some guys to use phrases like "no homo" to impress upon their friends that they're not gay. However, such language isn't necessarily an indicator of homophobia. It's telling what Ravi texted just hours before Clementi made his way to the George Washington Bridge:

> I've known you were gay and I have no problem with it. In fact, one of my closest friends is gay and he and I have a very open relationship. I just suspected you were shy about it which is why I never broached the topic. I don't want your freshman year to be ruined because of a petty misunderstanding, it's adding to my guilt.[9]

The webcam prank, therefore, wasn't merely an act of individual deviance but reflected conventional, albeit disturbing, norms within heterosexual male teen culture. The prank exhibited **toxic masculinity**, or the idea that being a "real man" hinges on acting abusively toward others, and often toward oneself too (see Chapters 8 and 9 for a fuller discussion of this concept).[10]

Just as norms about masculinity gave rise to the mean behavior Dharun Ravi displayed toward Tyler Clementi, the cyberbullying of Rebecca Sedwick was also rooted in teen culture. Recall that two girls sent her scathing Facebook messages after learning about her new boyfriend. While clearly they were jealous, their biting words also reflected the pressure heterosexual teen girls feel to fit in with their peers. Often a girl's most valuable asset for gaining and maintaining popularity is her attractiveness to boys. The girls who bullied Sedwick did so to defend their turf due to the threat her boyfriend posed to their status. The nasty Facebook posts underscored these

Bullying other boys is often a way for teenage guys to demonstrate their "masculinity" to their friends.
iStockphoto.com/SolStock

Girls' bullying of one another may reflect the competition they feel in trying to appear attractive to boys.
iStockphoto.com/FatCamera

girls' precarious position within the hierarchy of middle school. If girls like them do not continually engage in efforts to boost their popularity—including bullying—they risk becoming unpopular.[11]

Girls experience cyberbullying more often than boys, yet boys more frequently cyberbully others (see Figure 12.2). These trends pertain to all instances of cyberbullying, not just those that involve boys targeting boys or girls targeting girls. The reason for focusing on gender-segregated cases like the bullying of Clementi and Sedwick is to highlight how these cases reflect teen expectations about heterosexual gender roles. Think about ways that other cases of cyberbullying involving a mixture of boys, girls, and trans students might also be rooted in teen culture.

FIGURE 12.2 ● Cyberbullying by Gender

Whereas girls experience cyberbullying more often than boys, boys more frequently cyberbully others.

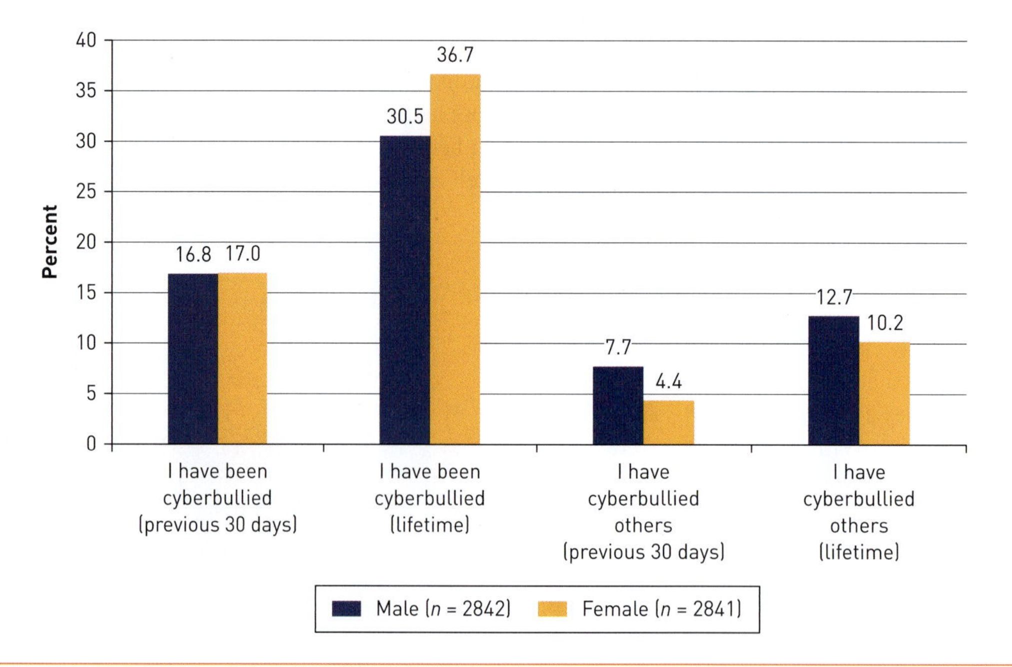

Source: Data are from a nationally representative sample of 5,683 12- to 17-year-olds in the U.S. Sameer Hinduja and Justin W. Patchin, "Cyberbullying Victimization," Cyberbullying Research Center, 2016, https://cyberbullying.org/2016-cyberbullying-data.

Understanding cyberbullying from a sociological perspective casts this behavior in an entirely different light than how we typically see it. The conventional wisdom is that the problem stems from popularity-obsessed teens using social media to prey on innocent victims. However, we're discovering that kids who appear as outliers for acting meanly toward their peers are exhibiting behavior that reflects *norms* within teen culture.

Whether they are on social media, at the mall, or hanging out outside, teens have the power to define their own markers of social status.
iStockphoto.com/FatCamera

One way to emphasize the significant influence of this culture is by recognizing how teens differ from both children and adults. Compared to children, teens spend more time with friends, and their friends exert a greater impact on them. As a result, teens often try to differentiate themselves from their parents by, for example, adopting tastes in new types of music or clothing styles. But relative to that of adults, teens' autonomy is still limited. Not only do they experience legal restrictions on drinking and voting but parents may also prohibit where they can go because of fears for their safety. However, there's one dimension of their lives over which teens have nearly total control: how they evaluate one another. We can understand bullying, gossiping, spreading rumors, teasing, and pranking as behaviors that enable teens to size up where people fit within the pecking order.[12]

Social forces that extend beyond teens' lives give them license to engage in these behaviors. Let's consider three of these forces:

Most teens are unlikely to become megastars like Liza Koshy did by posting comic videos online and creating a popular YouTube channel. Still, social media creates opportunities for attaining a degree of celebrity within one's network.
Presley Ann/FilmMagic

1. **The rewards attached to bullying.** In the business world, people who are willing to do whatever it takes to close the deal often get ahead. Likewise, politicians often muscle their way toward accomplishing what they want rather than engaging in deliberative dialogue with their adversaries. Donald Trump's supporters *and* his critics can agree on one thing: Bullying others to get what he wants is one reason for his political success.[13]

2. **The power of advertising.** Ads appear anywhere that people may pay attention to them, such as on billboards, television, websites, and smartphones. While advertisements promote

many different products, they all demonstrate that capturing others' attention can be extremely valuable. Cyberbullying has the same effect; people who behave outrageously online get noticed.[14]

3. **The cult of celebrity.** Teens often identify with people in the public eye whose fame hinges on winning others' approval. Instagram, Snapchat, and Twitter offer 24/7 platforms for accumulating influence, making it possible for a person to achieve a measure of celebrity among their friends and followers.[15]

Therefore, it's useful to see cyberbullying as a by-product of social forces that exist outside teen culture.

The Drama of Adolescence

Examining how cyberbullying is a by-product of teen culture uncovers another significant story. Whereas outsiders to this culture, such as parents, teachers, and lawmakers, may be quick to define instances of online meanness as cyberbullying, status-conscious teens are more inclined to view this behavior as "drama." This distinction matters since a person's role on social media is malleable from one situation to the next. They may be the target of meanness in one instance and the perpetrator in another. Even a teen who has been exploited may retaliate and assert power.[16]

To see how this can be, let's return to the events preceding Tyler Clementi's suicide. While media attention after his death focused on Dharun Ravi's homophobia, Clementi also exhibited bias. On their very first day together at college—and with Ravi unpacking his bags on the other side of their shared dorm room—Clementi messaged to a high school friend, "I'm reading his twitter page and umm he's sitting right next to me. I still don't kno how to say his name." The friend replied, "Fail!!!!! that's hilarious." Clementi then commented that his roommate's parents appeared "sooo Indian first gen americanish" and "defs owna dunkin."[17] This interaction between Clementi and his friend did not become part of the mainstream media story of what transpired between the two roommates prior to Clementi's death. Because he took his own life, for journalists to have mentioned his racial biases might have come across as blaming the victim and compromised Ravi's public image as a poster-child bully.

Focusing on teen culture exposes that bullies and victims may resemble one another in the emotional pain they've experienced throughout their lives.
iStockphoto.com/Tijana87

I'm certainly not suggesting Clementi was at fault for his tragic fate, but it's worth pointing out that what to adults looks like cyberbullying may be part of a wider teen drama. Victims may also participate in this drama, using social media in similar ways as perpetrators—to assert, maintain, and defend their status. Moreover, bullies can also be victims. Seeing them in this way certainly doesn't condone their malicious behavior; however, it does highlight that their meanness may reflect the pressure teen culture exerts on them to be self-conscious about their social status. Based on her comprehensive study of teen life online, sociologist Danah Boyd wrote,

> It's easy to empathize with those who are on the receiving end of meanness and cruelty. It's much harder—and yet perhaps more important—to offer empathy to those who are doing the attacking.[18]

Exposing the social forces that contribute to meanness in teens may lead you to feel a certain degree of compassion toward bullies, in a similar way that you likely do toward the people they victimize.

Not a Solitary Act of Desperation: Exposing Social Forces That Lead Teens to Die from Suicide

12.3 Explain how social forces, such as social class and social networks, influence a teen's susceptibility to developing suicidal thoughts.

The significant media attention given to teen suicides that occur after incidents of cyberbullying suggests these two social problems are linked. Indeed, middle and high

FIGURE 12.3 • Bullying and Suicide

Kids who've been bullied are more prone to consider taking their own life than kids who have not experienced such victimization.

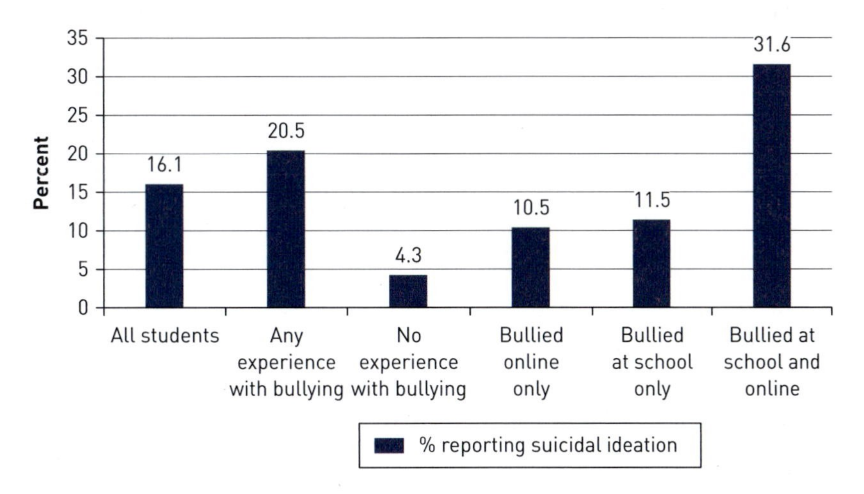

Source: Justin W. Patchin, "More on the Link between Bullying and Suicide," Cyberbullying Research Center, June 2, 2017, https://cyberbullying.org/more-on-the-link-between-bullying-and-suicide.

school students who've been bullied are 27.3 percent likelier to experience **suicidal ideation**, which involves giving serious thought to attempting suicide, than students who haven't been bullied (see Figure 12.3 on page 247). And bullying victims are 1.9 times more likely to have actually attempted suicide than nonvictims.[19]

Still, it's difficult to know how much cyberbullying, in itself, contributes to suicidal ideation or suicidal attempts. Since many kids who've been bullied are LGBTQ, disabled, overweight, or a member of another marginalized group, they typically have psychological risk factors for suicide that long predate being bullied (see Figure 12.4). There's certainly a **correlation**, or mutual relationship, between cyberbullying and suicide. However, showing that the two behaviors are correlated is hardly the same as proving one causes the other. Here's an analogy: Attending class on sunny days indicates your academic behavior is related to the weather; the two are correlated. However, sunshine doesn't cause you to attend class; if it's a warm day, sunshine may actually contribute to some people *not* going to class![20]

The details of Tyler Clementi's story lead us to question whether having been cyberbullied caused him to take his own life. Although initial news reports indicated

FIGURE 12.4 ● At Risk for Suicide

In a study of kids ages ten to seventeen who died from suicide, these are the percentages who had experienced various psychological risk factors.

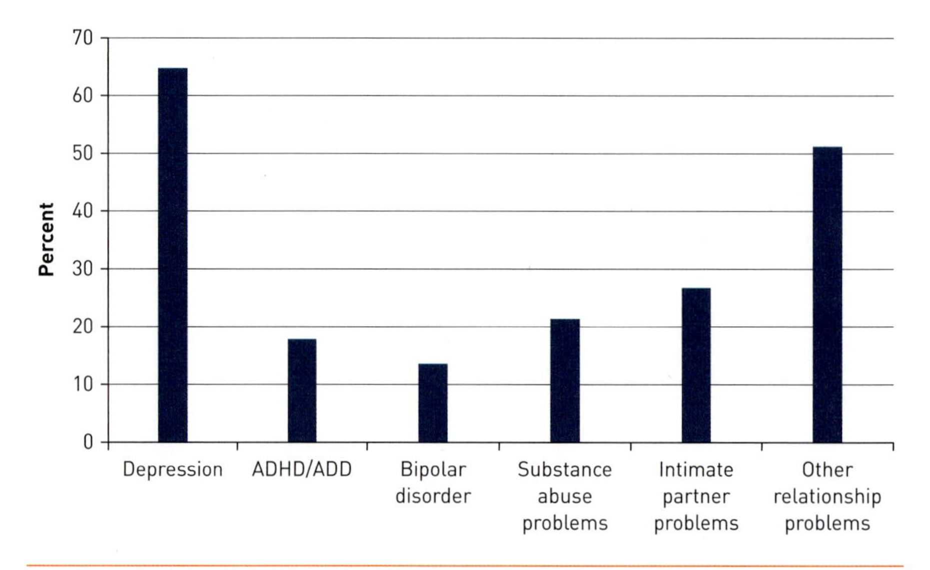

Source: Adapted from Debra L. Karch, J. Logan, Dawn D. McDaniel, C. Faye Floyd, and Kevin J. Vagi, "Precipitating Circumstances of Suicide among Youth Aged 10–17 Years by Sex: Data from the National Violent Death Reporting System, 16 States, 2005–2008," *Journal of Adolescent Health* 53, no. 1 (2013): S51–S53.

Note: Because some kids had multiple factors, these percentages add up to more than 100.

the webcam prank exposed him as gay, he'd actually come out to his parents three days before leaving for college. A file on his computer titled "Why is everything so painful" indicated that after sharing this news, he felt rejected by his mother. Her lack of support may have compounded the emotional challenges incoming residential college students often face. But even so, the webcam prank may not have been a significant factor in Clementi's death. After all, *he* was the one who'd asked his roommate, Dharun Ravi, to leave on a couple of occasions so he could be alone with another man. Since Ravi had never had a girlfriend, Clementi may have interpreted the prank as a sign that Ravi was envious of him. Even if the prank *did* play a role in the suicide, it's still presumptuous to conclude that it was a major reason why Clementi jumped to his death. Though media reports immediately following the suicide painted his death as resulting from cyberbullying, the extent of the causal link remains in question.[21]

Moreover, the tendency of media reports to play up this link diverts attention from the social forces that illuminate why suicide is the second most likely reason people ages ten to twenty-four die (the first is unintentional injury; for example, from car accidents).[22] Renowned sociologist Émile Durkheim explored these social forces in an 1897 book in which he argued that Catholics had a lower suicide rate than Protestants because Catholics did a better job integrating people into society and giving their lives meaning.[23] Although this difference may no longer hold true today, Durkheim's study was noteworthy for being the first to analyze suicide by focusing on factors external to the individual. The sociological perspective is valuable because it can identify which types of people are most at risk of dying from suicide—although it cannot predict who will actually *attempt* suicide. Let's now investigate two social forces that place certain teens at an elevated risk of experiencing suicidal ideation.

Social Class Pressures

Across many American suburban communities, the majority of kids grow up in middle- or upper-middle-class families and strive to do well in school. In these communities, there are also kids who come from working-class families. Whereas some of these working-class kids are invested in school, the ones who aren't are prone to suicidal thoughts. It's tempting to embrace an individual perspective to explain their lack of academic motivation—that it reflects an inborn shortcoming. But this is hardly the full story. We also need to explore what occurs within these communities that stifles their motivation.

Though they share many beliefs in common, Catholics tend to feel a stronger sense of purpose in their lives than do Protestants. This difference is critical in explaining the lower suicide rate among Catholics.
iStockphoto.com/FatCamera

High schools that tailor extracurricular activities to academically ambitious kids may be sending their lower-performing peers the message that the school is uninterested in their success.
iStockphoto.com/FangXiaNuo

Sociologist Donna Gaines investigated the suicides of four working-class teens in Bergenfield, New Jersey. None of these teens had been academically invested, which essentially excluded them from participating in constructive after-school activities. Additionally, the principal and other school administrators didn't believe there were organized clubs and sports that "nonconforming youth en masse might enjoy that would not be self-destructive, potentially criminal, or meaningless."[24] Because these kids lacked such opportunities, they didn't get the chance to have meaningful interactions with adults who might have motivated them to care more about school.

This lack of opportunities factored into why these kids turned to drugs, which put them in contact with police officers, substance abuse counselors, special education teachers, and other professionals who were constant reminders that they were academic outcasts. Being marginalized for failing to live up to community success standards eroded these kids' self-worth, making them more inclined than their higher-class peers to become susceptible to suicidal thoughts.[25]

In other suburban communities, teen suicide stems from growing up amidst affluence. Houses are spacious and often worth a million dollars or more. The public schools are well-funded and most kids are academically ambitious. These communities include places such as Winnetka, Illinois; West University Place, Texas; Clyde Hill, Washington; Dover, Massachusetts; Chevy Chase, Maryland; and Palo Alto, California.

Parents in wealthy suburban communities, who tend to be highly accomplished, often exert pressure on their children to excel in school.
iStockphoto.com/bmcent1

Palo Alto may seem like an ideal town to grow up, given that it's nestled in the Silicon Valley and home to Stanford University. Yet from 2005 to 2015 the teen suicide rate was five times the national average. During this period, there were two different **suicide clusters**, which occur when there are multiple suicides in the same area over a short time period. During nine months spanning 2009 and 2010, six teens died from suicide. Another five did so from October 2014 to February 2016. In a survey conducted during 2013–14, 12 percent of Palo Alto high school students reported having given serious thought during the previous year to taking their own lives.[26] On the opposite coast during this same school year, three teens living in one of Boston's priciest suburbs, Newton, died from suicide.[27] From 2011 to 2014, so did six students at W. T. Woodson High School in Fairfax, Virginia, an affluent town outside Washington, D.C.[28]

In affluent communities, many kids develop emotional scars from trying to earn top grades and make it look easy.
iStockphoto.com/PeopleImages

The tragic similarity Palo Alto, Newton, and Fairfax share is no coincidence. To understand why teens growing up in America's wealthiest communities are prone to suicidal thoughts, let's look at the culture surrounding success in these communities. Fitting in socially hinges on kids relentlessly competing with one another to meet their parents' high academic expectations. Students often take as many Advanced Placement courses as possible and become overcommitted with extracurricular activities.[29]

Not only must teens work hard to get good grades but they often feel they need to make doing so appear effortless and stress-free. Upon hearing that her classmate had died from suicide, one girl in Palo Alto thought it simply couldn't be true because he was popular and seemed unfazed by juggling a packed schedule of hard classes with sports practices. However, acting as if he did not have to work very hard to meet community standards for success took an unbearable psychological toll on this boy. Kids like him are keenly aware that appearing unable to manage a heavy workload risks appearing weak. Of course, other dynamics besides social class pressures likely factored in this Palo Alto teen's decision to take his own life. Nonetheless, the academic demands of growing up in a high-achieving family seem to have contributed to his suicidal thinking. For kids who internalize a sense of failure in meeting these demands, it may feel unbearable to keep living with such stress.[30]

It's striking how much the kids in affluent communities who die from suicide resemble their psychologically healthier classmates in terms of their academic ambitions.
iStockphoto.com/FatCamera

From a sociological perspective, kids growing up in communities like Palo Alto who die from suicide are *not* deviating from social norms even though their deaths are devastating both for their family and for the community. Rather, when one of these kids takes their own life, they are desperately expressing a desire to conform with the community's high expectations of them—and also tragically expressing their perceived failure to meet these expectations.[31]

The extensive media coverage following Marilyn Monroe's death produced a ripple effect, fueling a suicide cluster.
Alfred Eisenstaedt/Contributor/Getty Images

Network Effects

Sometimes when a person takes their own life, they leave evidence that their actions were a response to knowing other people who had recently died from suicide. Here we see the power of social networks. A famous illustration is the drug overdose in 1962 of actress Marilyn Monroe. The suicide rate in the U.S. jumped 12 percent during the month afterward, compared to the average over the same month in previous years. This was no mere coincidence. Although those who died from suicide soon after Monroe probably didn't know her directly, news reporting created a social network of people who tuned in to the circumstances of her death. For someone who already had suicidal thoughts, the publicity around Monroe's drug overdose may have tipped them toward actually deciding to take their own life.[32]

The impact of one person's suicide on others in their social network may be even more pronounced when the person who takes their own life is a close tie as opposed to a celebrity. Research indicates that even someone with no history of suicidal thoughts may be vulnerable to such thoughts after a family member or friend attempts suicide. The suicide of a friend can leave an especially pronounced impression on teens, given that they're at a stage of life when peers exert tremendous influence over their thinking and decision making. And because they spend a lot of time on social media, teens are often keyed into the thoughts and actions of their peers in ways that weren't possible in prior generations.[33]

Because Instagram, Snapchat, and other social media platforms provide constant updates about the people in one's network, these sites may act as conduits for contagious suicidal thoughts.
iStockphoto.com/alexsl

These network effects may be particularly strong when teens die from suicide after having been bullied. As researcher Deborah Temkin argues, media reports linking cyberbullying and suicide may give youth who already experience risk factors for taking their own life the message "that suicide is a normal reaction to bullying, and . . . suggest that if they do die by suicide, their name will be known across the country and perhaps the world—something any youth who feels alone and invisible could desire."[34] Temkin's words reveal that cyberbullying may contribute to teen suicide insofar that publicized cases like Tyler Clementi's and Rebecca Sedwick's make suicide more thinkable to others who have been bullied. The tragic irony is that news reporting about the supposed causal link between cyberbullying and suicide may actually strengthen this link. How do you think the media should portray teen suicide in order to generate constructive awareness and discussion about this very serious social problem while at the same time avoiding contributing to future suicides?

13 Reasons Why stirred controversy over whether it was contributing to suicidal ideation in teens. Shown here is the lead character Hannah who, before she died, made recordings explaining why she took her own life.
PictureLux/The Hollywood Archive/Alamy Stock Photo

Getting to the Root of Kids' Meanness toward One Another

12.4 Identify how focusing on the link between cyberbullying and suicide diverts attention from the root causes of both of these social problems.

Much of the conventional wisdom around cyberbullying focuses on teens' brute capacity nowadays to act meanly toward one another, anywhere, at any time. We can see this theme of kids having gotten meaner in news reporting about cases of cyberbullying in which the victim subsequently died from suicide. These cases suggest that growing up with mobile devices and social media has given teens license to behave in ways they never would in face-to-face interactions, carrying out malicious acts that can lead victims to take their own lives.

Teens' interactions online are certainly different than offline, if for no other reason than because everything they post is traceable. Nastiness is plainly visible for others to see and becomes encoded in one's digital footprint. But whether teens are meaner today than in prior generations is both impossible to know and beside the point. Focusing on particular mean kids distracts attention from two sociological perspectives we've developed in this chapter: (1) how teen culture contributes to these kids' egregious behaviors and (2) the social conditions that make some teens more prone to suicidal ideation than others.

Of course, cyberbullying can be very hurtful, perhaps even more so than the offline schoolyard bullying that was more common when I was growing up. That's because the threat cyberbullying poses is both constant and anonymous: The bully never has to look into the face of the person they're victimizing.

Given that for many teens social status reigns supreme, cyberbullying is often the norm rather than a deviation from it.
iStockphoto.com/AntonioGuillem

Yet we must tread carefully in how we think about this new social problem in relation to an old social problem—suicide. While cyberbullying may play a role when some teens take their own lives, focusing too much on this role clouds our understanding of the factors that lead to suicide. In order to better understand this public health crisis facing teens, we need to highlight how psychological and social forces jointly lead some of these kids to feel their lives are too burdensome to continue living.

What Do You Know Now?

1. How does paying attention to the influences of teen culture offer a deeper explanation for cyberbullying than focusing solely on particular kids' meanness?
2. Using the webcam prank at Rutgers University as an illustration, what evidence is there that cyberbullying is not necessarily a significant factor in teen suicide?
3. Why do you think media reporting reinforces the misconception that cyberbullying causes teen suicide?
4. What are the social conditions that put lower-class kids at an elevated risk of suicide compared to their peers? What are the different social conditions that put higher-class kids at a heightened risk?
5. Why are teens especially susceptible to suicide contagion?

Key Terms

Cyberbullying 240
Toxic masculinity 243
Suicidal ideation 248
Correlation 248
Suicide cluster 250

Visit **edge.sagepub.com/silver** to help you accomplish your coursework goals in an easy-to-use learning environment.

Notes

1. Lisa Foderaro, "Private Moment Made Public, Then a Fateful Jump," *New York Times*, September 29, 2010, http://www.nytimes.com/2010/09/30/nyregion/30suicide.html?_r=1?login=email.
2. Lizette Alvarez, "Girl's Suicide Points to Rise in Apps Used by Cyberbullies," *New York Times*, September 13, 2013, http://www.nytimes.com/2013/09/14/us/suicide-of-girl-after-bullying-raises-worries-on-web-sites.html?_r=0.

3. Kelly Wallace, "Police File Raises Questions in Rebecca Sedwick's Suicide," CNN, April 21, 2014, https://www.cnn.com/2014/04/18/living/rebecca-sedwick-bullying-suicide-follow-parents/index.html.
4. Felix Richter, Facebook Keeps on Growing, *Statista*, July 25, 2019, https://www.statista.com/chart/10047/facebooks-monthly-active-users/.
5. Sameer Hinduja and Justin W. Patchin, "Cyberbullying: Identification, Prevention, & Response," Cyberbullying Research Center, 2019, https://cyberbullying.org/Cyberbullying-Identification-Prevention-Response-2019.pdf.
6. Research documenting how rare it is that either suicide or suicidal thinking follows cyberbullying comes from Sameer Hinduja and Justin W. Patchin, "Bullying, Cyberbullying, and Suicide," *Archives of Suicide Research* 4, no. 3 (2010): 206–21.
7. Discussion of how many media reports of Tyler Clementi's death initially attributed it to the webcam prank come from Ian Parker, "The Story of a Suicide: Two College Roommates, a Webcam, and a Tragedy," *New Yorker*, February 6, 2012, https://www.newyorker.com/magazine/2012/02/06/the-story-of-a-suicide.
8. Parker, "The Story of a Suicide."
9. Ibid.
10. C. J. Pascoe, "Homophobia in Boys' Friendships," *Contexts* 12, no. 1 (2013): 17–8.
11. Neil Duncan and Larry Owens, "Bullying, Social Power and Heteronormativity: Girls' Constructions of Popularity," *Children & Society* 25, no. 4 (2011): 306–16; Don E. Merton, "The Meaning of Meanness: Popularity, Competition, and Conflict among Junior High School Girls," *Sociology of Education* 70, no. 3 (1997): 175–91.
12. Murray Milner, Jr., *Freaks, Geeks, and Cool Kids: American Teenagers, Schools, and the Culture of Consumption* (New York: Routledge, 2004); Danah Boyd, *It's Complicated: The Social Life of Networked Teens* (New Haven, CT: Yale University Press, 2014).
13. Charles Derber and Yale R. Magrass, *Bully Nation: How the American Establishment Creates a Bullying Society* (Lawrence: University Press of Kansas, 2016), 1–29.
14. Michael Schudson, *Advertising, The Uneasy Persuasion: Its Dubious Impact on American Society* (New York: Routledge, 2013), 3–13.
15. Boyd, *It's Complicated.*
16. Ibid.
17. Parker, "The Story of a Suicide."
18. Boyd, *It's Complicated*, 135.
19. Hinduja and Patchin, "Bullying, Cyberbullying, and Suicide."
20. Russell A. Sabella, Justin W. Patchin, and Sameer Hinduja, "Cyberbullying Myths and Realities," *Computers in Human Behavior* 29, no. 6 (2013): 2703–11.
21. Parker, "The Story of a Suicide."
22. Centers for Disease Control and Prevention, "10 Leading Causes of Death by Age Group, United States—2014," http://www.cdc.gov/injury/images/lc-charts/leading_causes_of_death_age_group_2014_1050w760h.gif.
23. Émile Durkheim, *Suicide* (London: Penguin Classics, 2006 [1897]).
24. Donna Gaines, *Teenage Wasteland: Suburbia's Dead-End Kids* (Chicago: University of Chicago Press), 1998.
25. Ibid.
26. Associated Press and Ashley Collman, "Teen Suicide Trend Rocks Wealthy Palo Alto," *The Daily Mail*, February 15, 2016, http://www.dailymail.co.uk/news/article-3448644/Teen-suicides-California-city-subject-US-study.html. For an in-depth investigative report about these suicides, see Hanna Rosin, "The Silicon Valley Suicides: Why Are So Many Kids with Bright Prospects Killing Themselves in Palo Alto?" *The Atlantic*, December 2015, https://www.theatlantic.com/magazine/archive/2015/12/the-silicon-valley-suicides/413140/.
27. Kathleen Burge, "A Newton Boy Left This Life without a Note or Clue," *Boston Globe*, March 2, 2014, http://www.bostonglobe.com/metro/2014/03/02/newton-mobilizing-after-suicides-teens/ddoqf9XlrCPhEl6fbd2KYI/amp.html.
28. Hillary Crosley Coker, "Why Have 6 Students Committed Suicide in 3 Years at Woodson High?" *Jezebel*, April 14, 2014, http://jezebel

.com/why-have-6-students-committed-suicide-in-3-years-at-woo-1563017918.

29. Anna S. Mueller and Seth Abrutyn, "Adolescents under Pressure: A New Durkheimian Framework for Understanding Adolescent Suicide in a Cohesive Community," *American Sociological Review* 81, no. 5 (2016): 877–99.
30. Rosin, "The Silicon Valley Suicides."
31. Mueller and Abrutyn, "Adolescents under Pressure"; Lucia Ciciolla et al., "When Mothers and Fathers Are Seen as Disproportionately Valuing Achievements: Implications for Adjustment Among Upper Middle Class Youth," *Journal of Youth and Adolescence* (2016): 1–19; Suniya S. Luthar, Samuel H. Barkin, and Elizabeth J. Crossman, "'I Can, Therefore I Must': Fragility in the Upper-Middle Classes," *Development and Psychopathology* (2013): 1529–49.
32. Margot Sanger-Katz, "The Science behind Suicide Contagion," *New York Times*, August 13, 2014, https://www.nytimes.com/2014/08/14/upshot/the-science-behind-suicide-contagion.html.
33. Seth Abrutyn and Anna S. Mueller, "Are Suicidal Behaviors Contagious in Adolescence? Using Longitudinal Data to Examine Suicide Suggestion," *American Sociological Review* 79, no. 2 (2014): 211–27.
34. Deborah Temkin, "Stop Saying Bullying Causes Suicide," *The Huffington Post*, September 27, 2013, https://www.huffpost.com/entry/stop-saying-bullying-caus_b_4002897.

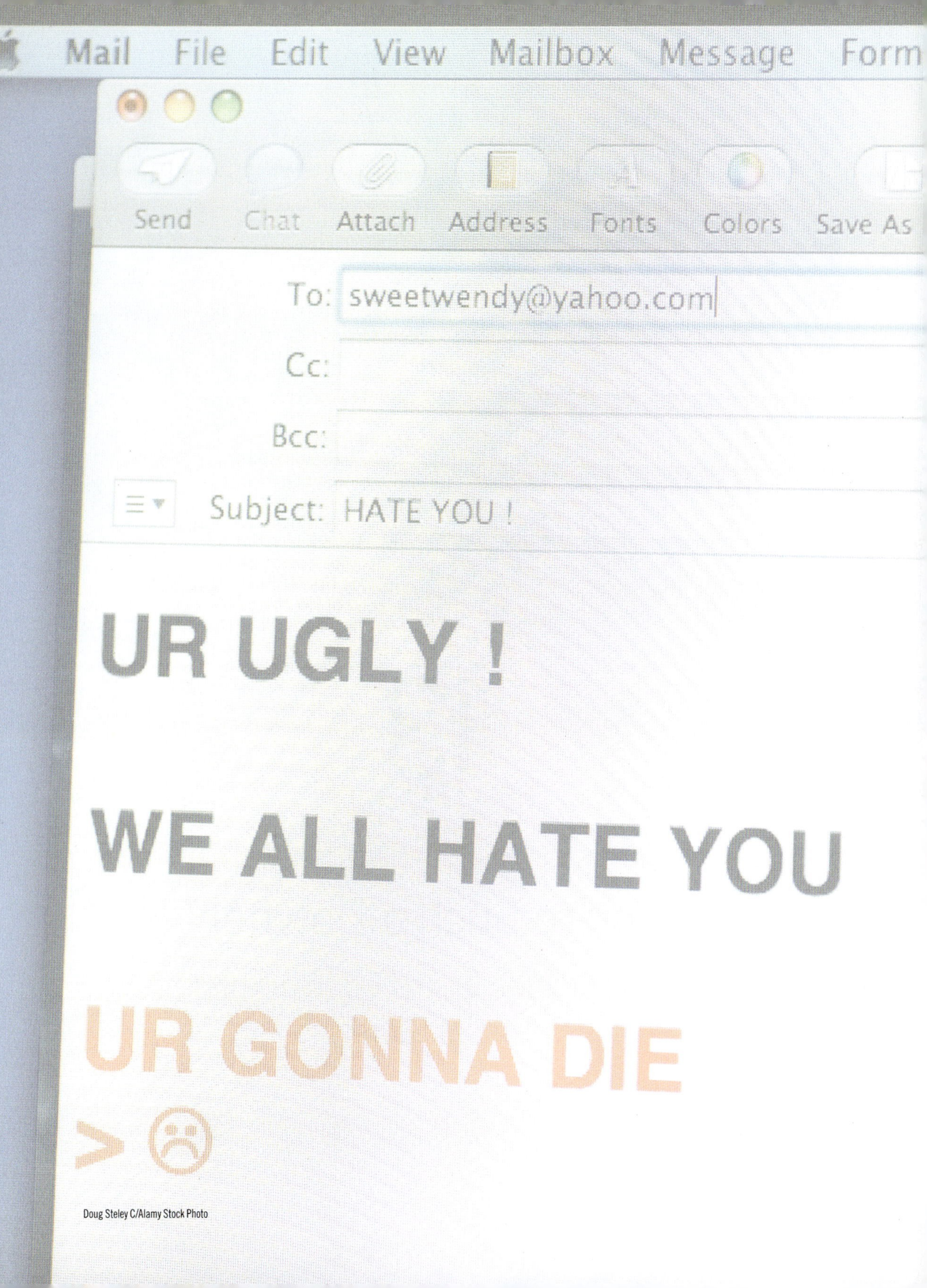

Doug Steley C/Alamy Stock Photo

13

"You're Such a Downer"

Why Mental Illness Goes beyond Personal Suffering

Learning Objectives

1. Describe the symptoms that characterize some of the most common types of mental illness.
2. Explain why changes over time in how we view certain mental health conditions don't merely reflect advances in scientific understanding of psychiatric disorders but the power of drug companies too.
3. Identify social forces that contribute to the misdiagnosis of ADHD in children.
4. Recognize how American culture contributes to the profound sadness felt by people with depression.
5. Describe how a person's gender influences their experience of mental illness.
6. Explain how the idea of neurodiversity can destigmatize mental illness.

Diseases of the Mind: Exploring the Wide Array of Psychiatric Disorders

13.1 Describe the symptoms that characterize some of the most common types of mental illness.

"I feel like I'm in a cage and I'm trapped, and I can't get out and it's night time and the daylight's never going to come. Because if the daylight came, I could figure out how to get out of the cage, but I can't," commented Ava Thompson, a nurse in her thirties, in an interview for a study about people's experiences with clinical depression. "Sometimes, I feel like I'm being smothered in that I can't breathe. I am being suffocated. . . . And it's like falling down a well, like I'm free-falling. That's what it is. And I have nowhere to grab onto to stop it."[1]

Depression is a paralyzing syndrome of loneliness and despair that affects approximately 6.7 percent of Americans over eighteen.[2]
iStockphoto.com/Carlo107

Scott Strossel, a journalist in his forties, recounted similar feelings of desperation: "My wedding was accompanied by sweating so torrential that it soaked through my clothes and by shakes so severe that I had to lean on my bride at the altar, so as not to collapse. At the birth of our first child, the nurses had to briefly stop ministering to my wife, who was in the throes of labor, to attend to me as I turned pale and keeled over."[3] Strossel went on to describe other occasions on which he experienced symptoms of severe anxiety, such as going out on dates, taking exams, interviewing for jobs,

iStockphoto.com/wildpixel

Roughly 18.1 percent of adults have one among an array of anxiety disorders, which are characterized by chronic worrying and feeling on edge.[5]
iStockphoto.com/SIphotography

ADHD is most prevalent in children.[7]
iStockphoto.com/kupicoo

and travelling by plane, train, or automobile. Even ordinary activities like walking down the street, talking on the phone, or playing tennis would often produce a sense of "existential dread" and bring on physical symptoms like nausea, vertigo, and shaking.[4]

During college, Kerri MacKay had trouble paying attention in class. She couldn't focus on her professor's words, even when she found the subject interesting. Her mind wandered among the various stimulants in the room—PowerPoint slides, classmates raising their hands and asking questions, and the sounds of people typing on their keyboards, coughing, or whispering to one another. Kerri has attention deficit hyperactivity disorder (ADHD), which is characterized by impulsivity and difficulty staying focused on a single activity. Doctors diagnose about 11 percent of Americans under age eighteen as having this condition.[6]

Depression, anxiety, and ADHD are three types of **mental illness** or *psychiatric disorder*—terms used interchangeably in this chapter to refer to a condition that impairs a person's thinking, emotions, or behavior. Globally, anxiety and depression are the most common types of mental illness; each affects close to 4 percent of the world's population (see Figure 13.1). Other types that may be familiar to you include bipolar disorder, autism spectrum disorder, schizophrenia, obsessive-compulsive disorder, anorexia nervosa, and bulimia nervosa. About one in five Americans suffers from a psychiatric disorder.[8] Therefore, it's likely you know someone who does. Perhaps that person is you.

Many experts, including psychiatrists, psychologists, social workers, neurologists, and primary care doctors, may diagnose mental illness. They base their determination on how closely a patient's behavior meets criteria listed in the *Diagnostic and Statistical Manual of Mental Disorders* (*DSM*). The dominant view within medical science is that psychiatric disorders are hereditary and triggered by chemical imbalances in the brain. Clinicians, for the most part, see mental illness from an individual perspective, viewing its causes as internal to the person suffering.

FIGURE 13.1 • Frequently Diagnosed

Among all psychiatric and substance abuse disorders, anxiety and depression are the most common worldwide.

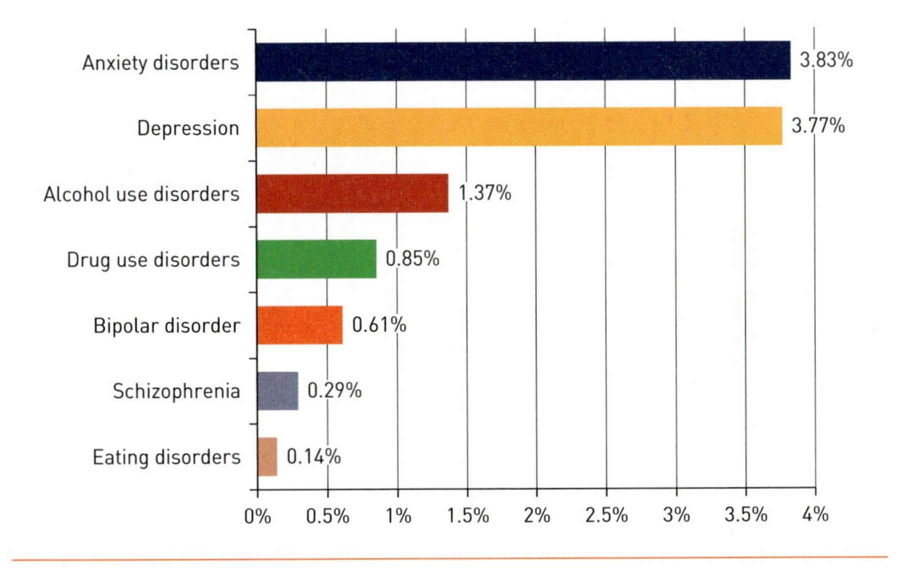

Source: Sean Fleming, "This Is the World's Biggest Mental Health Problem—and You Might Not Have Heard of It," World Economic Forum, January 14, 2019, https://www.weforum.org/agenda/2019/01/this-is-the-worlds-biggest-mental-health-problem.

Because the professionals who make *DSM* diagnoses influence how their patients and others in the general public understand these labels, it makes sense if you regard the individual perspective as the only possible way to think about mental illness. It's likely that you see these conditions as unique for each person, particularly if you suffer from one or know someone who does. And indeed, they are unique. At the same time, people who have a psychiatric disorder often share the commonality of experiencing **stigma**; their illness marks them as disreputable in the eyes of others. As a result, mentally ill people can feel inhibited from speaking openly about what they're feeling and therefore may not have an opportunity to think about their condition from a perspective other than their own suffering.[9]

This chapter offers such an opportunity. Embracing the sociological perspective enables you to recognize that the roots of mental illness extend beyond what a person personally feels. We'll explore how these feelings reflect social forces that, though often invisible to the people who are hurting, exert a significant impact on them. While millions suffer in profound ways from mental illness, their troubles are in part a reflection of the society around them. Even though psychiatric disorders don't

The *DSM* has been the most authoritative source on mental illness since the 1950s. Several updates have been made over the years; the most recent edition came out in 2013.
Amir Ridhwan/Shutterstock

directly afflict all of us, each of us plays a role in shaping how people experience and cope with these illnesses.

FIRST IMPRESSIONS?

1. What did it feel like to read the stories that opened this chapter?
2. Do you know anyone who has spoken to you about feeling the stigma of mental illness? If so, what has their experience been like?
3. Do a web search. How have the diagnoses listed in the *DSM* changed over time? Why are these changes significant?

Conditions Blood Tests Cannot Prove: The Social Construction of Mental Illness

13.2 Explain why changes over time in how we view certain mental health conditions don't merely reflect advances in scientific understanding of psychiatric disorders but the power of drug companies too.

Whereas in the U.S. two people of the same sex who love one another can now marry, medical science used to regard their romantic desires as a mental illness. Until the 1987 edition of the *DSM*, homosexuality was listed as a psychiatric disorder. While other conditions have also been removed over the years, different ones have been added and the book is now thicker than its original version. Moreover, entirely new diagnostic categories have come into being. When I was growing up, being "shy" simply meant feeling inhibited from interacting in social situations. Nowadays, some people who previously would have been characterized in this way receive a diagnosis of social anxiety disorder and take meds to become more outgoing.[10]

For some guys, the presence of others in a public bathroom can create such discomfort that they cannot urinate. The 2013 *DSM* for the first time included this condition, known as paruresis, or shy bladder syndrome, in its list of anxiety disorders.
Thierry Berrod, Mona Lisa Production/Science Source

These examples highlight the **social construction of mental illness**, which is the idea that people define which types of moods and behaviors characterize particular diagnostic categories. Categories of mental illness aren't set in stone. Health care providers make diagnoses based on observations of how closely patient behavior matches criteria in the *DSM*, which, as we've just seen with homosexuality and social anxiety disorder, can change over time. Moreover, as we'll see later in the chapter, a person's experience of a particular psychiatric disorder can vary widely depending on social forces in their lives.

Because insurance companies rely on *DSM* criteria in reimbursing for psychiatric evaluations, health care providers must regard these criteria as authoritative even if they hold reservations about their validity.
iStockphoto.com/SDI Productions

It's significant that there's no test to prove definitively that a person is mentally ill. This means that the manufacturers of drugs to treat ADHD, depression, anxiety, and other psychiatric disorders can exert significant influence over both the definition of diagnostic criteria and how people experience specific disorders. These companies have a financial incentive to encourage as many people as possible to believe that they suffer from a condition for which medication can make them more functional. Therefore, changes in the *DSM* from one edition to the next don't simply reflect the latest scientific research about psychiatric disorders but the profit motive of drug companies too.

These companies aggressively market the benefits of prescription drugs. It is interesting to note that the U.S. is one of just two countries that permit the pharmaceutical industry to advertise directly to consumers.[11] The industry also plays a hand in shaping the creation of *DSM* criteria. Over half the members of the committee that developed the most recent edition had industry ties. We can see the effects of these ties via changes in the definition of ADHD:

- *DSM-IV* (1994): Hyperactivity—often runs about or climbs excessively in situations in which it is inappropriate.
- *DSM-5* (2013): Hyperactivity and Impulsivity—often runs about or climbs in situations in which it is inappropriate.

The criteria are identical except for the deletion of one word: *excessively*. It's a consequential omission because it means that now a much larger population of kids behave in ways that fit the diagnosis. Because of this change and the aggressive marketing of ADHD drugs, prescriptions for, and sales of, these drugs have skyrocketed (see Figure 13.2).[12]

FIGURE 13.2 ● Profiting from Diagnosis

The pharmaceutical industry has substantially benefited from the marketing of ADHD medication.

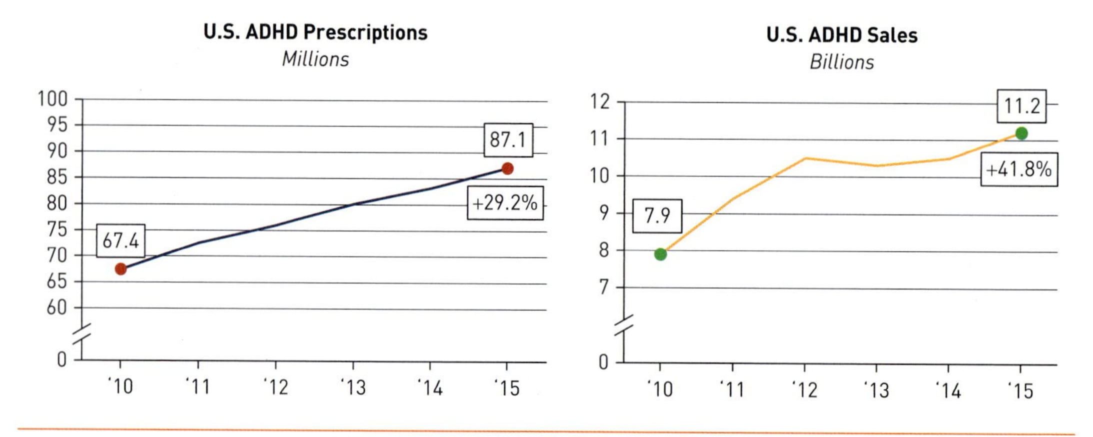

Source: Based on John Fauber, Matthew Wynn, and Kristina Fiore, "New Heroin? ADHD Drug Abuse Similar to Opioids," *App*, September 12, 2016, https://www.app.com/story/news/2016/09/12/new-heroin-abuse-adhd-drugs-similar-opioids/90272344.

Inexact Science: Explaining Cases Where ADHD Is Misdiagnosed

13.3 Identify social forces that contribute to the misdiagnosis of ADHD in children.

Kristin Parber was the sort of kindergartner who often leaped out of her chair with a hand raised high in the air, ready to share with her classmates all that she knew. Her inability to sit still irked her first grade teacher, who took away Parber's recess when she impulsively spoke out of turn. It was during second grade that her parents learned there was a problem. Because the girl's mind often wandered while she read or worked on math worksheets, her teacher recommended they consult a psychiatrist. At age seven, Parber became one of the roughly half a million kids in the United States who annually receive an ADHD diagnosis and a prescription for medication to improve focus and concentration.[13]

For Kristin Parber's parents, the spike in news reporting about ADHD that began shortly before Parber was born in the mid-1990s contributed to the sense of urgency teachers conveyed about their daughter's misbehavior in school.
iStockphoto.com/RapidEye

Even after Kristin Parber began taking meds, her troubles continued. She was among the roughly 20 percent of kids to experience severe side effects from ADHD drugs, which include anxiety and depression. Her psychiatrist prescribed Xanax for the former and Lexapro for the latter. In seventh grade, Parber and a friend started crushing and snorting their Ritalin pills. In addition to being more potent, the drug also became more addictive. After breaking her nose in a ski accident two years later, Parber began using opioid painkillers. As has so often happened to people prescribed Vicodin or OxyContin to manage pain, she became hooked (see Chapter 4 for a discussion of opioid addiction). Her behavior reflected a societal trend: the greater likelihood of people to abuse prescription drugs than illicit drugs. The other shoe dropped when her school gave her a two-week suspension for drinking.

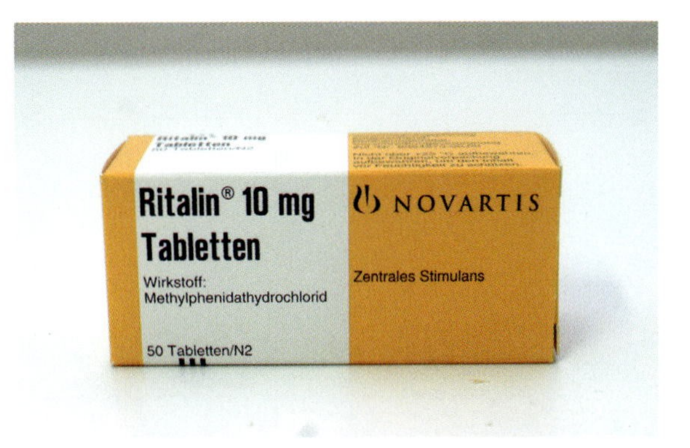

When Kristin Parber received her diagnosis in the 1990s, Ritalin was the main treatment for ADHD. Adderall and Concerta are now commonly prescribed as alternatives. Each of these drugs helps kids pay attention so they can perform better in school.
XAMAX\ullstein bild via Getty Images

Parber's parents were at a loss about what to do. After eventually deciding to send her to rehab, Parber was able to finish high school. She stayed sober throughout college, and now works at the same facility where she underwent treatment. However, rehab success stories are hardly the norm. Indeed, 40–60 percent of people treated for substance abuse disorders experience relapse. If you didn't know how often relapse occurs, you might take away from Parber's story the message that despite her bumpy childhood, all's well that ends well.[14]

Focusing on Parber's successful rehabilitation also overlooks a cautionary tale buried within her story: Having been given an ADHD label as a young child may have been a misdiagnosis of her behavior. Figure 13.3 highlights that the ADHD rate among children in the U.S. ages four to seventeen increased from 6.1 percent in 1997 to 10.2 percent in 2016. Certainly, some of this rise was the result of parents and teachers more carefully paying attention to, and seeking out help for, kids exhibiting behavioral problems. In prior years, such kids would have gone undiagnosed and been unable to do well in school. However, this doesn't fully account for the surge in ADHD diagnoses. There have also been many cases of kids like Kristin Parber, who might have been better off if she hadn't received this diagnosis.

Several social forces contribute to the misdiagnosis of ADHD in kids. First, there is the political influence of the pharmaceutical industry. At the time of Kristin Parber's diagnosis, drug companies had recently begun to spend heavily on elections and lobby Congress to vote against regulating their business practices. Such

FIGURE 13.3 ● Misdiagnosed

While the spike in ADHD since the late 1990s reflects better screening of kids who've needed help and benefited from medication, other kids should not have gotten this label.

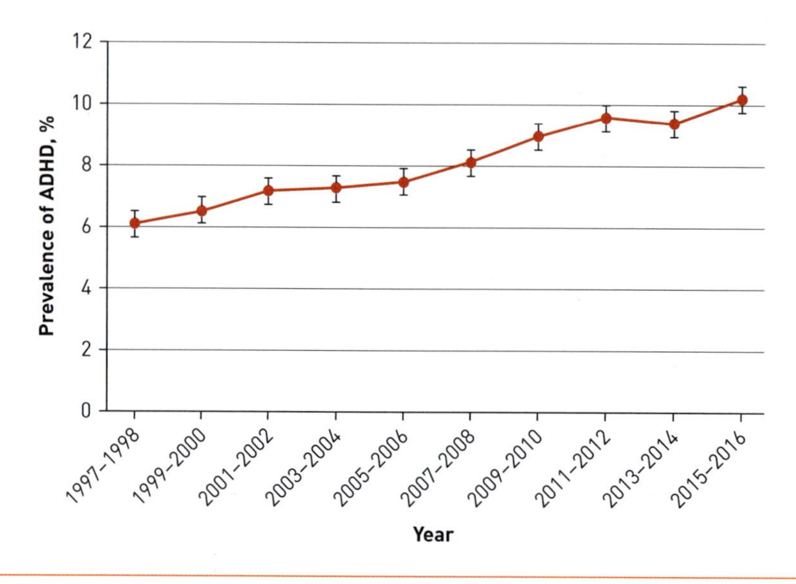

Source: Guifeng Xu, Lane Strathearn, and Buyun Liu, "Twenty-Year Trends in Diagnosed Attention-Deficit/Hyperactivity Disorder among U.S. Children and Adolescents, 1997–2016," *JAMA Open*, 2018, https://jamanetwork.com/journals/jamanetworkopen/fullarticle/2698633.

efforts earned these companies the nickname "Big Pharma." Second, drug companies send marketing reps to medical offices to tout the benefits of psychiatric medications and leave behind free samples. Drug reps particularly influence primary care physicians, who often make diagnoses of mental illness based on relatively limited information about these disorders. Third, Big Pharma aggressively markets drugs to consumers. Think about how common it is to see ads for meds that treat mental illness. The fourth social force is news reporting. Because the substantial media attention given to ADHD reinforces a view shared by doctors, parents, and teachers that this label explains a variety of misbehaviors in children, there's a potential for this diagnosis to be overused.[15]

Three words make drug ads so effective: *Ask your doctor.* Whereas consumers may harbor suspicions about an advertiser's motives, they usually trust their doctor's advice.

Hum Historical/Alamy Stock Photo

It becomes clear how these social forces contribute to misdiagnoses of ADHD when we look at which children within a particular grade get this label and which ones don't. Kids in the same grade can be nearly a year apart in age since some just make the cutoff and others just miss it. Research on children ages six to twelve found that boys born in December (the youngest in the grade) were 30 percent likelier to receive an ADHD diagnosis than boys born in January (the oldest in the grade). For girls, the difference was 70 percent. These findings underscore that a large segment of diagnosed kids do not have an underlying pathology: Their misbehavior in school simply reflects that they're less mature than their classmates.[16]

Of course, many kids, and adults too, significantly benefit from receiving an ADHD diagnosis. Kerri MacKay is one such person. Recall from the very beginning of this chapter that during college she had significant difficulties paying attention in class. The diagnosis enabled her to cease struggling in silence as she got the academic help and support she needed to complete her degree.[17]

People like MacKay who have been able to improve the quality of their lives after experiencing crippling difficulties focusing and concentrating owe gratitude to Keith Conners, often known as the "Father of ADHD." In the 1960s, he created a questionnaire with thirty-nine items that health care professionals continue to use in diagnosing this condition. His work ushered in a medical breakthrough, enabling millions of people to get crucially needed help and function comparably to their peers.

Toward the end of his life, however, Conners began to question the diagnostic criteria he'd promoted over his illustrious career. After visiting the rehab center where Kristin Parber underwent treatment and now works, he wrote in a letter to a colleague that

In the early elementary grades, some kids display a lack of impulse control, a short attention span, and other ADHD characteristics that may disappear over time without medical intervention.
iStockphoto.com/ranplett

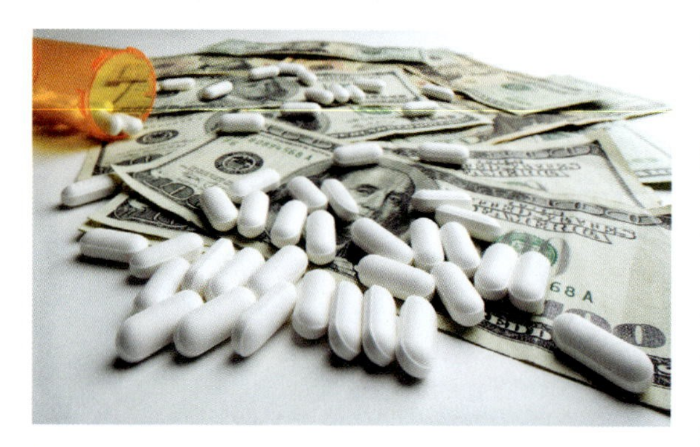

How do you look upon the pharmaceutical industry, which produces life-improving drugs yet is financially motivated to convince people who may not need help that medication can make their lives better?
iStockphoto.com/doug4537

> a constant theme [based on conversations with kids undergoing treatment] was the brief and casual way in which parents and physicians seemed to latch on to the diagnosis, with little attempt to sort out the complicating specific problems in their life. Similarly, no attempt to try a treatment other than medication. This small group experience was clear in showing how substance use appears as a risk not just of ADHD itself, which we already know, but the label of ADHD, whether the diagnosis is correct or not.[18]

Now in retirement, Conners was acknowledging what his life's work had previously prevented him from seeing: that the criteria for identifying and treating ADHD were easily subject to misapplication.

His words point to the dangers of **medicalization**, which refers to the process by which a group of people experiencing the same health problem come to be seen as having a treatable disease. Bolstered by the power of Big Pharma, clinicians may be inclined to affix the label *ADHD* instead of addressing deep-seated issues in children's lives that pills cannot fix. As a result, parents and teachers may buy into the wishful thinking that ADHD is a catch-all diagnosis for kids' misbehaviors and, as a result, not attempt to address underlying problems those kids face.

Being Sad When You're Supposed to Be Happy: How American Culture Influences the Experience of Depression

13.4 Recognize how American culture contributes to the profound sadness felt by people with depression.

Ruth Whippman and her husband moved from England to California after he received an offer from a Silicon Valley start-up. She left her job in television for the chance to be a full-time mother to their toddler. Whippman was immediately taken by her impressions of living in the United States. When with her son at the grocery store or the playground, she often overheard adults questioning whether they were following their passions, doing what they loved, and fulfilling their life's purpose. In short, she noticed Americans' obsession with being happy.[19]

Ruth Whippman's observations highlight that it sometimes takes a foreigner to notice significant characteristics of American culture that a person who's grown up in it may not fully recognize because these characteristics have become ingrained. Consider these trends:

American culture exerts pressure on people to wonder whether they are as happy as they could or should be.
iStockphoto.com/Guardiano5

- People frequently post photos on social media depicting themselves as the life of the party or blissfully enjoying some beautiful place.
- Customer service jobs require workers to put on a happy face whether or not they actually feel the way they're acting.
- Yoga, meditation, and mindfulness are popular pursuits for elevating one's satisfaction with life.
- The positive psychology movement promotes strategies for how a person can boost their mood.

Even if some of these examples are unfamiliar to you, seeing them listed together provides a glimpse into Americans' fixation with happiness. Interestingly, however, the U.S. is not among the happiest countries. The United Nations compiles an annual index based on an average of people's ratings of their quality of life across seven factors. In 2019, the U.S. ranked behind most of Western Europe, Australia, New Zealand, Israel, and Costa Rica (see Figure 13.4 on page 270).[20]

There are countless books written for American parents with advice about how to raise happy children.
CBW/Alamy Stock Photo

Although Americans on the whole are not as happy as they appear to be, advertising fosters the image that the opportunity to become happier is available to everybody. All one needs to do is buy products that will, supposedly, improve their quality of life. Visit any mall and you'll see this image in flying colors. Whether it's a new smartphone, pair of jeans, or piece of jewelry, the underlying message is the same: *This product will make you happier.* While the message is alluring, uncritically buying into it may instead foster the opposite feeling: an insecurity that arises because there's always another new product that supposedly will make you even happier.

FIGURE 13.4 ● If You're Happy and You Know It

Even though Americans place a lot of importance on happiness, the U.S. ranks at just #19 in worldwide measures of countries where people are the happiest.

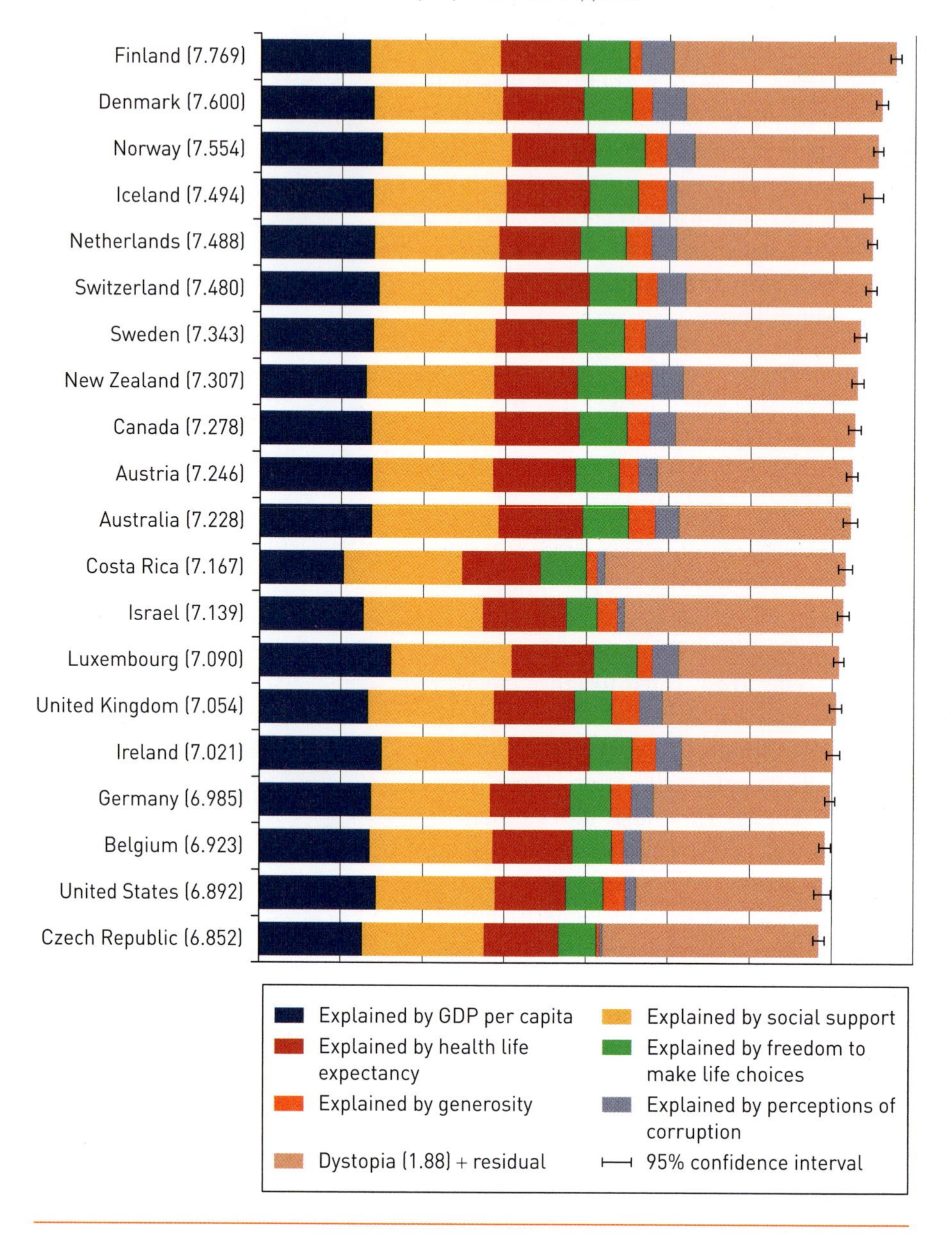

Source: John F. Helliwell, Richard Layard, and Jeffrey D. Sachs. *World Happiness Report 2019*, New York: Sustainable Development Solutions Network, https://s3.amazonaws.com/happiness-report/2019/WHR19.pdf.

Note: This ranking is based on data collected from 2016 to 2018.

The importance American culture places on the pursuit of happiness can adversely affect people who are clinically depressed and therefore unable to engage fully in this pursuit. Individuals take cues from their culture about how they are supposed to feel. Whereas a study found that people living in China reported feeling happy and sad at the same time, at any given moment Americans tend to be either one or the other. Moreover, whereas clinically depressed Chinese people are often bored, dizzy, fatigued, or uncomfortable, depressed Americans are prone toward a punctuated feeling of sadness. Since American culture continually reminds people that they should be happy, or at least be actively trying to become happier, those who suffer from depression may be particularly prone toward feeling badly about their inability to experience happiness.[21]

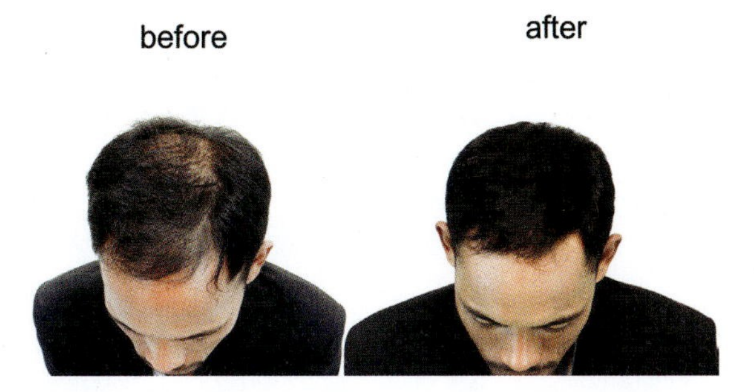

The popularity of antibalding medication highlights the belief in the U.S. that men who look younger are happier. Why might that belief not necessarily be true?
iStockphoto.com/saiyood

Consider Ruth Whippman's description of how living in the U.S. affected her: "The more conversations I have about happiness and the more I absorb the idea that there's a glittering happy ever after out there for the taking, the more I start to overthink the whole thing, compulsively monitoring how I am feeling and hyper-parenting my emotions."[22] These words reveal how living in American culture can cause depressed people to ruminate over their psychological impairments and feel shame for having them.[23]

The Declaration of Independence stipulates the pursuit of happiness as an inalienable right. Living in a society with this founding principle puts pressure on people to feel and appear happy —a pressure that can influence a person's brain chemistry, creating depression.
Tony Freeman/Science Source

Here's a different way to understand how the American obsession with happiness can shape the way depressed people feel. Consider that over 10 percent of the world gets by on less than $1.90 per day. Most poor people in the United States fare better; they earn several times that amount and have indoor plumbing, heat, electricity, and a phone. I don't say these things to sugarcoat the significant hardships low-income Americans experience. To the contrary, their struggles emphasize why poverty may be more debilitating in the U.S. than in other places around the world where people experience a much lower standard of living. Since America's poor live in a society that advertises any person can achieve a better life, they're susceptible to feeling like failures when they come up against obstacles they can't surmount (see Chapter 2 for a fuller discussion of poor Americans' internalization of self-blame).[24]

The latest *DSM* medicalizes grief by stipulating that people who still feel profound sadness just a few weeks after the loss of a loved one may be given a diagnosis of depression and prescribed medication.
iStockphoto.com/RichLegg

Depressed Americans may similarly experience a gulf between what the culture tells them they should feel and how they actually feel. The fact that poor people in the U.S. are disproportionately likely to be depressed underscores this point. One study found that 31 percent of respondents living in poverty had a diagnosis of depressive disorder, versus 15.8 percent of respondents who were not poor. The sociological perspective highlights that the suffering of those who have this menacing illness is in part a by-product of knowing they are supposed to be happy.[25]

Moreover, the criteria for diagnosing a person with depression have expanded over recent decades. Health care providers used to make a clear distinction between sadness brought on by loss or other life stresses and prolonged sadness without an immediate cause. The former was normal while the latter met *DSM* criteria for depressive disorder. However, in recent years there has been a trend, heavily influenced by Big Pharma's marketing of Prozac and other antidepressants, of diagnosticians classifying normal sadness as depression. This example is similar to what we saw earlier with ADHD: Drug companies use their power to persuade as many people as possible that they suffer from a psychiatric disorder requiring medication.[26]

"Feeling Crazy": How Gender Shapes the Ways People Think about and Cope with Mental Illness

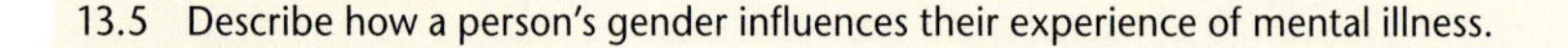

13.5 Describe how a person's gender influences their experience of mental illness.

Have you ever noticed how often people use the words *crazy* and *psycho* to describe how girls and women sometimes act when they're having their periods?[27]
iStockphoto.com/pxhidalgo

Imagine that your friend invites you to their parents' house for Thanksgiving. Halfway through dinner, the person sitting on the other side of the table starts talking about how psycho their ex is. They mention that the ex was often a nervous wreck while they were dating and acted hysterically during the breakup. As you're listening to this description, what sort of picture of the ex is running through your mind?

It wouldn't be surprising if you assumed the person to be female. Indeed, much of the stigma surrounding mental illness is a by-product of the **social construction of gender**, which is the idea that a society views certain roles, but not others, as acceptable for individuals with particular types of bodies. Whereas males are not supposed to express emotion openly, females are encouraged to show others how they feel. Yet girls and women who act too emotionally risk encountering ridicule. Others may characterize them in the ways the person at the Thanksgiving dinner described their ex.

The musical sitcom *Crazy Ex-Girlfriend* satirizes the gendered stigma associated with mental illness. The show offers a realistic picture of the struggles with depression and anxiety suffered by the lead character, played by Rachel Bloom.
Robert Landau/Alamy Stock Photo

Consider that until 1980, hysteria was included in the *DSM*. This condition gets its name from the Greek word for *uterus*, which, of course, only females have. Hysterectomy is the surgical procedure to remove the uterus. Although hysteria is no longer considered a psychiatric disorder, its stigma persists in our language. This word disparages girls and women for having emotions that, in the eyes of those doing the labeling, seem irrational.[28]

Belittling females by associating them with mental illness stigmas can, within the context of intimate relationships, become psychologically manipulative. **Gaslighting** refers to situations where one person in an intimate relationship continually dubs and treats the other as mentally unstable, to the point where the belittled partner begins to question their own sanity. Being the brunt of such abuse can fuel feelings of insecurity that foster a **self-fulfilling prophecy**, which occurs when a person acts in ways that confirm how others label them. In this case, the person internalizes others' biases about them and starts believing they're damaged. Women who don't have a *DSM* diagnosis but who've been gaslighted by a male partner are prone to becoming convinced that they need psychiatric help.

In a Nike ad, Serena Williams appropriates the term *crazy* in describing women's amazing sports feats, like running a marathon, coaching a professional basketball team, or, in her case, winning over twenty Grand Slam tennis tournaments. The ad ends with Williams saying, "So if they want to call you crazy, fine. Show them what crazy can do."
Henk Koster/Alamy Stock Photo

The social construction of gender also influences how people cope with mental illness. Let's focus on the most devastating effect of depression: suicide. Whereas the individual perspective focuses on the psychological struggles that can lead someone to feel their life is not worth living, the sociological perspective widens the lens to consider a person's varying susceptibility to suicide based on how social forces influence them (see Chapter 12 for a fuller discussion). One key source of variation is gender. Research highlights a paradox:

FIGURE 13.5 ● Gender and Suicide

The male suicide rate in the U.S. has consistently been three to four times higher than the female suicide rate. However, the numbers in this graph overlook an important story: Girls and women *more frequently attempt* to take their own lives.

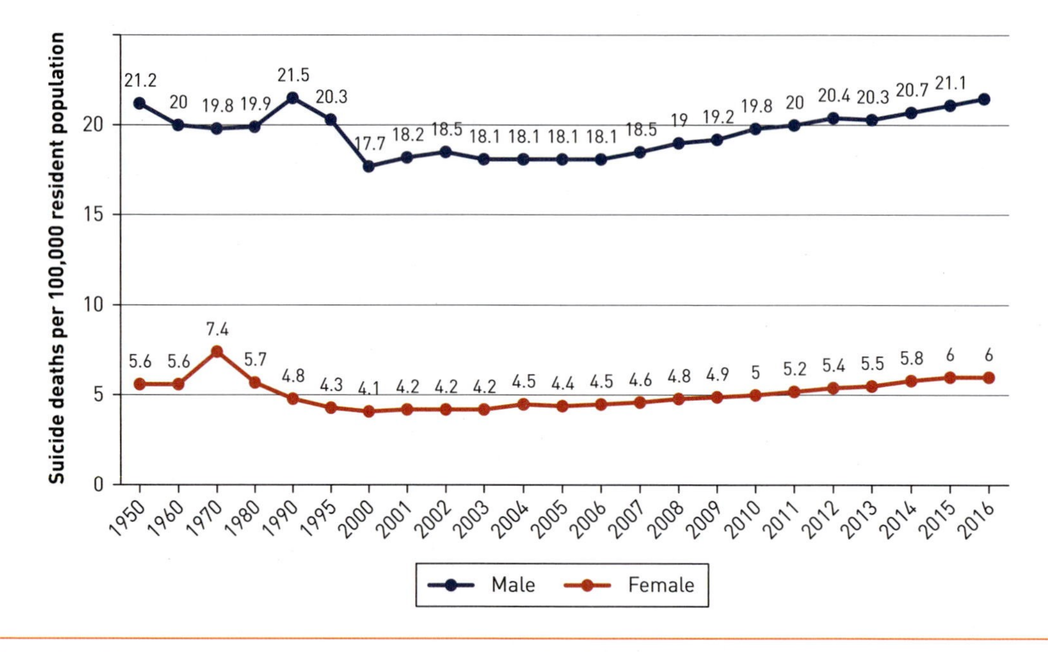

Source: John Elflein, "Death Rate for Suicide per 100,000 Resident Population in the U.S. from 1950 to 2016, by Gender," *Statista*, https://www.statista.com/statistics/187478/death-rate-from-suicide-in-the-us-by-gender-since-1950.

Whereas girls and women more often attempt to take their own lives, boys and men are more likely to die from suicide (see Figure 13.5).[29]

We can begin to understand this paradox by recognizing that girls and women are continually socialized to believe that seeking others' help is an acceptable "feminine" behavior. Therefore, females who have depression and attempt suicide are inclined toward methods that do not quickly lead to death but instead allow time for people to discover and respond to their pain. These methods include cutting oneself or overdosing on pills. Depressed boys and men, on the other hand, are more apt to pick up a firearm or hang themselves—tactics that immediately cause death. Males choose these tactics because they reflect a recurring message reinforced by parents, coaches, friends, the media, the military, and elsewhere: being "masculine" hinges on pushing away pain and masking weakness.[30]

"One of the ironies about men's depression," according to journalist Terrence Real, "is that the very forces that help create it keep us from seeing it. Men are not supposed to be vulnerable. Pain is something we are to rise above."[31] Real's words offer one explanation for why women are twice as likely to be diagnosed with

A woman who feels suicidal may overdose on drugs. Since this method doesn't immediately cause death, others may discover that she needs help, which is an acceptable characteristic for American females to project.
iStockphoto.com/spukkato

The males who carry out mass shootings often end their rampage by turning the weapon on themselves. This finale to their mayhem conveys "manliness" in that the shooter addresses his pain through self-destruction rather than vocally admitting it to others.
iStockphoto.com/turk_stock_photographer

depression as men. According to this view, men experience the symptoms of this illness about as often as women. However, men may be disinclined to seek out treatment for depression due to a belief that it's not "manly" to expose vulnerability by asking for help.[32]

Pete Davidson is one of several celebrities in recent years to speak openly about the particular challenges of male depression. It is noteworthy that these celebrities have come forward, given the widespread belief that boys and men are supposed to muscle through pain.
Ava Williams/NBC/NBCU Photo Bank via Getty Images

While there's variation from person to person in how depression affects them, the sociological perspective enables us to see that there are also significant gender differences in how people experience and cope with this illness. Because boys and men are often discouraged from talking openly about their own emotional distress, depressed males with suicidal thoughts have a motive not to seek help. They may instead opt to end their pain in the quickest way possible: self-inflicted violence. The sociological perspective, therefore, enables us to see why suicide rates are higher in males even though females more frequently attempt to take their own lives.

People Whose Minds Are Merely Different: Challenging the Stigma Associated with Mental Illness

13.6 Explain how the idea of neurodiversity can destigmatize mental illness.

Because I have a visual impairment, I've experienced firsthand what is common for many people with mental illness: Others negatively view them through the narrow prism of their disability. Having dealt my entire life with the stigma of seeing poorly has contributed to my continual attraction to sociology. The sociological perspective gives me a language for understanding insecurities I used to internalize as simply my own problem. I have come to see these insecurities as stemming from how others respond to my poor sight. I've also been able to recognize that the similar feelings of stigma I share with people who are mentally ill are not unique to those who have a disability. Anyone whose behavior or appearance deviates from the socially constructed norm, including LGBTQ and fat people, may experience insecurities because others deem them as "abnormal" and hence unworthy.

Temple Grandin, a scientist and proponent of neurodiversity, has written extensively about how being autistic gives her a perception of animals' thoughts and feeling that most people do not have.
ZUMA Press, Inc./Alamy Stock Photo

Understanding psychiatric disorders sociologically opens the door to seeing how people who have these disorders might live without the burden of others' negative judgments. That's the aim of those who promote **neurodiversity**, or the idea that everyone deserves dignity regardless of the different ways their brains work. The neurodiversity movement regards a person's mental state as just one aspect—rather than the totality—of who they are, challenging the notion of one single type of "normal" brain functioning that's morally superior to others.[33]

This movement doesn't minimize the impairments of mental illness nor the unbearable pain that people with a diagnosed condition often endure. Rather, neurodiversity proponents emphasize the talents that sometimes stem from having a particular psychiatric disorder. Here are some examples:

- Those with ADHD can be highly creative.
- Dyslexic people often have keen perceptual abilities.
- Some individuals on the autism spectrum may be able to recognize tiny details in complex situations that many others cannot discern.

Because people with these talents can be valuable workers, employers are increasingly recognizing the assets that may accompany certain mental illnesses.[34]

Neurodiversity is an important idea because it highlights the possibility that a person with mental illness may experience less suffering if others view them as "merely different" rather than as "abnormal." Abnormality is not a quality inherent to people diagnosed as meeting *DSM* criteria but rather a label often assigned to them. It's crucial to distinguish between how a person's brain functions and the meanings given to that level of functionality. Whereas the former may be genetically based, the latter is socially constructed. Biology need not be destiny.

Therapy offers an opportunity for people to see themselves as worthy despite the stigma others may attach to their mental illness.
iStockphoto.com/tommaso79

This chapter has highlighted the sociological perspective toward mental illness by telling stories about (1) social forces that produce ADHD misdiagnoses, (2) how living in the U.S. shapes the way a person experiences depression, and (3) the influence of gender expectations on how people think about and cope with psychiatric disorders. This perspective encourages you to pay attention to the roles you play in defining how people understand what's going on inside their own heads and the heads of others. If you have a psychiatric disorder, you can now see why medical labels don't have to define how you see yourself. If you do not personally suffer from mental illness but know someone who does, the payoffs of the sociological perspective are equally rich. This lens illustrates how you can develop greater compassion toward people who not only experience profound suffering but may also feel deep shame because of it.

What Do You Know Now?

1. What does it mean to view mental illness as a social construction? What is significant about viewing mental illness in this way?
2. How does Kristin Parber's story expose the power of Big Pharma to contribute to the misdiagnosis of mental illness?
3. How does the pursuit of happiness in American society adversely affect people who are clinically depressed and therefore impaired in their ability to experience happiness?
4. Why are boys and men likelier to die from their suicide attempts than girls and women? What do these differences reveal about societal messages about how acceptable it is for males and females to experience emotional pain?
5. What prospects does the neurodiversity movement have for fueling greater compassion toward those who not only suffer from mental illness but also feel ashamed about their impairments?
6. Many people your age have symptoms of a particular *DSM* diagnosis yet don't get the help they need. Use the sociological perspective to make an educated guess about why this may be so.

Key Terms

Visit **edge.sagepub.com/silver** to help you accomplish your coursework goals in an easy-to-use learning environment.

Notes

1. Quoted in David A. Karp, *Speaking of Sadness: Depression, Disconnection, and the Meanings of Mental Illness* (New York: Oxford University Press, 1996), 29. Ava Thompson is not her real name but one I created. Karp quotes her anonymously.
2. National Institute of Mental Health, "Major Depression," February 2019, https://www.nimh.nih.gov/health/statistics/prevalence/major-depression-among-adults.shtml.
3. Scott Strossel, "Surviving Anxiety," *The Atlantic*, January/February 2014, https://www.theatlantic.com/magazine/archive/2014/01/surviving_anxiety/355741/?utm_source=masthead-newsletter&utm_medium=email&utm_campaign=member-newsletter-20171025-78&silverid=MzU1MTU0MDQ2OTA5S0.
4. Ibid.
5. Data about the prevalence of anxiety disorders come from the Anxiety & Depression Association of America, "Facts & Statistics," https://adaa.org/about-adaa/press-room/facts-statistics.
6. Kerri MacKay, "What ADHD Feels Like to Me," *The Inside Track Blog*, January 12, 2016, https://www.understood.org/en/community-events/blogs/the-inside-track/2016/01/12/what-adhd-feels-like-to-me.
7. Centers for Disease Control and Prevention, "Data and Statistics about ADHD," September 21, 2018, https://www.cdc.gov/ncbddd/adhd/data.html.
8. National Alliance on Mental Health, "Mental Health by the Numbers," https://www.nami.org/Learn-More/Mental-Health-By-the-Numbers.
9. Robert Rigo, "Let's Talk about Suicide," *New York Times*, July 24, 2017, https://www.nytimes.com/2017/07/24/opinion/chester-bennington-linkin-park-suicide.html?_r=0.
10. Neel Burton, "When Homosexuality Stopped Being a Mental Disorder," *Psychology Today*, September 18, 2015, https://www.psychologytoday.com/blog/hide-and-seek/201509/when-homosexuality-stopped-being-mental-disorder; Allen Frances, *Saving Normal: An Insider's Revolt against Out-of-Control Psychiatric Diagnosis,* DSM-5, *Big Pharma, and the Medicalization of Ordinary Life* (New York: William Morrow, 2013), 152–3.
11. The other country that permits direct marketing of drugs to consumers is New Zealand; see John Marshall, "Why You See Such Weird

Drug Commercials on TV All the Time," *Thrillist*, March 23, 2016, https://www.thrillist.com/health/nation/why-are-prescription-drug-advertisements-legal-in-america.

12. Tara Parker-Pope, "Psychiatry Handbook Linked to Drug Industry," *New York Times*, May 6, 2008, https://well.blogs.nytimes.com/2008/05/06/psychiatry-handbook-linked-to-drug-industry/.
13. Kristin Parber is a pseudonym. Her story comes from Alan Schwarz, *ADHD Nation: Children, Doctors, Big Pharma, and the Making of an American Epidemic* (New York: Simon and Schuster, 2016), 78–80, 148, 232.
14. Frances, *Saving Normal*, 95; Schwarz, *ADHD Nation*, 148–56, 211. Data about the relapse rate among people treated for substance abuse disorders come from The National Institute on Drug Abuse, "Drugs, Brains, and Behavior: The Science of Addiction," https://www.drugabuse.gov/publications/drugs-brains-behavior-science-addiction/treatment-recovery.
15. Discussion of these social forces comes from Schwarz, *ADHD Nation*, 53–4 and Frances, *Saving Normal*, 92, 101–5, 168.
16. Richard L. Morrow et al., "Influence of Relative Age on Diagnosis and Treatment of Attention-Deficit/Hyperactivity Disorder in Children," *Canadian Medical Association Journal* 184, no. 7 (2012): 755–62.
17. MacKay, "What ADHD Feels Like to Me."
18. Quoted in Schwarz, *ADHD Nation*, 218.
19. My knowledge of Ruth Whippman's story is from her essay "America Is Obsessed with Happiness—and it's Making Us Miserable," *Vox*, December 22, 2016, https://www.vox.com/first-person/2016/10/4/13093380/happiness-america-ruth-whippman.
20. William Davies, *The Happiness Industry: How the Government and Big Business Sold Us Well-Being* (New York: Verso, 2016), 4–5.
21. Tamara Sims et al., "Wanting to Maximize the Positive and Minimize the Negative: Implications for Mixed Affective Experience in American and Chinese Contexts," *Journal of Personality and Social Psychology* 109, no. 2 (2015): 292–315; Arthur Kleinman, "Culture and Depression," *New England Journal of Medicine* 351 (2004): 951–3.
22. Ruth Whippman, *America the Anxious: How Our Pursuit of Happiness Is Producing a Nation of Nervous Wrecks* (New York: St. Martin's Press, 2016), 8.
23. Lucy McGuirk et al., "Does a Culture of Happiness Increase Rumination over Failure?" *Emotion*, 2017, https://ppw.kuleuven.be/okp/_pdf/McGuirkInPressDACOH.pdf.
24. The World Bank, "Poverty," April 3, 2019, http://www.worldbank.org/en/topic/poverty/overview.
25. These findings come from a Gallup poll and are reported in Lindsay Abrams, "People Living in Poverty Are Twice as Likely to be Depressed," *The Atlantic*, October 31, 2012, https://www.theatlantic.com/health/archive/2012/10/study-people-living-in-poverty-are-twice-as-likely-to-be-depressed/264320/. Also see Karp, *Speaking of Sadness*, 171.
26. Allan V. Horwitz and Jerome C. Wakefield, *The Loss of Sadness: How Psychiatry Transformed Normal Sorrow into Depressive Disorder* (New York: Oxford University Press, 2007), 3–4; Frances, *Saving Normal*, 186–7.
27. Ingrid Johnston-Robledo and Joan C. Chrisler, "The Menstrual Mark: Menstruation as Social Stigma," *Sex Roles* 68, no. 1/2 (2011): 9–18.
28. Carol S. North, "The Classification of Hysteria and Related Disorders: Historical and Phenomenological Considerations," *Behavioral Sciences* 5, no. 4 (2015): 496–517.
29. Silvia Sarah Canetto and Isaac Sakinofsky, "The Gender Paradox in Suicide," *Suicide and Life Threatening Behavior* 28 (1998): 1–23.
30. Konstantinos Tsirigotis, Wojciech Gruszczynski, and Marta Tsirigotis. "Gender Differentiation in Methods of Suicide Attempts," *Medical Science Monitor* 17, no. 8 (2011): PH65–PH70; Anne Cleary, "Suicidal Action, Emotional Expression, and the Performance of Masculinities," *Social Science and Medicine* 74, no. 4 (2012): 498–505.

31. Terrence Real, *I Don't Want to Talk about It: Overcoming the Secret Legacy of Male Depression* (New York: Scribner, 1997), 22.
32. Harvard Health Publishing, "Women and Depression," *Harvard Mental Health Letter*, May 2011, https://www.health.harvard.edu/womens-health/women-and-depression.
33. Mark Peters, "Neurodiversity: When You're Not Flawed, Just Mentally Different," *Boston Globe*, June 15, 2017, http://www.bostonglobe.com/ideas/2017/06/15/neurodiversity-when-you-not-flawed-just-mentally-different/PURSa1XmQdjzYPOejcThAN/story.html.
34. Thomas Armstrong, "The Myth of the Normal Brain: Embracing Neurodiversity," *AMA Journal of Ethics* 17, no. 4 (2015): 348–52; Robert D. Austin and Gary P. Pisano, "Neurodiversity as a Competitive Advantage," *Harvard Business Review*, May/June 2017, https://hbr.org/2017/05/neurodiversity-as-a-competitive-advantage.

iStockphoto.com/wildpixel

14

Eyes Wide Open

The Benefits of Seeing Social Problems from Multiple Perspectives

Learning Objectives

1. Explain how the sociological perspective expands upon the individual perspective toward social problems.
2. Identify patterns in how the sociological perspective explains the social problems discussed in earlier chapters.
3. Describe the root cause of a social problem explored in this book and explain why people rarely connect that cause with the problem.
4. Recognize how all individuals bear some responsibility for a given social problem, even if they have not directly experienced or perpetrated it.

People who like to avoid shocking discoveries, who prefer to believe that society is just what they were taught in Sunday school, who like the safety of the rules . . . should stay away from sociology.

—Sociologist Peter Berger[1]

These words are on a poster outside my office. It's an invitation to students that comes with a warning: Sociology offers an amazing adventure, but it's not for the faint-hearted—kind of like a thrill ride. As this adventure winds down, let's think about its many shocking discoveries. This concluding chapter encourages you to consider the significance of Peter Berger's words by highlighting four benefits of the sociological perspective toward social problems. After a brief discussion of each, there's a list of questions meant to reinforce and further develop ideas presented in earlier chapters.

Benefit #1: Recognizing Different Realities

14.1 Explain how the sociological perspective expands upon the individual perspective toward social problems.

There's an ancient story about six blind people who hear that someone has brought an elephant to their village. They're unfamiliar with this animal and curious to learn about it, so they each reach out their hand to touch it. The first person feels the trunk and comments that an elephant is like a thick snake. The second touches the ear and says an elephant is like a fan. The third, after reaching toward the leg, believes this animal is like a tree. The fourth touches its side and remarks that an elephant is like a wall. The fifth person feels the tail and describes the animal as akin to a rope. After touching the tusk, the last person describes an elephant as hard, smooth, and spear-like.[2]

Though people often see the social world differently from one another, their perspectives may be compatible.
iStockphoto.com/tintin75

All of them understand what an elephant is, and yet none completely understands. Likewise, whereas before reading this book you knew certain truths about social problems, you only had partial knowledge. In the preceding chapters, you've discovered the value in building upon what you already know. Just as the six blind people's understandings of the elephant are compatible with one another, so too are the individual and sociological perspectives. They offer complementary ways to explain social problems. Recognizing that you can gain a greater understanding of these issues by looking at them from both perspectives is one of the major takeaways of this book.

The individual perspective explains the problems plaguing our society by focusing on particular people's irresponsible behavior. For example:

- Opioid abuse reflects a person's lack of willpower; they prefer getting high over saying no to drugs.
- A poor mother is uncaring if she spends her limited income on chips, soda, and other junk food for her children.
- An athlete is unscrupulous for taking performance-enhancing drugs to get an edge on the competition.

The case studies in this book illustrate that focusing exclusively on the individual limits your understanding of these issues. That's why it's also important to recognize the social forces that lead a person to behave in deviant ways:

- Many people get hooked on opioids because they trust their doctor's recommendation to fill a prescription for drugs heavily marketed by pharmaceutical companies as remedies for chronic pain. (Chapter 4)
- Low-income women who feed their kids junk food do not have the money for more nutritious alternatives and may live in a food desert where such alternatives are not even available. (Chapter 5)
- Athletes who use performance-enhancing drugs are, like students who write exam answers on their palms, responding to the pressures of trying to be the best in a winner-take-all setting. (Chapter 7)

Looking at social problems sociologically gives you reason to resist the inclination to focus solely on an individual's decision to abuse drugs, eat poorly, or cheat—and instead recognize what underlies that decision. This book has highlighted the sociological perspective because, unlike the individual perspective, this way of understanding social problems is rarely accessible based on your everyday experiences.

Questions

1. Pick a social problem discussed in this book. How does the individual perspective enable you to understand this problem? How does the sociological perspective contextualize the explanation raised by the individual perspective?
2. Think of a social problem not discussed in this book. From the individual perspective, what are its causes? Then, do some educated guesswork. How does the sociological perspective build upon the individual perspective?
3. Think about a person you know. How is the story about the blind people and the elephant a useful tool to better understand this person?

Benefit #2: Expanding Your Focus

14.2 Identify patterns in how the sociological perspective explains the social problems discussed in earlier chapters.

When I was in college and my mind wandered during class, I'd sometimes find a blank page in my notebook and draw tiny dots. My doodling usually amounted to nothing more than a way to pass the time. But every once in a while when I moved my head away from the page, I'd notice something I hadn't intended to create. Whereas up close the dots looked random, from further away they formed a picture of a person's face, the body of an animal, or something else familiar.

Had I discovered something no one else had? Should I change my major from sociology to art? These questions occasionally ran through my mind whenever I doodled in class. In time, I learned that there was nothing original about my creativity. The artistry I'd stumbled upon by happenstance is a genre called pointillism that emerged in France during the 1880s. Painters etch dots with an eye to how they

From far away, George Seurat's *A Sunday on La Grande Jatte* depicts people enjoying a beautiful day by the water. But when you move the image closer to your face, it becomes a random array of dots.
Everett Collection Historical/Alamy Stock Photo

will collectively appear from afar as a recognizable image. Pointillism is popular because it demonstrates that the perspective a person takes toward an object shapes what they notice about it.

The same is true for our society. I didn't have to switch majors to discover fascinating patterns submerged within the seemingly disconnected details of everyday life. Indeed, this is the very essence of sociology. The sociological perspective enables you to notice that many distinct trees comprise one single forest. Embracing this perspective reveals that every part of your life has a broader story behind it. This book has shown you how to see patterns among cheating, obesity, mental illness, cyberbullying, and many other social problems. Whereas up close these issues often seem unrelated, taking a step back enables you to recognize deeper connections among them.

Questions

1. Pick two social problems in this book that appear up close to be unrelated. From the sociological perspective, what are the connections between them?

2. Think of a social problem not discussed in this book. Use the sociological perspective to make an educated guess about how its root causes are similar to the root causes of the issues you picked for Question 1.

Benefit #3: Getting to the Heart of the Matter

14.3 Describe the root cause of a social problem explored in this book and explain why people rarely connect that cause with the problem.

When our son Benjamin was nearly two, Nancy and I learned he was on the autism spectrum. We were devastated to contemplate what this might mean for his future. We went searching for answers about how this could've happened and soon heard about Dr. Andrew Wakefield. Four years earlier, he'd published a study linking autism to the MMR (measles, mumps, and rubella) vaccine, which babies typically get around eighteen months of age. Frequent media reports had bolstered the legitimacy of his findings. Our daughter Arielle had been born just ten days before Benjamin got his diagnosis, and, based on these reports, we decided to join a growing movement of parents who either delayed vaccinating their babies or refused to do so. This movement grew amidst reports that autism spectrum disorder among children was becoming more and more common each year.[3] According to data from the Centers for Disease Control and Prevention, the number of children in the U.S. diagnosed with this disorder has risen steadily in recent years, from 1 in 166 children in 2004 to 1 in 59 children in 2018.[4] Although in 2011 a respected medical journal published an investigation showing that Wakefield had falsified his data, many parents continue to believe that vaccines cause autism.[5]

These parents, like Nancy and I had done years earlier, are acting out of fear. Although anxiety about vaccinating kids isn't a subject discussed earlier in this book, many of the issues explored in the case studies similarly revolve around fear. It's a common thread among seemingly disparate topics, such as the bleak opportunities of a baby born to a teen mother (Chapter 6), the terror a gunman creates by opening fire in a classroom (Chapter 8), and the exploitation a child experiences from an anonymous online stranger (Chapter 11). From a sociological perspective, fear isn't merely a response to the reality of danger; moreover, fear stems from how powerful groups like the government and news media manipulate people's perceptions of reality.

"One of the paradoxes of a culture of fear," wrote sociologist Barry Glassner, "is that serious problems remain widely ignored even though they give rise to precisely the dangers that the populace most abhors."[6] Pause for a moment to reread these intriguing words and consider their significance. Glassner suggests that if you want to thoroughly understand a social problem, you need to pay attention to serious, yet possibly unfamiliar, issues that lie at the root of the behaviors that alarm you. For example, the root cause of gender violence is the messages our society gives boys and men about what it means to "be a man" (see Chapter 9). Once you understand how fear can blind you from seeing a social problem's underlying story, new ways of understanding the problem become possible. That's one of this book's most eye-opening ideas.

Questions

1. Choose two of the case studies in this book. Identify the root issue for each and explain how it causes the publicized danger.
2. Why do you think the root issue receives little public attention compared to the problem the public fears?

Benefit #4: Looking Inward

14.4 Recognize how all individuals bear some responsibility for a given social problem, even if they have not directly experienced or perpetrated it.

Because you now recognize that social problems are rooted in the very structure of our society, it would be easy to throw your hands up in the air and believe there's nothing you can do to address these problems. However, that's not your only option. The sociological perspective has continually motivated me to explore how taking a close look at myself is a crucial step in addressing these problems. I encourage you to do the same. Because this perspective shifts attention from the wrongdoer to the social

Taking a close look in the mirror reveals inconvenient truths about the underlying causes of social problems that may motivate you to change your attitudes or behaviors.
iStockphoto.com/Marco_Piunti

influences on that person's behavior, often you and I contribute to those influences—for example, by participating in the winner-take-all society that produces cheating or affirming the masculine norms that lie at the heart of gender violence.

Cops who kill unarmed suspects and teens who get pregnant are certainly responsible for the troubles they cause our society. But the responsibility for these and other social problems also extends more widely. All of us need to look at the roles we may play in reinforcing the beliefs that give rise to these problematic behaviors. Developing this kind of introspection is crucial, yet it is not easy. Even after reading the preceding chapters that highlight our collective responsibility for social problems, you still may be inclined to resist this way of thinking. Adhering to the individual perspective may be more appealing because it places blame on those whom you may see as distinctly unlike you, such as people who abuse drugs or bully their peers.

You may never before have thought about your responsibility for behaviors that, when viewed from the perspective of your personal experiences, do not involve you. Beginning to think about this responsibility doesn't mean that you deserve blame the next time there is a mass shooting. The takeaway, instead, is that you risk overlooking the root causes of this social problem by focusing solely on the gunman. You can expose these root causes by paying attention to how and where he learned to see his behavior as acceptable. Doing so will enable you to recognize the part you may play in validating to this behavior.

The point of looking inward isn't to make you feel guilty about the harms that afflict our society but to highlight that all of us can contribute to fixing social problems once we begin to look at them from a sociological perspective. This perspective illuminates the importance of the phrase "No person is an island," calling on each of us to imagine our world differently than how we typically see it. Even though social problems afflict only certain people, we all bear responsibility for them. The solutions, therefore, lie with each of us.

Questions

1. Pick two of the case studies in this book. Even if you've never engaged in either type of deviance, in what sense do your actions or beliefs play a role in contributing to them?

2. Why is it significant to acknowledge your responsibility for social problems? What duties accompany that responsibility?

3. Some critics of sociology incorrectly see it as offering excuses for wrongful behaviors. In truth, how does sociology explain a person's faults and failures without justifying them?

Visit **edge.sagepub.com/silver** to help you accomplish your coursework goals in an easy-to-use learning environment.

Notes

1. Quoted in Peter Berger, *Invitation to Sociology: A Humanist Perspective* (New York: Anchor, 1963).
2. This story is thousands of years old and can be traced back to ancient India. A common version of it comes from "The Blind Men and the Elephant," a poem written by John Godrey Saxe in the nineteenth century.
3. For discussion of the parental frenzy over vaccines causing autism, see Barry Glassner, *The Culture of Fear* (New York: Basic Books, 2009), 174–9. Research documents that there is no credible evidence behind these fears and that not vaccinating elevates kids' chances of contracting a number of other preventable diseases; see Frank DeStefano, "Vaccines and Autism: Evidence Does Not Support a Causal Association," *Clinical Pharmacology and Therapeutics* 82, no. 6 (2007): 756–9.
4. Autism Speaks, "CDC Increases Estimate of Autism's Prevalence by 15 Percent, to 1 in 59 Children," April 26, 2018, https://www.autismspeaks.org/science-news/cdc-increases-estimate-autisms-prevalence-15-percent-1-59-children.
5. For an exposé of how Andrew Wakefield falsified his data, see Brian Deer, "How the Case against the MMR Vaccine Was Fixed," *British Medical Journal*, January 6, 2011, https://www.bmj.com/content/342/bmj.c5347.
6. Glassner, *The Culture of Fear*, xxvi.

• Glossary •

American Dream: the belief that hard work and self-discipline enable people to get ahead.

Anabolic steroids: drugs that promote muscle growth by mimicking the effects of testosterone.

Animal cruelty: when humans inflict pain, suffering, physical injury, or death on members of other species.

Black Lives Matter: an organization that has been a leading voice exposing racial injustices in policing.

Blue Lives Matter: an organization formed in response to cop killings that honors the important work of law enforcement.

Body mass index: a medical chart that labels a person's size relative to what doctors consider healthy.

Boomerang effect: how aggressive efforts to protect kids from danger can hurt them in unforeseen ways.

Broken windows: a crime-control strategy in which police target highly visible nonviolent offenses in order to preclude people from committing more serious crimes.

Bystander effect: public situations where a person needs help, yet someone who witnesses their distress chooses not to offer assistance because they believe that others will instead.

Canary in the coalmine: an alarm signaling even greater danger than there appears to be.

Case study: an example that exposes and illustrates a broader theme.

Catfishing: creating a fake online identity in order to deceive others.

Cognitive dissonance: the uncomfortable feeling a person gets when their beliefs are out of sync with their behaviors.

Columbine effect: the tendency for mass shooters to model their attacks on the 1999 rampage at Columbine High School in Colorado.

Comprehensive sex education: school curricula that promote birth control as a way to avoid pregnancy and sexually transmitted infections.

Conventional wisdom: what you think is true about a topic based on information you've acquired over the course of your life.

Correlation: a mutual relationship between two or more things.

Crack: a smokable crystal created by cooking powder cocaine with baking soda and water.

Criminalization: the view that because a person has broken the law, they should be incarcerated.

Cultural capital: the particular types of knowledge one needs in order to achieve upward mobility.

Culture: the beliefs, values, and behaviors that a particular group of people share in common.

Cyberbullying: when one or more people repeatedly ridicule another person via text or e-mail, post mean content about them on social media, or use technology in some other way to hurt them.

Deviance: behavior that violates the rules for acceptable conduct.

Disordered eating: a condition characterized by an irregular meal schedule, limited food intake, binging, or self-induced vomiting after eating.

Drug addiction: a physical and/or psychological dependence on a dangerous substance.

Economic inequality: gaps among people based on their financial worth.

Elephant in the room: a major cause of a social problem that people often ignore.

Epidemic: a harm that spreads rapidly across a larger and larger segment of people.

Euphemism: a sanitized way of speaking about something dirty or unpleasant.

Explicit bias: the overt prejudices a person holds toward a particular group of people

Factory farming: the method of raising massive quantities of livestock under tightly controlled conditions until slaughter.

Fat shaming: when a person receives insults because they have a large body.

Feminization of poverty: women's greater likelihood than men to be poor.

Food deserts: neighborhoods where the only available foods are at convenience stores or fast food restaurants, both of which have few nutritious options.

Gaslighting: when one person in a romantic relationship psychologically manipulates the other to the point where the victim starts questioning their own sanity.

Gender binary: a social norm that a person may legitimately identify as either male or as female, and the valued characteristics of each identity lie in opposition to one another.

Gender violence: harm inflicted by people in powerful positions that reinforces norms about appropriate male and female behavior.

Grooming: when an online predator offers gifts or flattery as a way of forging an emotional connection with the person they intend to victimize.

Heteronormativity: the view that heterosexuality is the only legitimate expression of desire.

Images: the pictures people have in their heads about a particular idea or group of people.

Implicit bias: the deep-seated, often unconscious prejudices a person holds toward a particular group of people.

Incel: an online subculture of heterosexual males who see themselves as victims because of their involuntary celibacy.

Income inequality: the disparity in earnings between the richest and poorest people in a society.

Individual perspective: a way of understanding social problems that focuses on the person who commits wrongdoing.

Intersectionality: the idea that a person's identities are interwoven.

Intimate partner violence: when one person in a romantic relationship uses their power to weaken, shame, and humiliate their lover.

Jim Crow laws: statutes enacted after the U.S. Civil War that permitted segregation in many aspects of everyday life, including busing, schooling, and public accommodations.

Latent function: a person's or organization's hidden intentions for engaging in a given behavior.

Man Box: a way to visualize the prevalent expectations in American society about what it means to be "manly."

Manifest function: a person's or organization's stated intentions for engaging in a given behavior.

Mass incarceration: the surge in the percentage of Americans behind bars as a result of the War on Drugs.

Mass shooting: the murder of four or more people in succession with a firearm.

Medicalization: the process by which a group of people experiencing the same health problem come to be seen as having a treatable disease.

Mental illness: a condition that impairs thinking, emotions, or behavior. This term is used synonymously with *psychiatric disorder*.

Middle-class mindset: an understanding that there are certain things a person needs to do to achieve what our society defines as success.

Neurodiversity: the idea that everyone deserves dignity regardless of the different ways their brains work.

Obesity: a medical condition afflicting people with a body mass index exceeding 30.

Opioids: a class of drugs that relieve pain by dulling the senses.

Opportunity divide: the inequalities between people born into higher-income families who have lots of chances to better their lives and those born into lower-income families who have comparatively few chances.

Police brutality: cops unnecessarily using force against criminal suspects.

Poverty: a minimal standard of living that the federal government defines as having a household income below a given amount that is annually adjusted for inflation.

Public mass shooting: a mass shooting that receives significant publicity because it occurs in a place where anyone may happen to be at a given moment, such as a school, mall, workplace, restaurant, airport, theater, or house of worship. This term is used synonymously with *rampage* and *massacre*.

Racial discrimination: the systemic disadvantaging of people based on the color of their skin.

Rape: sexual assault that involves oral, anal, or vaginal penetration.

Recidivism: when an ex-convict commits another crime.

Redlining: a practice where Black people interested in purchasing a home were barred from doing so in certain communities.

Revictimization: reliving the pain and humiliation of a terrifying experience.

Scapegoat: someone or something that is unfairly given the blame for a social problem simply because they're an easy target.

Self-fulfilling prophecy: when a person acts in ways that confirm how others label them.

Sexting: using mobile devices to exchange sexually explicit photos and videos.

Sextortion: when one person threatens to share provocative images of another unless they offer sexual favors.

Sexual assault: when a person asserts their power by physically making sexual advances on someone else without their consent.

Sexual double standard: the idea that girls and women deserve criticism for sexual behavior that is praiseworthy if carried out by boys or men.

Sexual harassment: when a person asserts their power over someone else by making sexual innuendos or unwanted advances.

Sexualization of childhood: kids' continuous exposure to provocative images as they're growing up.

Sexual revolution: the loosening of public attitudes toward sex that began during the 1960s.

Shotgun marriage: when a young heterosexual couple quickly decides to marry after the girl becomes pregnant, as a way to avoid embarrassment over her having had premarital sex.

Silencing: when males ignore, censor, or reprimand females simply because of who they are.

Size acceptance activists: a group of people who aim to diminish the importance of weight to a person's feelings of self-worth.

Size discrimination: unequal treatment people encounter simply for being overweight.

Social class: a person's financial standing and how it relates to other types of resources and opportunities they have or don't have.

Social construction of drugs: the idea that people decide whether the possession, use, and sale of a particular drug are crimes and, if so, the penalty for breaking the law.

Social construction of gender: the idea that a society views certain roles, but not others, as acceptable for individuals with particular types of bodies.

Social construction of mental illness: the idea that people define which types of moods and behaviors characterize particular diagnostic categories.

Social construction of weight: the idea that people assign bodies of varying sizes and shapes unequal amounts of social worth.

Social forces: group influences that shape how a person behaves or thinks.

Social inequality: the disparities among people in their amount of money, power, or status.

Social norms: the rules defining appropriate behavior.

Social problem: a harm in our society that people believe should and can be fixed.

Sociological data: evidence about how people behave or think.

Sociological perspective: a way of understanding social problems that highlights the social forces shaping individual behavior.

Sociology: the study of the intricate workings of a society.

Stigma: a characteristic that marks a person as disreputable in the eyes of others.

Stop and frisk: a crime-control strategy that gives cops the authority to question and search anyone they suspect of carrying weapons or drugs.

Street harassment: when a person receives unwanted sexual attention in public from unfamiliar men.

Subsidy: government money given to a recipient to reduce their out-of-pocket price for goods or services.

Suicidal ideation: giving serious thought to attempting suicide.

Suicide cluster: when multiple suicides occur in the same area over a short time period.

Toxic masculinity: the idea that being a "real man" hinges on acting abusively toward others, and often toward oneself too.

Upward mobility: a long-term increase in a person's income and status.

War on Drugs: the government's substantial expansion, beginning in the 1970s, of funding for the policing and punishment of drug crimes.

Wealth: a person's total assets.

Weight anxiety: the unease many people have that their bodies are too big.

Welfare: government assistance to low-income people to ensure that they have a basic standard of living.

White supremacy: the belief that Whites belong to the superior race and are therefore justified in subordinating people of other races.

Winner-take-all society: the idea that the educational system and many occupations follow the same competitive structure that's so apparent in sports.

• Index •